The global greenhouse regime

Who pays?

NOTE TO THE READER FROM THE UNU

The United Nations University's programme on Human and Policy Dimensions of Global Change is concerned with the complex interlinkages between human activities and the environment. It is recognised that in order to devise effective responses to combat global environmental change it is essential to understand the human and societal causes underlying the transformation of the physical environment. The objectives of the programme are:

1 to increase awareness of the complex dynamics governing human interaction with the Earth as a whole system;
2 to strengthen efforts to anticipate social change affecting the global environment;
3 to analyse policy options for dealing with global environmental change; and
4 to identify broad social strategies to prevent or mitigate undesirable impacts of global environmental change.

After the United Nations Conference on Environment and Development in Rio de Janeiro in 1992, a central issue in the Climate Change Convention relates to the amounts and sources of the greenhouse gases emitted from the various countries and regions, both industrialized and developing, and their relation to international governance. To date, the lack of agreed principles has stalled agreement as to what concrete and practical steps should be taken to meet the needs for stabilizing climate change.

The present book is the outcome of the UNU international collaborative research carried out under the Human and Policy Dimensions of Global Change programme. It is aimed at presenting the state of the art in greenhouse indices, and related international policy making and governance, clarifying key technical issues relating to greenhouse gas emissions, and outlining the economic responsibilities of various countries based on the emissions. It makes an argument for the necessary North–South resource transfers.

The global greenhouse regime

Who pays?

Science, economics and North-South politics in the Climate Change Convention

Edited by
Peter Hayes and Kirk Smith

earthscan
from Routledge

Dedication

This book is dedicated to the two Nadias
May their parents' generation learn in time that accepting liability for past errors is the other side of undertaking responsibility for the future

First published by Earthscan in the UK and USA in 1993

This edition published 2013 by Earthscan

For a full list of publications please contact:

Earthscan
2 Park Square, Milton Park, Abingdon, Oxfordshire OX14 4RN
Simultaneously published in the USA and Canada by Earthscan
711 Third Avenue, New York, NY 10017

First issued in hardback 2017

Earthscan is an imprint of the Taylor & Francis Group, an informa business

A catalogue record for this book is available from the British Library

ISBN 13: 978-1-138-16337-9 (hbk)
ISBN 13: 978-1-85383-136-2 (pbk)

United Nations University Press is the publishing division of the United Nations University

Contents

List of Illustrations

Figures

List of illustrations

Tables

Boxes

List of illustrations

List of contributors

Dilip Ahuja is a Senior Analyst with the Bruce Company which provides technical support to the USEPA's Climate Change Division: #215, 1100 6th St, SW, Washington DC, 20024 USA; fax 1 202 4791009

Mark Cherniack directs IIEC's regional office in Bangkok and helped to develop the Demand Side Management (DSM) Master Plan for Thai utilities: International Institute for Energy Conservation, Racquet Club Building, 8 Sukhumvit Soi 49/9, Bangkok 10110 Thailand; fax 66 2 381 0815

Ogunlade R Davidson is Director, Research & Development, University of Sierra Leone, Freetown, Sierra Leone; fax 232 22 224439

Peter du Pont works with the Asia Office of the International Institute for Energy Conservation (IIEC) in Bangkok to implement a major energy efficiency initiative in Thailand: International Institute for Energy Conservation, Racquet Club Building, 8 Sukhumvit Soi 49/9, Bangkok 10110 Thailand; fax 66 2 381 0815

Peter Hayes is a researcher on security, resource, and environmental issues related to Asia Pacific, at Nautilus Pacific Research, 746 Ensenada Ave, Berkeley, CA 94707, USA; fax 1 510 526 9297

Stanislav F Kolar is a consultant specializing in energy efficiency and greenhouse gas reductions in Eastern Europe: Kolar Associates, 1527 Q St, NW, Washington DC, 20009, USA; fax 202 3877701

Jose Roberto Moreira is Executive Director of the Biomass Users Network, an NGO dealing at the international level with rational use and production of biomass: BAN, Rua Francisco Dias Velho, 814, Brooklyn Novo, 04581-001-S. Paulo-SP-Brazil; fax 55 11 5435494

Somthawin Patanavanich is a research associate in the Natural Resources and Environment Program at the Thailand Development Research Institute. She has done extensive research on the link between deforestation, energy use, and climate change: TDRI, Rajapark Building, 163 Asoke Rd, Bangkok 10110 Thailand; fax 66 2 3810815

Michael Philips recently completed a major report on the energy-lending activities of the multilateral development banks for the International Institute for Energy Conservation, Racquet Club Building, 8 Sukhumvit Soi 49/9, Bangkok 10110 Thailand; fax 66 2 381 0815

Alan Douglas Poole is an independent consultant for energy, transport and environmental planning in developing countries: Rua Marques de Sao Vicente 86, apt 211, Gavea, 22451, Rio de Janeiro, RJ Brazil; fax c/o WEE, 552 11233 4334

Amulya Reddy is President of the International Energy Initiative 25/5 Borebank Road, Benson Town, Bangalore, 560 046, India; fax 91 812 346 234

Hugh Saddler led Australia's first major study of the economic impact of energy efficiency and greenhouse gas reductions at: Energy Policy Analysis Pty Ltd, Suite 105, 55 Townshend St, Phillip ACT 2606, Australia; fax 61 6 2853583

Jayant Sathaye is Co-Leader of the International Energy Studies Group at the Lawrence Berkeley Laboratory, Bldg 90, Berkeley, CA 94720, USA; fax 1 415 4866294

Kirk R Smith works on environmental risks in Asian developing countries, especially those relating to unprocessed biofuels for cooking and spaceheating. He is Senior Fellow at the Program on Environment at the East West Center, 1777 East West Rd, Honolulu, Hawaii, 96848, USA; fax 1 808 944 7298

Susan Subak coordinates the climate change work at Stockholm Environment Institute's Boston office, and is a technical reviewer and modeller for the IPCC's Task Force on Greenhouse Gases; SEI, 89 Broad St, Boston, MA 02110, USA; fax 1 617 4267692

Joel Swisher is an engineering economist specializing in resource planning and greenhouse economics at RISO National Laboratory, PO Box 49, DK 4000 Roskilde, Denmark; fax 45 46 321999

Michael Wilford is a solicitor with experience in maritime law and insurance law. He is a consultant to Clyde & Co, an international law firm, and a Senior Associate of the Foundation for International Environmental Law and Development, King's College, Manresa Road, London SW3 6LX, United Kingdom; fax 44-351-6435

Preface

The United Nations University (UNU), an autonomous organ of the United Nations, is mandated in its Charter to conduct 'research into the pressing global problems of human survival, development and welfare that are the concern of the United Nations and its agencies.'

As the 1992 Rio UNCED ('the Earth Summit') made clear, the interrelated issues of the environment and development, and the search for sustainable development, are among the most urgent problems on the international political agenda.

The global community must now face directly the problem of accommodating in the twenty-first century a likely doubling of population and a five-fold economic growth without destroying its global life support system.

A key problem to be resolved is the limitation of the build-up of the greenhouse gases with their potential for altering global climate patterns. The Climate Change Convention agreed to in Rio is still largely symbolic, and must be succeeded by protocols and firm abatement agreements yet to be negotiated. The realization of such agreements is obviously of enormous importance for the future of society. The needed greenhouse gas regime will require a contract between rich and poor nations, and will only follow arduous negotiations.

A central issue in these negotiations will be the allocation of costs. A unique contribution of this book is its proposed composite index to determine who should pay for creating a global greenhouse gas regime, an index that incorporates both ability to pay – that is, economic realism – and historical contribution to climate change – that is, equity based on the polluter pays principle.

This, however, is only one of the contributions of the book to environmental diplomacy debates, which inevitably involve issues of science and technology, politics and economics, and, not the least, ethics. Environmental diplomacy must therefore be based on comprehensive interdisciplinary viewpoints as illustrated in this work. It must also be sensitive to varying regional approaches and views, as also reflected in this work, itself the product of a global network of scholars drawn from Australia, North America, India, Africa, South America, the Pacific Islands, and Europe.

This is a timely, important book. It deserves to be read by a wide audience of policy makers, academics and a public interested in the future of the earth.

Roland J Fuchs
Vice-Rector, United Nations University

Acknowledgements

The editors are grateful to the many people who assisted in the research, writing, and editing, and production of this book, including: Tom Athanasiou, Harriet Barlow, Marcy Darnovsky, Sandic Dela Cruz, Howard Geller, Peter Green and colleagues at Mac Advice, Allen Hammond, John Holdren, Evelyn Klinkmann, Rob Lindgren, Alan Miller, Irving Mintzer, Frank Muller, Cassandra Morrow, Pahdma Muthiah, Toufiq Siddiqi, Helen Takeuchi, John Topping, Chieko Umetsu, Mark Valencia, Megan Van Frank, David Victor, Isabel Wadc, Lyuba Zarsky, and three anonymous reviewers.

We are especially grateful for support and encouragement received from Dr Juha Uitto and Dr Roland Fuchs, Academic Officer and Vice Rector respectively of United Nations University; to United Nations University Press; to Jonathan Sinclair Wilson and Jo O'Driscoll of Earthscan Publications; and to Ian Carruthers of the Australian Department of Arts, Sports, Environment, Tourism and Territories for permission to draw in Chapters 5 and 6 on a study commissioned by the Department.

As always, any errors in this book remain the responsibility of the editors and authors.

Part I

Measuring responsibility

1

Introduction

Peter Hayes and Kirk R Smith

At the 1992 Earth Summit in Brazil, which was attended by more heads of state than any other meeting in human history, the UN Framework Convention on Climate Change (included as the Appendix to this volume) was opened for signature. By mid-October 1992, 158 nations had signed it. To become law, it must be ratified by national legislatures of at least fifty countries, a process that may take two years.[1]

Unfortunately, the Convention did not contain any specific provisions for funding its implementation. This lack is a major obstacle to its realization. The questions of how to decide who should pay and how much it might cost are the central topics of this book.

The participation of developing countries in a Climate Change Convention will determine whether the world responds prudently to the greenhouse effect. Even if the wealthy states radically reduce their greenhouse gas emissions, the poorer states will replace and eventually surpass them as major contributors to the greenhouse effect.[2] Action by members of the Organization for Economic Co-operation and Development and other industrialized countries can significantly slow the rate and reduce the magnitude of global warming. But unless the developing countries also act, the threat remains to everyone. Based on current trends, big poor countries like China, Indonesia, India, and Brazil will become major carbon dioxide contributors. They are already big methane gas emitters even though their per capita output is small.

As is argued below, the Climate Change Convention itself is still mostly symbolic. Unresolved issues include the practical implementation of the Convention in protocols to the Convention on technology and resource transfer; obtaining commitments from parties to limit carbon emissions; and the design and implementation of abatement strategies. All this and much more remains to be settled in protocols to be negotiated now that the Convention itself has been signed.

In this book, we do not tackle all these important issues. Instead, we postulate that the major determinant of developing country participation will be the terms offered by the developed world. The need for the rich and poor nations to work together to respond to the greenhouse effect could create a new political–economic interdependence between them. Alternatively, as

Norwegian analyst Anne Kristin Sydnes warns, it could portend 'another twenty years of fruitless North–South bargaining.'[3] The authors of this book examine the grounds for, the scale of, and possible conditions on possible resource transfer agreements from rich to poor states that will be central to any successful greenhouse management regime.[a]

In this chapter, we undertake four tasks. First, we review the basic scientific understanding of the greenhouse gas effect that gave rise to the Climate Change Convention. Second, we describe the content of the Convention and note its limitations. Third, we review the novel negotiating difficulties that will arise in the course of developing effective protocols under the Convention. Fourth, we summarize the key issues for the ongoing negotiations under the rubric of the Convention as presented in this book. In the latter section, we also provide a synopsis of each chapter of the book.

The greenhouse effect

Planet Earth's capacities for dispersing, diluting, and degrading most human-generated pollutants are large, but limited. As pollution rates increase, the natural processes that absorb and assimilate pollutants are eventually overwhelmed, leading to rising concentrations of pollutants in the environment. Depending on the pollutants, this overloading can create local disruptions in human health and ecosystem sustainability or, eventually, even global effects, such as climate change.

The authors of this book focus on the largest and oldest of all human pollutant releases, carbon into the atmosphere. Since their mastery of fire, human beings have disrupted the global carbon cycle by burning wood and other biomass at greater rates than occur naturally. Being about half carbon, biomass upon combustion releases carbon dioxide, methane, carbon monoxide, and other carbon-containing pollutants, which must be transported and broken down by natural processes.

Some of this release has been the direct result of using wood and other biomass forms such as crop residues as fuel. Another part is due to the clearing of biomass so that the land could be used for farming or other human purposes. Throughout most of human history, however, it is thought that human biomass combustion did not create large disturbances in the atmosphere. That is, natural processes such as regrowth replaced sufficient portions of the burned biomass to prevent significant build-up of these carbon-containing pollutants in the atmosphere.

The industrial revolution in what are now the economically developed countries increased the combustion of fossil fuels, which are mostly carbon. Today, fossil fuel combustion is the major source of atmospheric carbon

a An international regime is a set of norms, procedures and institutions that define the boundaries of legitimate behaviour in a given issue area.

releases although biomass burning has also probably increased. The rates of release became such that in the second half of the twentieth century, it became clear that atmospheric levels of important carbon-containing gases, particularly carbon dioxide and methane, were steadily increasing over their natural levels.

The atmospheric concentrations of these gases are far from those thought to be toxic or otherwise of much acute concern. Their impact is more subtle, for they act to blanket Earth, keeping in more of the sun's warmth than otherwise would be the case. This effect is indisputable because humans have observed the warming due to natural levels and variations of these same gases. Indeed, without the existence of these natural amounts of 'greenhouse gases', Earth would be too cold for life.

The higher-than-natural rates of greenhouse gas releases resulting from human fossil fuel and biomass combustion are boosting carbon dioxide and methane levels at greater rates than has occurred in recent Earth history. Earth's natural systems may not be able to cope with the extra heat being absorbed except by an overall increase in temperature, that is, global warming. This warming may in turn be associated with significant disruptions in local weather, such as patterns of precipitation and cloudiness. It may also have global impacts through thermal expansion of the oceans and melting of glaciers to cause sea-level rise. It may even disrupt ocean current patterns and marine and terrestrial ecosystems.

The nature and magnitude of global warming and associated climate change and sea-level rise resulting from greenhouse gases released by human activities are not known with certainty. The global atmospheric/ocean/climate systems are extremely complex, so much so that even the largest computers can only model a small portion of them at one time. Some natural processes seem to reduce the effect. Extra heat, for example, leads to more evaporation, which leads to more clouds, which leads to more sunlight being reflected, which leads to less heat. In contrast, other processes may enhance the effect. Extra heat, for example, leads to more snow melting, which leads to less sunlight reflection, which leads to more heat. Thus, contemporary estimates of the global effects are imprecise and so uncertain as to be not usable for predicting effects at any one place and time.

Nevertheless, a growing number of scientists believe that there is a significant chance that damaging levels of global warming will occur sometime before the middle of next century if existing trends of greenhouse gas emissions are continued. The most authoritative source for this view is the scientific report of the Intergovernmental Panel on Climate Change (IPCC).[4]

The IPCC consists of scientists from many countries working together under the aegis of the World Meteorological Society and United Nations Environment Programme and who have reviewed and summarized the available knowledge and uncertainties about greenhouse gases and climate change. As part of its conclusion, the IPCC predicts that unless emissions patterns change, there will be:

> a rate of increase in global mean temperature during the next century of 0.3°C per decade (with an uncertainty range of 0.2–0.5°C per decade); this is greater than that seen over the past 10,000 years. This will result in a likely increase in global mean temperature of about 1°C above the present value by 2025 and 3°C before the end of the next century.[5]

Its collective judgement is that:

> Rapid change in climate will change the composition of ecosystems; some species will benefit while others will be unable to migrate or adapt fast enough . . . The effect of warming on biological processes . . . may increase the atmospheric concentrations of natural greenhouse gasses.[6]

IPCC's impact report concluded that this warming could cause major and mostly negative local impacts on agriculture, forestry, water resources, natural ecosystems, air quality, and coastal zones among other sectors important to humanity.[7]

Recognizing that there are still significant uncertainties, particularly about the timing of impacts, and that some countries, notably the United States, have not yet officially acknowledged a need to take immediate steps, this book nevertheless starts with the premise that the world community decided when it signed the Climate Change Convention that a serious effort must be made to reduce the probability and magnitude of adverse impacts from global warming. According to the IPCC, a successful response strategy is likely to be one which recognizes that:

- 'Climate change is a global issue, effective responses . . . require a global effort';
- '[Effective responses] may have considerable impact on humankind and individual societies';
- 'Industrialized countries and developing countries have a common responsibility in dealing with problems arising from climate change'.[8]

Although the developed countries have emitted far more greenhouse gases to date, based on the current relationships between economic development and greenhouse-gas emissions, and on the sorely needed increases in economic welfare that are required, large poor countries such as China, India, Indonesia, Nigeria, and Brazil will also turn into major greenhouse-gas contributors.[b] Whether developing countries are engaged adequately in the implementation of a Climate Change Convention will be a major determinant of how quickly and completely the world responds to the dangers presented by human-enhanced greenhouse warming. Action in the Organization for Economic Co-operation and Development (OECD) and other developed countries alone can slow the rate and magnitude of global warming in the short term, but unless action is also started in developing countries, significant threat remains. The

b As discussed in some detail in Chapters 2–4, when making international comparisons it is important to distinguish carefully between cumulative past emissions and present emission rates as well as between emissions per nation and per capita.

fact – often discussed in this book – that there are attractive greenhouse amelioration projects in developing countries, however, does *not* necessarily mean that developing countries should pay for them.

The challenge facing humanity, therefore, is to find ways that the many benefits accompanying economic development can be attained by the world's poor without simultaneously emitting the amounts of greenhouse gases that have accompanied such economic development in the past; and to reduce dramatically the emissions from rich countries at the same time.

What was decided at Rio?

The Climate Change Convention signed at Rio sought:

1 to stabilize 'greenhouse gas concentrations in the atmosphere at a level that would prevent dangerous anthropogenic interference with the climate system';
2 to do so quickly enough 'to allow ecosystems to adapt naturally to climate change';
3 'to ensure that food production is not threatened';
4 'to enable economic development to proceed in a sustainable manner'.[9]

To paraphrase, the world's leaders committed themselves to work together in first slowing and then stopping the growth of greenhouse gases in the atmosphere at concentrations that would not be so greatly above natural levels as to significantly threaten human welfare and natural ecosystem balance. In this process, other important goals must not be sacrificed, including provision of adequate food supplies and eradication of poverty.

Abatement commitments

Developed countries committed themselves to a vaguely worded 'aim of returning individually or jointly to their 1990 levels [of] these anthropogenic emissions of carbon dioxide and other greenhouse gases not controlled by the Montreal Protocol.'[10] Developing countries which signed the treaty undertook to provide an inventory of greenhouse gas emissions and sinks and a national climate change response strategy. They undertook no specific commitments to reduce their greenhouse gas emissions.

Indeed, developing countries asserted their right to increase their greenhouse gas emissions. As the preamble to the treaty states:

> All countries, especially developing countries, need access to resources required to achieve sustainable social and economic development . . . In order for developing countries to progress towards that goal, their energy consumption will need to grow taking into account the possibilities for achieving greater energy efficiency and for controlling greenhouse gas emissions in general, including through the application of new technologies on terms which make such an application economically and socially beneficial.[11]

In effect, parties from developing countries made any future possible abatement commitments on their part contingent upon the developed world demonstrating its intent to make available atmospheric space for future growth in emissions from the poor countries; and upon financing and technology transfer that would enable the poor countries to fulfil more stringent commitments in the future. This refusal to commit to reductions enabled the developed world to leave vague the exact scope and scale of their commitments to provide funds and technology.

The Climate Change Convention addressed four key issues that determine who will pay for the costs of incremental abatement in developing countries. These are the provision of financial resources, technology transfer, an interim financial mechanism, and the definition of incremental abatement cost. The following sections briefly outlines the agreement reached on each of these concerns.

New and additional financial resources

In Article 4 of the Convention, developed country parties committed themselves to 'provide new and additional financial resources to meet the agreed full costs incurred by developing country Parties in complying with their obligations under Article 12.' They also pledged to 'provide such financial resources, including for the transfer of technology, needed by the developing country Parties to meet the agreed full incremental costs of implementing measures.' The flow of financial resources and technology, states the treaty, is to be accomplished at a level adequate to the task, and in a predictable fashion.[12]

The Convention also declares that developing countries can claim extra assistance if they are particularly vulnerable to the adverse effects of climate change and related adaptation costs. In addition to small island, land-locked and transit states, countries are also eligible for extra help if they have: low-lying coastal areas; arid and semi-arid areas, forested areas and areas liable to forest decay; areas prone to natural disasters, drought and desertification; high urban atmospheric pollution; fragile ecosystems, including mountainous ecosystems; or are highly dependent on income generated from the production, processing and export, and/or on consumption of fossil fuels and associated energy-intensive products.[c,13]

Technology transfer

The developed country parties also undertook to 'take all practical steps' to

c Although many landlocked and transit states contain vulnerable mountainous or arid ecosystems, many do not; thus, their inclusion is a concession to politics. A transit state (for example, India) is one through which a landlocked state (for example, Nepal) must move its imports/exports, thereby increasing a transit state's transport sector greenhouse gas emissions.

'promote, facilitate and finance, as appropriate, the transfer of, or access to, environmentally sound technologies and know-how' to developing country parties, as well as to support the development of endogenous capacities and technologies within developing countries.[14]

Not only did the developed countries recognize that the ability of developing countries to fulfil their commitments will depend on provision of financial resources and transfer of technology; but they also recognized that developing countries must develop at the same time. In effect, developed countries admitted the obvious: that emissions from developing countries must and will increase and that reductions in the rich countries must offset this inevitable increase in emissions by the poor.[15]

Financing mechanism

In Article 11, the signatories to the Convention declared that they would create a mechanism whereby grants or concessional financing would be achieved. Moreover, they decided that this mechanism would be accountable to the conference of parties to the Convention. Thus, although it was agreed that the World Bank's Global Environment Facility (GEF) would be entrusted with this responsibility as an interim measure, ultimate control over the mechanism does not rest with the Bank but with the parties to the treaty. As the treaty states that the financial mechanism shall have an 'equitable and balanced representation of all Parties within a transparent system of governance', the signatories effectively defined a reform agenda for the governance of the GEF. GEF is given four years to work effectively, at which time the parties reserved their right to review and redefine the financial mechanism used to implement the Convention. To reinforce this point, Article 14 declares that the GEF must be 'appropriately restructured and its membership made universal'.[16]

Full incremental cost

The treaty did not define what constitutes the 'full incremental cost' of abatement in developing countries, except to refer to the costs of reporting on national emission sources and sinks and in conducting research on climate change, and for costs incurred on such activities as are agreed between developing countries with the international financial mechanism created under the Convention.

The latter step, however, implies that the international community will adopt guidelines as to what are admissible costs. The mandate of the new Working Group III of the Intergovernmental Panel on Climate Change covers these technical and economic issues. Almost certainly, the restrictive criteria adopted by the Global Environment Facility will be widened as it would not support many sound greenhouse abatement measures under its current guidelines.[d,17]

In short, the signatories to the Convention did not define the specific levels of abatement nor the scale and content of the effort required to achieve the overarching goals of the treaty. While some found this disappointing, most analysts viewed it as inevitable that, as a 'framework,' the Convention would only outline a set of general principles and obligations in various areas. Subsequent negotiations are to produce specific targets and quantitative reductions which – if agreed to – will be added as protocols to the framework Convention.

Protocol negotiating difficulties

Protocols to the Convention that deal with carbon dioxide, methane and other greenhouse gases must address and resolve much more difficult and complex issues than the Vienna Convention that covers ozone depleting gases. Relative to ozone depleting gases, for example, these gases are far more integral to lifestyles. Take, for example, methane emitting rice paddies in Indonesia, carbon dioxide spewing automobiles in cities such as Melbourne, or slash and burn agriculture in the highlands of Papua New Guinea. Moreover, the number of producers and consumers of these greenhouse gases is far greater than was the case for ozone destroying gases which created a trading cartel devoted to eliminating its major product. In comparison, climate change presents many novel negotiating difficulties to the international community (see Table 1.1) which have not been overcome in the Convention.

First, free riding on a greenhouse gas reduction regime is likely and attractive at all levels of human society – international, regional, national, and local. Second, a successful agreement will be based on measures that are in national self-interest, are normatively self-policing or are economically self-regulating. Third, greenhouse polluters are separated in time (cross-generationally) and space (due to global mixing rates relative to mean residence time of greenhouse gases) so that liability is difficult to determine. Fifth, responsibility is clouded further by uneven regional climate impacts. Sixth, institutional change within states to implement greenhouse reductions will also be major compared with those entailed by past environmental agreements.

Finally, the economic costs of reducing greenhouse gases may be large, concentrated on existing interests at the national or subnational level, and

d Even to be considered, an abatement project must not only reduce greenhouse gas emissions, but among other criteria, also be replicable internationally; contain an incentive design; be unlikely to be funded without GEF funds; the subsidiary criteria assume that projects are defined around specific 'emission reduction technologies' rather than broad technological capabilities and incremental learning; and take no account of non-technological components of cost such as intersectoral adjustment costs, trade impacts, human resource development costs, and institutional change costs.

Table 1.1 *Greenhouse gas negotiating novelties*

1	The atmosphere is a true global commons precluding appropriation
2	The dispersed users of the atmosphere and the huge number of dispersed sources of GHGs mean that: • monitoring is difficult • free riding is easy • self-policing is based on self-interest
3	Liability for damages is difficult to allocate
4	There are big costs now, potent blocking coalitions versus uncertain benefits later, weak promoting coalitions
5	There are unconventional negotiating axes
6	Discounting of GHG damage is controversial
7	Prudence may delay validation of models
8	There is uncertainty about the benefits of GHG abatement due to frequent, rapid, and unforseeable changes in scientific assessment of climate change
9	Treatment of sinks
10	GHG equivalencies are controversial; GHGs are largely non-substitutes

GHG: greenhouse gases

may involve restrictions on existing resources rather than the allocation of new resources as in the Law of the Seas negotiations. These costs are given a great deal of attention in this book due to their importance in determining who should pay what to whom in a global greenhouse regime (see Chapters 5–13).

Admittedly, the economic benefits of curtailing greenhouse gases may be also large because damages from climate change may be immense. But the realization of the benefits of avoiding climate change is uncertain, will likely come later rather than sooner, and will be distributed diffusely. Moreover, the benefits of using current emissions are widespread; and stakeholders in the status quo are well organized and powerful.

As was evident in the negotiations leading up to the Climate Change Convention (see Chapter 14), the size and ranking of greenhouse gas polluters (depending on how emissions are measured) cut across virtually all prior axes of interstate negotiation on security, economic, or environmental grounds. Simple targets make little sense as the energy intensities of economies vary internationally by an order of magnitude. Other simple criteria such as population, per capita GDP, fuel mix, energy reserves, and industrial patterns greatly complicate emission reduction or energy efficiency targets.[18]

Determining the *net* emissions of greenhouse gases is also more difficult than for ozone depleting gases. Ozone depleting gases come from a relatively small number of human sources, and the gases remain in the atmosphere for hundreds of years before they decompose. In contrast, the major greenhouse gases have large natural sources and sinks, and have much shorter lifetimes in the atmosphere. States may claim that nationally controlled sinks for greenhouse gases should be subtracted from national emissions of greenhouse gases in determining emission quotas. Others may object strongly on grounds of scientific uncertainty (nearly a quarter of the carbon sink is currently unexplained by scientific models) or to the allocation of sink property rights. The concept of sink itself is a shifting sand on which to base target emissions and allocations (see Chapter 2).

The IPCC has already produced an index of heating equivalence across greenhouse gases and normalized to carbon dioxide, as was done in the Montreal Protocol across ozone depleting gases. However, many of the ozone depleting gases were close technical substitutes. It may be more difficult to apply the scientific equivalencies that might be used to evaluate control activities within an overall weighted emission quota for greenhouse gases than it was in the Montreal Protocol. Either a CO_2-only or a separated, gas-by-gas protocol is therefore more likely under the Convention than an integrated, multiple-gas protocol implied by the ozone precedent.

Faced with such vast uncertainty, many scientists suggest that a 'no regrets' policy should be implemented now by incurring short run costs of emission reduction in anticipation of uncertain, long-run benefits. Incontrovertible validation of scientific simulations of climate change may not be available until (if the models are right) massive climate change may be irreversible.[e,19] By reducing climate change, prudent behaviour now may deny positive evidence that the scientific models were correct. Relatedly, frequent, rapid, and unforeseeable changes may occur in scientific assessments of climate change, making negotiations on protocols to the Convention crisis-ridden and fraught with uncertainty.

In this book, the authors explore the implications of a 'no regrets' policy in which emissions-reduction measures are chosen along a 'least-cost' pathway. These measures consist of energy efficiency projects and other actions that will have many other benefits even if present climate change concerns should turn out to be unwarranted.[f,20] This policy option entails radical reductions in carbon emissions to about 50–60 per cent less than those in 1990.[21] This stringent reduction goal therefore poses an unambiguous and measurable challenge to today's decision makers that must be met if they are to fulfil their obligations to future generations.

Key issues for climate change negotiations

The major greenhouse gas emitters have not yet committed themselves to major emission reductions. Nor have the international donors backed up their words with money and action. The authors of this book explore a way that the

e Predictions of the rate of global climate change may improve about 1995; of regional climate differences about 2005; and predictive capabilities of climate change variation about 2010.

f A 'no regrets' policy refers to the implementation of those abatement measures that are justified on other, non-greenhouse, grounds (such as energy efficiency), and contrasts with 'wait and see' (do nothing except more research), and 'buying insurance' (that is, undertake preventive and adaptive measures now even if these cannot be justified on current cost-benefit grounds).

greenhouse management contract between rich and poor states can be constructed in an open, efficient, and equitable manner.

These issues are not solely political and economic, however. They are also technical and scientific in nature, relating as they do to complex and poorly understood issues of climate and ecology.[g] In Part I, four authors explain these issues and produce a technical and scientific foundation for determining who is responsible for climate change.

In Chapter 2, Kirk Smith explains the science that underlies comparisons of different greenhouse gas emissions from different nations and time periods, that is, the indices that can be used. Implicitly or explicitly, an index of some sort must be used so that choices among options can be made. He concludes that there are some hidden value judgements in choosing indices and that the choice of index depends strongly on the particular policy question being asked.

In Chapter 3, Susan Subak compares the results of applying five different indices to the question of relative national contribution to global greenhouse gas emissions. These are:

1 cumulative carbon dioxide emissions from fossil fuel combustion only;
2 cumulative carbon dioxide from fossil fuel and land use changes;
3 current annual carbon dioxide emissions;
4 an expanded list of current emissions including methane from landfills and fossil fuel production;
5 a comprehensive range of current greenhouse gas emissions from energy, land use change, and agricultural sources.

When allocated on a per capita basis, each measure produces a different distribution of national responsibility for past and present contributions to climate change. Subak shows how the scientific and technical dimensions of building indices of responsibility have significant political and economic implications for different polluters. In short, depending on which index is chosen for the protocols that allocate responsibility, the Convention will impose differential burdens on states of widely varying characteristics. The outcome will be widely varying incentives and disincentives for future action under the Convention. It is incumbent on policy makers, therefore, to pay careful attention to these technical issues.

Drawing on these technical and scientific foundations, Dilip Ahuja, Kirk Smith and Joel Swisher present a simple, transparent method in Chapter 4 to determine who should pay the cost of creating a global greenhouse regime. A composite indice is proposed that includes both ability to pay on the one hand, and historical contribution to climate change on the other. The former index confronts the issues of equity and economic realism that will affect participation rates of the poor. The latter index embodies the polluter pays principle and reflects the practical politics that the poor, small polluters are not likely to

g Indeed, some of the thorniest political and economic issues stem from the considerable remaining uncertainty about the science.

constrain their behaviour unless the wealthy, big polluters recognize that they have occupied the available 'ecological space' and must compensate latecomers for this pre-emption.

Smith, Swisher and Ahuja's approach provides a powerful philosophical and practical underpinning for discussions of the distribution of cost associated with managing climate change. If accepted, it would influence the outlook of key parties even if they are unable to accept specific numbers based upon it in actual negotiations. Such indices can be recomputed to investigate alternative yardsticks of ability to pay and historic responsibility for climate change. Whatever the final numbers, what is crucial to creating an effective global greenhouse regime is that it rests on these twin principles of equity and polluter pays responsibility.

In Part II, Peter Hayes confronts directly the deceptively simple question: who should pay? In Chapter 5, he introduces a method to calculate the likely costs to developing countries of complying with a global convention of climate change. He calculates the incremental cost of abating carbon dioxide emissions from the use of fossil fuels by the following procedure.

He begins by estimating projected emissions and required reductions of carbon dioxide that meet stringent IPCC emission targets which would restrain the growth of realized temperature and sea level to 0.1°C and 3 cm per decade respectively.[h,22] To achieve this goal, the global permitted emission in 2025 is about 2.7 gigatonnes of carbon as carbon dioxide.[i,23] This target is about 60 per cent of projected global emissions in that year, or about 50 per cent of emissions in 1990. Each country is required to reduce its emissions from projected 1995 levels so that it eventually reaches its fraction of this global permitted emission in 2025. This fraction is set to equal current national sink rights distributed to nations on the basis of current population and land area (although some other allocational criteria could and probably should be used to avoid problems associated with defining sinks, as noted in Chapter 4). High, medium and low marginal abatement cost curves are applied to these profiles of carbon abatement over time. In this way, the method generates a stream of annual incremental abatement costs for each country.

In Chapter 6, Hayes applies the quantitative allocational rules developed in Chapter 4 to the range of numerical estimates of the cost of carbon emission abatement and coastal protection from Chapter 5. He presents two rules that have been proposed to allocate the cost to various parties of meeting emission

h These twin criteria are based on scenario D of the IPCC's 1990 report to policy makers on the climate change threat.

i Based on IPCC estimates of a box diffusion carbon cycle model relating carbon dioxide emission rates to carbon dioxide atmospheric concentrations that provide estimates consistent with scenario D referred to in the previous footnote. Two 'low emission' cases examined by the IPCC resulted in atmospheric carbon reaching an (arbitrary) level of 150 per cent of its pre-industrial level, or stabilizing the concentration at current levels. The mid-point of these two cases in 2025 corresponds with the emissions level below that in scenario D in the same year. The subsequent analysis is therefore based on a stringent reduction requirement that is conservative with respect to the IPCC's scenario D.

targets imposed by a climate change agreement. These rules are 'obligation to pay,' based on each nation's historic emissions and ability to pay; and the UN scale of payments.

Under the UN scale, the OECD (North) pays about 77 per cent, the former Soviet Union and Eastern Europe (East) about 14 per cent, and the developing world (South) about 9 per cent of total UN cost. In the obligation to pay (OTP) index, the North's OTP is about 73 per cent; the East's about 20 per cent; and the South's about 7 per cent. Hayes argues that negotiations are likely to proceed therefore by countries making bids to vary their contribution relative to the UN scale until consensus is reached. The obligation to pay index provides a sound, transparent baseline against which to measure the fairness of departures from the UN scale.

In Chapter 6, Hayes treats the South's incremental cost, minus the South's obligation to pay, as the responsibility of the wealthy countries of the North and transfers it to the North's account. The annual transfer from North to South is estimated at $29–34 billion in the medium and high marginal abatement cost cases respectively. (There is no economic case based on incremental cost for transfer from the North to the South in the low cost case, although Hayes cites the need to finance front end costs of abatement measures and to increase scientific and technical capability in developing countries as reasons for providing funds in any case.) He concludes by examining how the substantial funds required might be collected and transferred by carbon taxes, or earned by the sale of tradeable permits or abatement services that would also push countries toward equalizing their marginal abatement costs at a global level.

An alternative approach would have been to construct a global marginal abatement cost curve from national marginal abatement cost curves; and to apportion reduction activity to each country up to the marginal cost that delivers the desired total emission each year. The reduction activity would be paid for according to the relative obligation to pay for each country (as updated periodically). On this basis, no country would be asked to reduce emissions more than the then-current marginal cost level, and no nation would remain unpenalized for failing to undertake reductions found to be cheaper than the current marginal cost criterion. While attractive in principle, it remains difficult to put this approach into computational practice due to the lack of meaningful global and national marginal abatement cost curves as well as difficulties associated with allocating and quantifying emission rights over time.

The small island states most vulnerable to the impacts of climate change have been among the most vocal proponents of a strong Convention. In Chapter 7, Michael Wilford enumerates a proposal emanating from the Alliance of Small Island States that an insurance fund be established to cover the costs of adaptation to sea-level rise. Precedents exist for this approach in the oil and nuclear industries, but neither approaches the scale or scope of a fund that would cover losses implied by the greenhouse effect. The moral influence of these small states whose very existence is at stake is evident in the

Convention article cited above which declares that vulnerable states are eligible for extra assistance.

A major difficulty that hampers calculations of the required funding by the rich countries of the incremental costs of developing countries is our ignorance as to the shape of the latters' emission abatement curves at various levels of required reduction. Part III contains the work of eleven authors who are deeply immersed in the empirical calculation of incremental costs associated with climate change and greenhouse gas abatement. These authors present abatement cost curves at the national or regional level in Asia, Africa, Australia, and Eastern Europe/Russia. All conclude that significant cost savings will likely accrue at the outset of carbon reduction programmes, although the absolute cost levels vary widely.

In India, (Chapter 8), Jayant Sathaye and Amulya Reddy show that while emissions are likely to grow, there are substantial opportunities to abate emissions or to fix additional carbon at low costs or a net saving. They demonstrate that India could largely offset its carbon emissions by fixing carbon in forestry reservoirs, thereby emphasizing the importance of the sink issue in determining responsibility for cost. However, they also identify public and private institutional and informational obstacles to the realization of this potential. They argue that a basic needs economic strategy will itself enhance development and reduce emissions. They conclude optimistically that for the first time ever, fundamental interests of the rich and poor countries are aligned in the climate change area – provided that these multiple barriers to abatement can be overcome.

In West Africa (Chapter 9), Ogunlade Davidson notes that although Africa contributes only a small fraction (about 3 per cent) of total global carbon emissions, its energy usage and greenhouse gas emissions will grow substantially. He estimates that emissions can be reduced in this region by between 13 and 36 per cent (depending on the country) by the year 2025, simply by introducing economically justified carbon conservation measures. As in other developing countries, he finds that a significant number of financial, institutional and technical obstacles exist which block the region from implementing these abatement options.

Because the generally poor management of energy and related institutions also hinders the effective implementation of these measures, he concludes that institutional reform is an essential ingredient of a carbon abatement strategy. He also finds that lack of investment finance is a major obstacle to energy development in the region. Weak capital markets and heavy indebtness require major economic reforms. These steps alone, however, will be inadequate to the task unless supplemented by external financing.

In Brazil (Chapter 10), Jose Moreira and Alan Poole present an aggregate cost curve for abating carbon emissions from Brazil's fossil fuel and biomass energy use which incorporates eighteen categories of abatement technology. Brazil is unusual in that a large fraction of its electricity is generated by hydropower and a large amount of alcohol from biomass is used in the transport sector. Moreover, the authors did not include steps related to re- or de-forestation in Amazonia in determining the potential for Brazil to abate or to offset its energy-related emissions. Nonetheless, they identified abatement potential that amounts to about 16 per cent of projected energy-related emissions, much of which can be obtained at negative cost (that is, at a savings).

In Thailand (Chapter 11), Peter du Pont, Somthawin Patanavanich, Mark Cherniack, and Michael Philips demonstrate that the most rapidly growing source of carbon emissions, the electric power sector, can be curbed significantly at a low or negative cost. They project that Thailand's carbon dioxide emissions from fuel combustion will double over the next decade, from 24 to nearly 50 million tonnes annually. They estimate that an aggressive demand side management effort in the power sector could reduce emissions by 2.5 million tonnes annually by the year 2001 at an average cost of conserved carbon of about US$190/tonne. While still nascent, Thailand's electric utility has an aggressive programme to tap this potential saving and may offer a good model for other countries to emulate.

In Central and Eastern Europe (Chapter 12), Stanislav Kolar produces aggregate cost curves based on detailed local research to demonstrate that carbon abatement can be achieved with major economic savings until high levels of carbon abatement are reached. Kolar points out that all the countries of the former Soviet Union are grossly energy inefficient. Equally, they also offer massive and relatively cheap carbon abatement. He concludes that energy efficiency is their most effective means of reducing carbon dioxide emissions and can achieve the twin goals of economic development and environmental protection. He concludes that these states have considerable flexibility as to which combination of price reforms and regulations would serve best to realize this potential.

Finally, in Australia (Chapter 13), Hugh Saddler examines estimates of the cost and scope of emission abatement measures and reviews estimates of the impact on the Australian economy of achieving various levels of abatement. Relative to a business-as-usual scenario, he reports that about 20 per cent of projected emissions can be abated with economically justified carbon abatement steps. He also notes that the local manufacturing-versus-import content of equipment needed to implement this strategy is a crucial determinant of the macroeconomic impact of carbon abatement – a variable that many developing countries may do well to examine carefully.

Generally, these studies indicate that abatement is possible at negative cost or a savings at the outset of the abatement strategy, but that costs will become positive fairly quickly. The studies also point to the obvious and urgent need for demonstration carbon abatement programmes and additional research into costs to obtain much better cost estimates necessary for the formulation of sound policy.

Politics will not end once robust analysis and widely accepted estimates of the cost are available. Part IV returns to these *realpolitik* considerations which will intrude into future negotiations over protocols to the Convention. For example, big wealthy countries anticipate a new wave of technological innovation associated with greenhouse gas abatement. As the major donors, they will seek to tie resource transfers to exports of their own equipment and services. For their part, recipient states will seek minimal ties on these funds. Aid flows justified on the grounds of greenhouse abatement will be no more or less susceptible to mismanagement, waste and corruption than existing development assistance.

In Chapter 14, Peter Hayes outlines some of the practical political issues that will arise in negotiations over resource transfers from the North to the South on the scale justified by the earlier chapters. Simply moving money across the North–South divide may only worsen existing development difficulties by creating an ongoing technological dependency. Studies of energy efficiency and related carbon abatement show that a wide range of scientific, technological and managerial capabilities must be created in developing countries to achieve effective carbon conservation. Financial shortages are a critical obstacle to the emergence of an endogenous technological capacity needed to reduce greenhouse gases – but they are only one of a range of issues that must be resolved before such programmes can be realized.

In Chapter 14, Hayes also analyses the potential for regional greenhouse initiatives in Pacific Asia as a precursor to a global climate change convention. He concludes that demonstration abatement projects in developing countries of Asia and the Pacific are needed urgently to demonstrate the viability of schemes such as tradeable permits and trade in abatement services. He emphasizes that 'first in, first served' will dominate the emerging markets for greenhouse abatement markets and related technological competitiveness.

In summary, the authors of this book believe that signing the Climate Change Convention was only the first step on a long path to creating a greenhouse gas regime. It remains to be seen whether the parties to the Convention can muster the domestic political will needed to meet the commitments contained or implicit in the Convention.

In this respect, the 'review and pledge' procedure implied by Articles 4 and 10 of the Convention is particularly important. In Article 4, countries undertook to prepare and to communicate national greenhouse gas inventories using common methodologies and to implement mitigation measures on sources and sinks of emissions. In Article 10, they created an international body that will 'assess the overall aggregated effect of the steps taken by the Parties in the light of the latest scientific assessments concerning climate change.' Implementing these two commitments will create an iterative dynamic that will lead to stronger action under the Convention in the future.

Action, however, will require that resources be allocated to match the rhetoric of the Convention. The parties to the Convention must adopt a transparent method of calculating obligations and cost if the commitments on funding and technology transfer are to be fulfilled. The studies reported in this book illustrate the complex and difficult issues that must be addressed in such a method. At this stage, however, it is not the specific method nor its results that are important. Rather, what is vital is that negotiators of protocols to the Convention develop parallel ways of thinking that facilitate communication and agreement on these concerns. We hope that this book contributes to this task.

References

1 UN Conference on Environment and Development, *Convention on Climate Change, Final Text,* Rio de Janeiro, Brazil, June 3, 1992; from Department of Public Information, Room S-845, UN, New York, New York, 10017, USA, October 1992
2 See D Lashof and D Tirpak (eds), *Policy Options for Stabilising Global Climate;* report to Congress, Environmental Protection Agency, Washington DC (Office of Policy, Planning and Evaluation), February 1989
3 A Kristin Sydnes, 'Global Climate Negotiations, Another Twenty Years of Fruitless North-South Bargaining', *International Challenges,* volume 11, no 1, 1991, pp 58–66
4 J T Houghton *et al, Climate Change, The IPCC Scientific Assessment,* Cambridge University Press, New York, 1990
5 *Ibid,* p xi
6 *Ibid,* p xii
7 G Tegart *et al, Climate Change, The IPCC Impact Assessment,* Australian Government Publishing Service, New York, 1990
8 Intergovernmental Panel on Climate Change, *Climate Change, IPCC Response Strategies,* Island Press, Washington DC, 1991, p xxvi
9 Intergovernmental Negotiating Committee for a Framework Convention on Climate Change, *Climate Change Convention, op cit*
10 Article 12, *ibid*
11 Preamble, *ibid*
12 *Ibid*
13 *Ibid*
14 *Ibid*
15 *Ibid*
16 *Ibid*
17 Scientific and Technical Advisory Panel, *Criteria for Eligibility and Priorities for Selection of Global Environment Facility Projects,* World Bank/Global Environment Facility, May 1992, pp 2–6
18 M Grubb, 'The Greenhouse Effect: Negotiating Targets,' *International Affairs,* volume 66, no 1, 1990, p 71
19 G McBean and J McCarthy, 'Narrowing the Uncertainties: A Scientific Action Plan for Improved Prediction of Global Climate Change', in J Houghton *et al, Climate Change, op cit,* p 328
20 See R Pachauri and M Damodaran, '"Wait and See" versus "No Regrets": Comparing the Costs of Economic Strategies', in I Mintzer (ed) *Confronting Climate Change, Risks, Implications and Responses,* Cambridge University Press, 1992, p 238
21 J T Houghton *et al, Climate Change, op cit,* p xxxiv
22 IPCC Working Group I, 'Policymakers Summary', in *ibid,* pp xxii, xxxi, xxxiv
23 R Watson *et al,* 'Greenhouse Gases and Aerosols', in J T Houghton *et al, Climate Change, op cit,* Figure 1.8, p 15

2

The basics of greenhouse gas indices

Kirk R Smith

Deciding which greenhouse-gas emissions reduction or absorption projects to fund and which countries should contribute to the cost implies the use of indices to weigh the comparative net greenhouse gas (GG) implications of potential projects and the net emissions of nations. These indices should be composed of individual indicators that are deemed to be relevant according to the criteria of scientific validity, economic efficiency, political equity, ease of use, and flexibility.[a] The application of the appropriate index should not only rank but, preferably, also give a quantitative indication of how much better one project is over another or how much more one country should contribute than another.

Most of the indices that are used to determine accountability contain the structure, 'net greenhouse gases emitted per unit', where the unit is nation, population, income, energy use, etc., depending on the intended application. There are several important considerations and implications in choosing these various index denominators as will be discussed in Chapters 3 and 4.

Nearly all the indices also require the careful choice of appropriate numerator, the method by which the different greenhouse gases are weighted so that they can be compared or aggregated.

Apples and oranges

As late as the mid-1980s, policy discussions of global warming induced by greenhouse gases focused almost entirely on carbon dioxide (CO_2), with relatively little discussion of the other greenhouse gases.[b] Since then, however, it has become well recognized that the others play important roles. Indeed, recent conventional wisdom is that, in total, these other gases

a 'An index is a single number derived from two or more indicators. In computing an index, the first step is often to compute the individual indicators' (Ott, *Environmental Indices*, 1978:8).

b Only 7 of 496 pages, for example, in a major National Academy of Sciences review in 1983 (USNAS, 1983).

together account for an amount of warming comparable to that due to CO_2, and, consequently, greatly shorten the time until an effective doubling of atmospheric CO_2 content (CO_2 equivalent) occurs (from 2075 to 2030, for example, in WRI 1990). Figure 2.1 duplicates one of the most commonly reproduced illustrations of the current relative contributions of the different gases.

Understanding the relative contribution of the different gases is vital to developing an appropriate index for comparing and ranking greenhouse-gas mixtures. This is because the different gases are produced in different relative amounts by different activities, which in turn are undertaken in different degrees by different countries. These differences can be seen in Figures 2.2 and 2.3 which are commonly reproduced illustrations of the relative importance of different activities and countries implied by the greenhouse-gas weightings of Figure 2.1.

Agreeing on the relative importance of the gases is a crucial first step in determining national accountabilities and the relative value of different greenhouse-gas reduction projects. Carbon dioxide, for example, is released

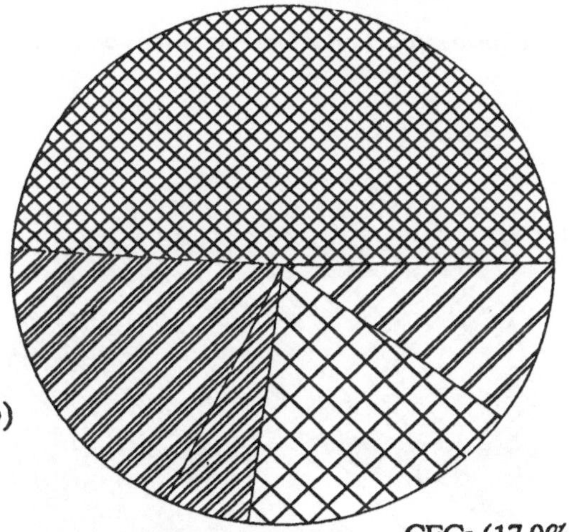

One commonly cited estimate of the relative contributions of different greenhouse gases to global warming during the 1980s (data originally from Hansen et al. (1988), but widely spread as a pie chart in the USEPA draft report to Congress (Lashof and Tirpak 1989) and popularized in such places as the *New York Times* (1989)). The USEPA final report to Congress (Lashof and Tirpak, 1990: ii, 36–39) contained a brief discussion of the need to consider indices and of the assumptions inherent in choosing one over another.

Figure 2.1 *Relative contributions of different greenhouse gases to global warming during the 1980s*

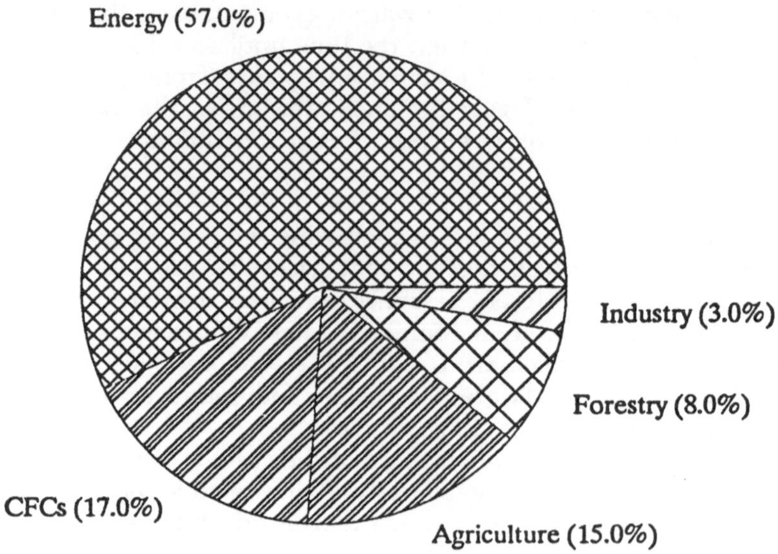

Based on Figure 2.1 and inventories of sources and emission factors (Lashof and Tirpak 1990).

Figure 2.2 *Relative impact of different activities on global warming*

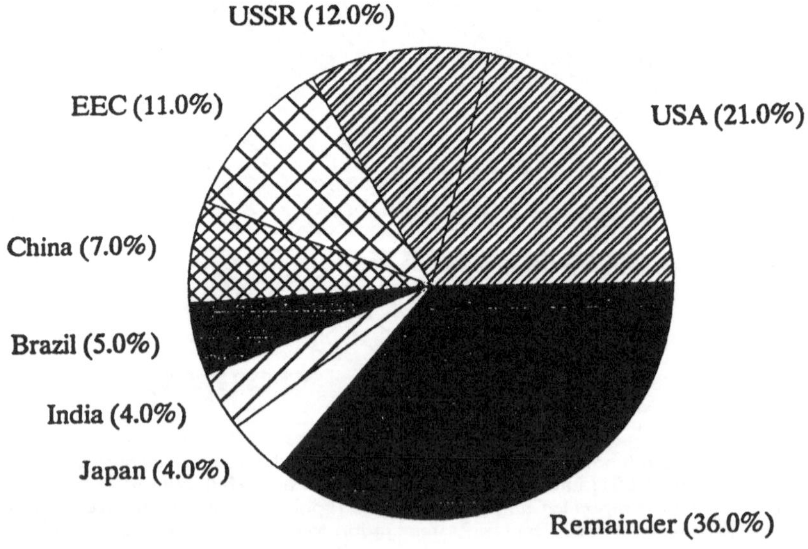

Based on Figures 2.1 and 2.2, this represents the relative contribution of different countries to global warming during the 1980s (Lashof and Tirpak, 1990).

Figure 2.3 *Relative contribution of different countries to global warming during the 1980s* mainly by fossil fuel use and land-use changes, while methane is released by livestock, wetland agriculture, and landfills. An index that weighted methane more heavily, therefore, would tend to make more attractive those projects addressing emission reductions within these agricultural activities and, conversely, make more accountable those countries, such as many developing countries, that are more engaged in such activities. An index weighting CO_2 more heavily, on the other hand, would tend to focus attention on fossil fuel combustion, the bulk of which occurs in developed countries.

Much of the discussion of policy alternatives has been based on indices of greenhouse gases that have not been carefully thought through (Smith and Ahuja 1990). The resulting confusion has led to a number of proposed indexing systems by which the different gases can be aggregated so that there is consistency both with physical reality and the needs of policy (for example, Krause et al. 1989; Fujii 1990; Ellington and Meo 1990; Lashof and Ahuja 1990; Agarwal and Narain 1991; Hammond et al. 1990; Rodhe, 1990; Shine et al. 1990; Smith et al. 1991; Gurney 1991; Grubler and Fujii 1991). The complexity of the problem has led to considerable controversy, not only about particular indexing alternatives (*Environment* 1991; McCully 1991; Mitchell 1992; Pachauri et al. 1992; WRI 1992) but even about the need to index at all (Victor 1990).

We argue here, however, that it is not possible to avoid indexing. To ignore all greenhouse gases except CO_2, for example, is to implicitly give them a weight of zero. Alternatively, to aggregate them by weight or number of molecules creates implicit indices with little physical meaning or policy relevance. It is far better to meet the problem head on, to choose indices that address the issues of concern in a way that reflects both physical and policy realities. Indeed, it is such indices that act as the interface between the science and the policy, and thus those chosen must be sufficiently robust to survive the inevitable complaints that will occur on both sides that they are not ideal.

Radiative forcing

The release of a GG results in increased warming because of what is called the 'radiative forcing' of the gas molecules in the atmosphere. To quite varying degrees, the different GG molecules act to make the atmosphere retain additional amounts of solar energy reradiated from Earth's surface, thereby leading to warming. Relative to CO_2, for example, a methane molecule in the atmosphere has a radiative forcing about 21 times higher, and a molecule of CFC-12 has nearly 16,000 times higher forcing. By weight, they are, respectively, 58 and 5700 times more effective. Clearly, with such widely ranging radiative forcings, neither the total weight nor number of molecules in a GG mixture is a good index of relative importance.

The relative radiative forcings, as shown in the second column of Table 2.1

Table 2.1 *Parameters for important greenhouse gases*

Trace gases	Radiative forcing relative to CO_2	Estimated atmospheric residence times (years)	*Global warming potential* — *Direct effects* Integration time horizon, years			*Direct + indirect efforts*[c] Integration time horizon, years		
			20	100	500	20	100	500
1 CO_2	1 (1)	(120)[b]	1 (1)	1 (1)	1 (1)	1 (1)	1 (1)	1 (1)
2 CH_4	58 (21)	10.5	35 (13)	11 (4)	4 (1.5)	60 (22)	21 (7.5)	9 (3.2)
3 CFC-11	4000 (12000)	55	4500 (13500)	3400 (10200)	1400 (4200)	-	-	-
4 CFC-12	5700 (16000)	116	7100 (20000)	7100 (20000)	4100 (11500)	-	-	-
5 N_2O	210 (210)	132	260 (260)	270 (270)	170 (170)	-	-	-
			Conversion to CO_2					
6 CO	weak	<1	1.6 (1)	1.6 (1)	1.6 (1)	7 (4.5)	3 (1.9)	2 (1.3)
7 NO_x	weak	<1	-	-	-	150 (130)	40 (35)	14 (12)
8 NMHC[a]	-	<1	2.6 (1)	2.6 (1)	2.6 (1)	31 (12)	11 (4.1)	6 (2.3)

Numbers refer to ratios by weight: numbers in brackets refer to ratios by molecule (or carbon atom for NMHC).

a NMHC = Non-methane hydrocarbons, assumed to have a mean molecular weight of 17 per carbon atom.

b The duration of an increase in CO_2 in the atmosphere is described only approximately by a single exponential decay (l/e) time of 120 years. The integrated value (infinite integration time) results in a period equivalent to about 320 years. (See Siegenthaler 1983.)

c Indirect effects refer to the impact of the gas on atmospheric chemistry, particularly with regard to the concentrations of ozone and water vapour, two powerful greenhouse gases. The original IPCC report (1990) listed indirect effects for all gases shown here. These are shown for CO, NO_x, and NMHC (nos. 6–8). Those for CFCs and N_2O (nos. 3–5), however, are not shown because of new evidence indicating possible indirect cooling as well as warming effects (IPCC 1992). It should be noted, however, that the 1992 IPCC Supplement considers all indirect effects to be so uncertain as to be unusable for policy purposes at present. For CO, NO_x, and NMHC, this uncertainty is partly because there may be large variations in GWP depending on the local conditions where emissions occur. Recent recalculations of methane's indirect effects are shown here for the situation in which moderate emissions controls are implemented in the next decade (Lelieveld and Crutzen 1992).

The radioactivity analogy

To explain the need to choose appropriate indices for comparing different mixtures of greenhouse gases, consider an analogous situation with an entirely different kind of hazard, radioactive waste.[a]

Imagine two containers of nuclear waste containing different mixtures of radioactive substances. Just as with GGs, the total weight of waste would not be a good index of relative hazard, since some substances are thousands of times more radioactive than others by weight. The amount of radioactive disintegrations per second measured in curies or becquerels is, in radiation, the rough analog to radiative forcing in GGs, useful as an index of the immediate relative hazard of radioactive mixtures.

To compare the total or long-term hazard of such nuclear waste mixtures also requires knowledge about the half-lives of the different radioactive substances and whether they change into other radioactive substances (daughter products) as they decay. One mixture could have a high initial radioactivity but be composed mainly of iodine-131, which has a half-life of 8 days. The other, however, might contain significant amounts of cesium-137 and plutonium-238, which decay with half-lives of 30 and 24,000 years, respectively. Clearly the total hazard would not be represented well by immediate radioactivity alone, which ignores that the iodine -131 would be essentially gone in a few months while the others would be nearly unchanged.

In some cases, radioactive material of one kind decays into radioactive material of another, uranium to radium, for example. Failure to take these 'daughter products' into account can lead to misrepresentations of the actual hazard represented by a mixture of radioactive substances.

Thus, both lifetime and physical transformation into other hazardous materials are taken into account when, for example, the long-term hazard represented by nuclear waste is calculated. So, too, must both residence time in the atmosphere and chemical transformation into other important materials be considered when comparing the global warming potential of different mixtures of greenhouse gases.

This analogy should not be taken too far, however. Unlike the half-lives of radioactive materials, the atmospheric residence times of greenhouse gases are not fixed, but are affected by many factors in the atmosphere, including concentrations of other gases. For example, carbon monoxide is not a greenhouse gas itself, but is thought to affect the amount of ozone and methane in the atmosphere, both powerful greenhouse gases. This makes the calculation of the relative warming produced by different mixtures of greenhouse gases much more complicated and uncertain than is the determination of relative radioactivity in different mixtures of nuclear waste.

a Please note that 'radiative forcing' and 'radioactivity' are entirely separate processes, even though the words are similar. Greenhouse gases are not 'radioactive'. The two terms should not be confused with one another.

for the most important GGs, are commonly used to index GGs. Used this way, they portray the relative radiative forcings of the GG at any one moment. If all the GGs acted the same in the atmosphere, including having the same lifetime, then the relative radiative forcings would also be appropriate measures of their relative total impacts. This, however, is not the case, because of two other factors: atmospheric residence times and chemical interactions (see box).

Atmospheric residence times

Some GGs are removed from the atmosphere in a few years, while others remain for hundreds of years. Thus, to compare different GG mixtures, for example those emitted from different countries, it is important to consider the relative atmospheric lifetimes. These are listed in the third column of Table 2.1.

Figure 2.4 shows that as a result of having different lifetimes, the relative importance of different GGs can change dramatically over time. Here, the curves start from the relative radiative forcings of current global anthropo-

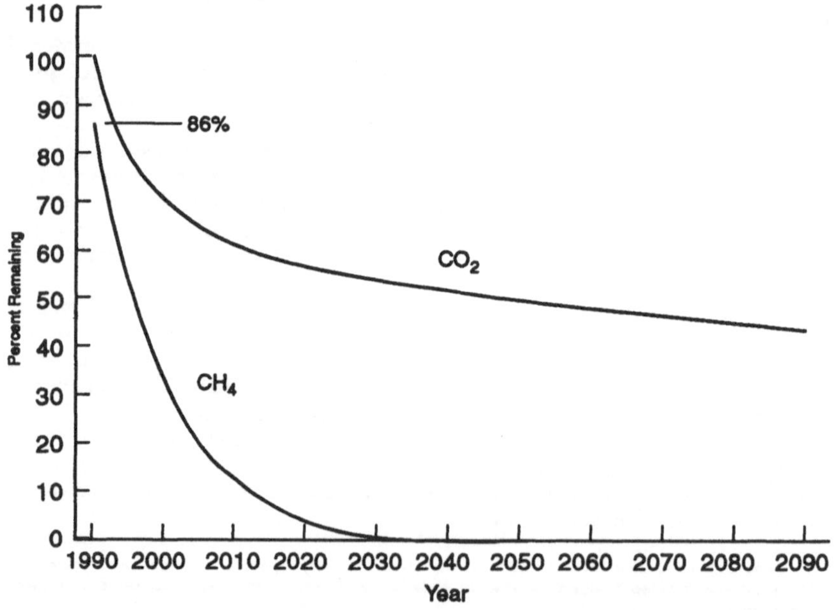

Starting from the instantaneous radiative forcing from their global emissions in 1990, this figure traces the change in forcing over time of the two gases (data from Subak et al., 1992). Note how much faster methane is removed, as indicated by the relative lifetimes in Table 2.1. See Table 2.2 for quantitative comparisons of the two removal curves.

Figure 2.4 *Relative rates of removal from the atmosphere for CO_2 and methane* genic CO_2 and methane emissions, shown on the vertical axis where t = 0. This point corresponds to an instantaneous index, one not considering lifetimes, where total methane emissions are about 86 per cent as important as CO_2. This index is based solely on radiative forcing at the moment of release. The downward sloping curves after this point represent the change in relative radiative forcing that occurs as the GGs are removed from the atmosphere over time. Because methane has such a short lifetime compared to CO_2, its relative importance decreases quickly. After 20 years, the methane released at the beginning of the period has become less than 11 per cent as important as the CO_2. After one hundred years, less than one in ten thousand of the original methane molecules are left, while more than two-fifths of the CO_2 remains. From most perspectives, therefore, the relative direct impact of current atmospheric methane concentrations is actually much less than the 86 per cent indicated by an instantaneous index.

It is not exactly true to say that an instantaneous index is wrong, for it does represent a systematic means to compare mixtures, albeit with a peculiar time horizon. Such an index is not suitable for addressing most

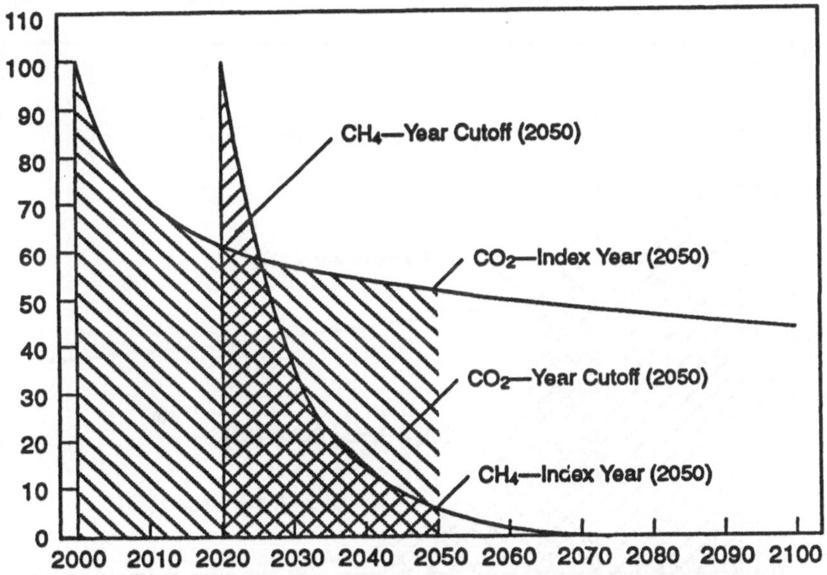

This and the next figure illustrate how the relative warming of two different gases released at two different times can be compared. Here, a sufficient amount of each is assumed to have been released so that their initial radiative forcing is equal. As in Figure 2.4, however, the two gases are removed from the atmosphere at quite different rates. The two methods of comparison are: Index Year, where the impact is taken as the amount of warming left at a certain year, here 2050; and Year Cutoff, in which the area under the curve (integrated warming) up to a particular year is used as the index.

Figure 2.5 *The index year and year cutoff methods of comparing the contributions of CO_2 and methane*

policy questions, however. Societal concern about GGs clearly extends beyond the immediate year that emissions occurred, and an index should be chosen accordingly. The time horizon could take various forms:

- Index years. As shown in Figure 2.5 present GG mixtures could be weighted on the basis of their relative impact on radiative forcing in a particular future year, say 2050.
- Index periods. Instead of a particular year, a particular period could be chosen, say the relative forcing in the thirtieth year after release (Figure 2.6).
- Year cutoffs. Mixtures could be weighted according to the total radiative forcing from the date of emissions through to the year of cutoff, say 2050. These weightings are represented by the areas under the curves to the left of the chosen year in Figure 2.5.
- Period cutoffs. Instead of a particular year, a particular period could be chosen, say the integrated radiative forcing for 30 years beyond the year of emissions (Figure 2.6).
- Discounting. Radiative forcings in future years could be weighted according to how distant they are from the present, as is standard in

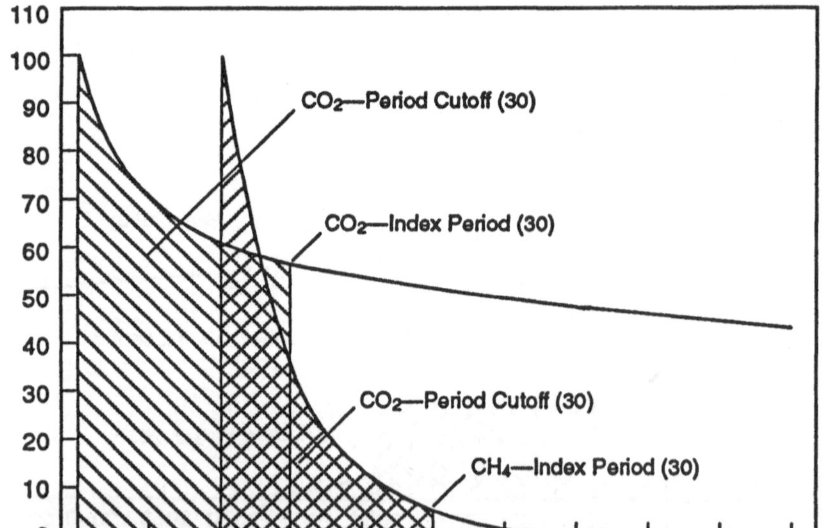

This figure illustrates two additional methods: Index Period, when the impact is taken as the remaining warming at the end of a particular number of years after the emissions, here after 30 years; and Period Cutoff is where the area under the curve up to a particular number of years after the emission is taken as the index. The last method has been adopted by the IPCC (1990) and by this book.

Figure 2.6 *The index period and period cutoff methods of comparing the contributions of CO_2 and methane*

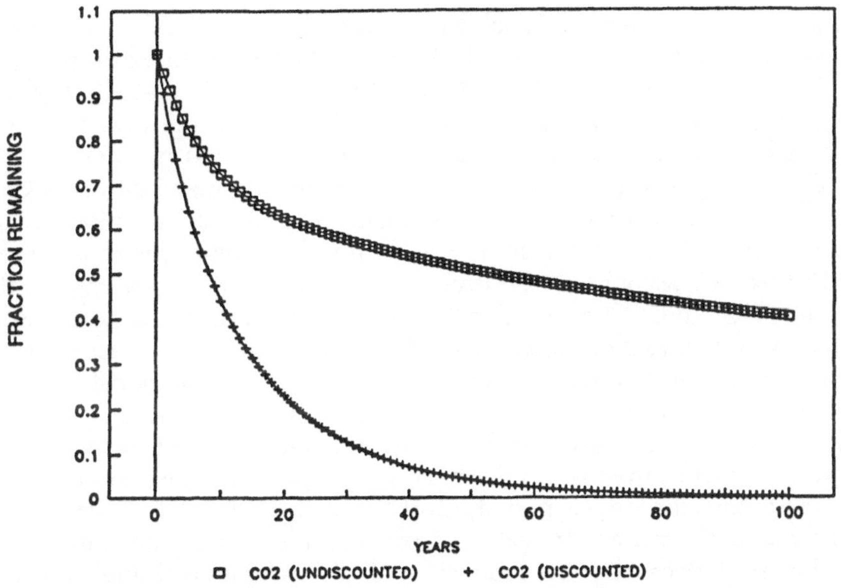

This shows the relative area under the discounted and undiscounted atmospheric decay curves for CO_2. At a 5 per cent discount rate, the ratios of the two areas are 1.5 at 20 years, 3.8 at 100 years, and 11.4 at 500 years. The ratio with integration to infinity is about 23.

Figure 2.7 *The effect of discounting on CO_2 decay*

economics for comparing monetary flows over time. As shown in Figure 2.7, at a 5 per cent discount rate, radiative forcings next year would only count about 95 per cent as heavily as those this year. The radiative forcing due to this year's emissions in the twentieth year would count less than 40 per cent as much.[c] The discount factor is applied *in addition* to the decay factor due to GG removal from the atmosphere over time.[d]

Direct effects

We focus on period cutoffs in this book, which is consistent with the reports

c A 5 per cent discount rate for CO_2 is roughly equivalent to an 18.6-year time horizon (period cutoff). Both approaches lead to the same relative total weight for the global warming produced by an equal amount of emissions, although with quite different temporal patterns.

d Although discount rates are usually taken to be positive, that is, impacts in the future are weighted less than those near the present, there are occasions when society seems to apply negative discount rates, that is, happenings in the future are weighted more heavily than those in the present. Decades of suppressing consumer demand by the Soviet Union in order to keep total investment high and thus build future industrial power is one example. More recently, societal expenditure on nuclear waste disposal may be another.

of the Intergovernmental Panel on Climate Change (IPCC 1990, 1992). Although not without problems, the best available way to make numerators for GG indices is to weight the different GGs by their respective global warming potentials (GWP), which is the estimated ratio of total warming produced by each gas over a particular period compared to an equal amount of CO_2 released at the same time. This allows the impacts of different GGs to be aggregated or compared in units of CO_2 equivalents. The middle set of columns in Table 2.1 show direct GWPs for three time horizons. Direct GWPs take into account the relative lifetimes and radiative forcings of the different GGs (IPCC 1992). Table 2.2, which applies to the same emission pattern as in Figure 2.4, shows that although instantaneously representing 86 per cent as much warming as CO_2, over 20 years, 1990 global methane emissions represent just 52 per cent as much warming as the CO_2 released and, over 100 and 500 years, only 16 per cent and 6 per cent, respectively.[e]

Except for CO_2, the major GGs have atmospheric lifetimes less than 150 years and, thus, the longest period cutoff shown in Table 2.1, 500 years, gives nearly the same warming for each as would complete integration (an infinite time horizon). Because some proportion of CO_2 releases is thought to have a rather long residence time (at least 800 years), however, a 500-year time horizon accounts for something like four-fifths or less of what would be indicated by a complete integration (Seigenthaler 1983; Lashof and Ahuja

Table 2.2 *World anthropogenic emissions of CO_2 and methane and global warming potentials*

Year emitted	*Carbon (kT)*		*Percentages*		*Per cent of CO_2 GWP*	
	CO_2	CH_4	CO_2	CH_4	Direct	Total
1990	6430	264	100	100	86	86
			Remaining in atmosphere			
2010	3980	39	62	15	52	89
2090	2780	0.018	43	0.07	16	31
2490	2180	5.5×10^{-19}	34	2.1×10^{-19}	6	14

Emissions data from Subak et al. 1992

GWP = global warming potential
kT = 1000 tonnes

e Note that the relative GWP over the period is much larger than the relative amounts of each gas at the end of the period, for example, 16 compared to 0.018 per cent at 100 years. This is because the GWP considers the warming over the entire period, not just during the last year.

1990). Thus, the GWPs, which are ratios of the warming of each gas compared to CO_2, would continue to fall slowly, at longer time horizons than the 500 years shown.[f]

Indirect effects

In addition to radiative forcing and atmospheric lifetime, which are used to determine direct effects of each GG compared to CO_2, indirect effects through chemical reactions create additional complications in determining GWPs and constructing accurate indexing schemes. Several types of chemical interactions are important:

- Some GGs and non-GGs change into other GGs, as methane (a GG) and carbon monoxide (essentially a non-GG) can eventually change into CO_2.
- Some non-GGs act to increase the atmospheric lifetime of GGs, as carbon monoxide does for methane, giving them GG equivalence even though they are not GGs themselves.
- Some GGs, as well as non-GGs, affect the creation of ozone and water vapor – important natural GGs.

This book is not the place to discuss these interactions in detail, partly because knowledge is still rapidly developing (IPCC 1992). Estimates of important indirect effects, which take account of various chemical interactions, are shown in the last set of columns in Table 2.1 for illustration. With the possible exception of methane (Lelieveld and Crutzen 1992), the magnitude of these indirect effects should be treated as extremely tentative.

The indirect effects are known with much less confidence than lifetimes, which, in turn, are less well understood than radiative forcing. In addition, local conditions have the largest influence on indirect effects, intermediate on lifetimes, and least on radiative forcing. Both indirect effects and radiative forcing depend on actual concentrations reached by the various gases over time; in other words, the GWPs change over time (Penner et al. 1989). Consequently, as more research is done, the magnitude of changes in estimates of these parameters can be expected to be roughly according to these relative uncertainties.[g]

The impact of adding indirect effects is quite important for methane, as indicated in the last three columns of Table 2.1. As listed in the last column of Table 2.2, including indirect effects actually increases the 20-year GWP slightly above the instantaneous level (89 compared to 86 per cent). Including

---

f These dynamics are due to the differential time periods over which CO_2 is captured first by the surface layers, then by the deeper parts of the ocean.

g Between IPCC (1990) and IPCC (1992), for example, it was determined that the indirect effects of the CFCs are probably negative rather than positive, that is, tend to produce cooling rather than warming.

indirect effects means that, at time horizons of 100 and 500 years, the 1990 methane emissions are 31 and 14 per cent, respectively, as important as 1990 CO_2 emissions.[h]

Which time horizon?

There is no unassailably objective way to choose the appropriate time horizon for GGs. It is not simply a technical question, but is related more to social and moral values and conventions as well as to the particular policy issue being addressed. Most people seem comfortable with a 100-year horizon. (For CO_2, this is roughly equivalent to a discount rate of 0.84 per cent.) This period can be justified on the basis that society clearly has concerns extending beyond a few years and yet is not likely to feel obligated to make large sacrifices to protect generations hundreds of years into the future, the capacities and needs of whom are so uncertain today. Intermediate time horizons thus seem most logical, but there is nothing sacred about the precise figure of 100 years.[i]

In addition, as pointed out by the IPCC (1990), each type of potential impact from global warming has its own timescale. General warming might be most appropriately indicated by a 500-year horizon, sea-level rise by 100, and rates of temperature change by 20, for illustration. Thus, gases that have different atmospheric residence times will tend to have a different pattern of influence on the impacts of concern. Too little is known at present, however, to apply such subtleties to policy-relevant indices.[j]

Past, current, and future emissions

The preceding discussion focused on greenhouse gas releases from a single year. Often, however, we will need to consider emissions over a number of years. Figure 2.8 shows how the individual contributions from single years combine in what look to be a pile of annual warming commitments. The

h We argue later that the three-quarters of methane and one-sixth of CO_2 released from biological processes and biomass burning (Subak et al. 1992) should not be included in indices designed to compare national accountabilities, although changes in these flows should be included when comparing alternative GG-control measures and judging each country's progress toward meeting its obligations. Taking out these flows reduces the relative GWPs of 1990 methane emissions in Table 2.2 (last 2 columns) by a factor of 3.4.

i For some comparisons, as explained in the next chapter, the choice of time horizon makes little difference.

j In addition to having effects over different time periods, the different greenhouse gases also exert their effects in different parts of the atmosphere (Wang et al. 1991). When understood more fully, these differences, too, may become part of the indices.

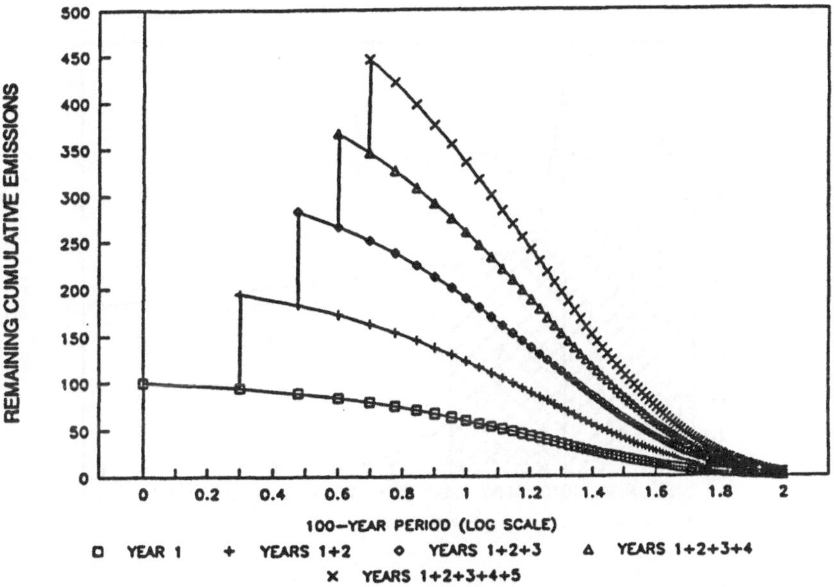

This illustrates the interaction of the incremental warming added year by year (for 5 years only in this example). The area under each year's warming increment represents the total contribution to global warming, here represented as one instantaneous release at the beginning of each year integrated out for 100 years. The warming in any particular year is the total of the remaining warming from all previous years (less than 100 years ago) plus the release that year. Note the logarithmic scale on the horizontal axis, which is needed to fit the curves onto one legible graph.

Figure 2.8 *Incremental warming from annual emissions*

warming commitment from each year's emissions extends to the 100-year cutoff (time horizon). At any year, the total radiative forcing is due to the total contribution remaining that year from each of the past 100 warming commitments. Between now and any point in the future, the total radiative impact is the area under all the warming commitments from now until that year.

In some cases, one may want to examine past cumulative emissions as part of an index. With a 100-year period time horizon, no present effect is attributed to emissions previous to 100 years ago. For emissions during the last 100 years, radiative forcing impact could be indexed in three ways:

1 that expressed to date;
2 that committed out to 100 years after the original emissions, but not yet expressed;
3 the total of (1) and (2).

These periods are illustrated in Figure 2.9. We do not have space to compare the use of these different indices and have chosen here to rely on option 3 in

Figure 2.9a

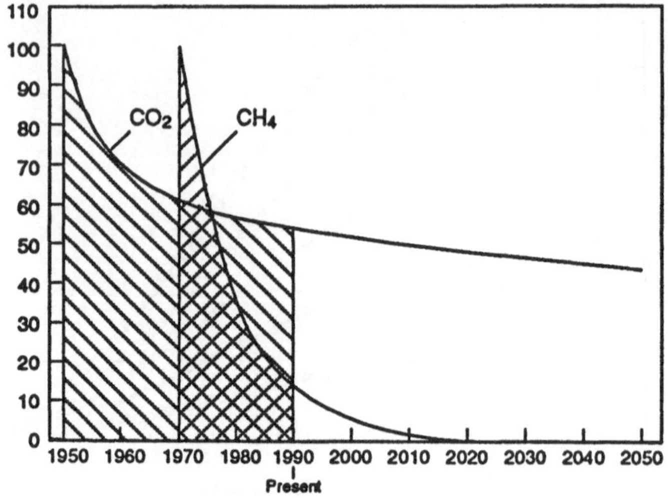

Figure 2.9b

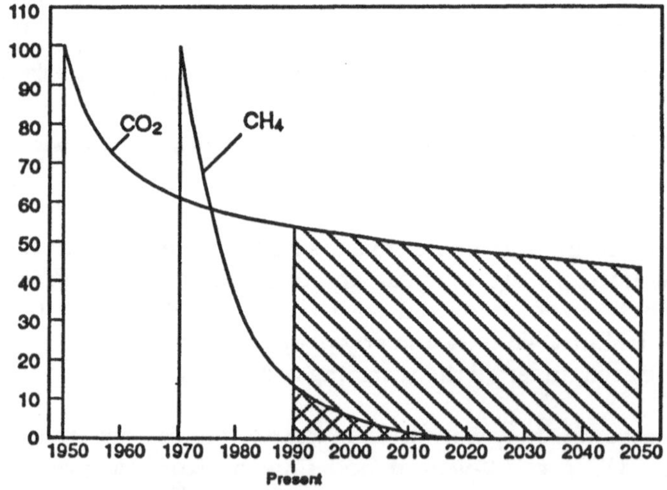

Methods of indexing past emissions using a 100-year time horizon. In 1990, emissions previous to 1890 are no longer considered to be affecting present decisions. Emissions during the past 100 years can be counted in three ways as illustrated in the figure, which shows hypothetical emissions from CO_2 in 1950 and methane in 1970. Two choices are: 2.9a, count only the warming up to the present, which is the area under the two curves up to 1990; 2.9b, count only the warming under the curves from the present forward out to 100 years from the original emissions. The third choice, which is supported here, is to add these two together, in other words, to treat past emissions the same as future emissions (Figure 2.5).

Figure 2.9 *Methods of indexing past emissions using a hundred-year time horizon* our calculations. This option seems most consistent with our choice for present and future emissions, that is, period cutoffs.[k]

Implications

With this brief background of the constraints imposed on indices by physical reality, let us examine the implications concerning relative national responsibilities for choosing different kinds of indices. The entire landscape of nations and gases - historical, present, and future - is complex. Before addressing the total picture in the next chapter, therefore, we can take a cue from atmospheric scientists, who often make use of greatly simplified models of the world (only one-dimensional or two-dimensional, for example), in their quest to understand and predict the behaviour of extremely complex systems.

The one-nation, one-pollutant model

The most obvious measure of a nation's responsibility for greenhouse gas emissions is simply its present emissions. This measure has the clear benefit of being relatively easily determined and being the most responsive to control efforts. For these and other reasons, it has many advantages as an index.

A problem with this measure, however, is that it does not completely reflect physical reality. The extra greenhouse warming that occurs at any time is actually due to the cumulative amount of greenhouse gases remaining at that time, rather than to the emissions that year. That year's emissions are important only to the extent that they add to the accumulation.

The amount of greenhouse gases remaining in the atmosphere at any one year due to a nation's emissions has been termed the 'natural debt' (Smith 1989b, 1991). A national debt is built by borrowing financial resources from the future, but the natural debt is built by borrowing assimilative capacity of the atmosphere from the future, through the release of greenhouse gases faster than they can be naturally removed. Just as with the national debt, borrowing on the natural debt has allowed nations to build up their infrastructure and economic wealth faster than would have occurred otherwise. Like the national debt, however, if the natural debt becomes too

k Alternative GG-emitting actions will also have different temporal patterns; for example, cutting down a forest will result in CO_2 emissions (1) immediately from material burned on site, (2) on an intermediate scale, e.g., loss of soil carbon, and (3) over the long term, e.g., decay of lumber made from the trees (see Chapter 8). Thus some methods may also have to be developed to compare emissions committed but not yet expressed by actions today (see Chapter 3).

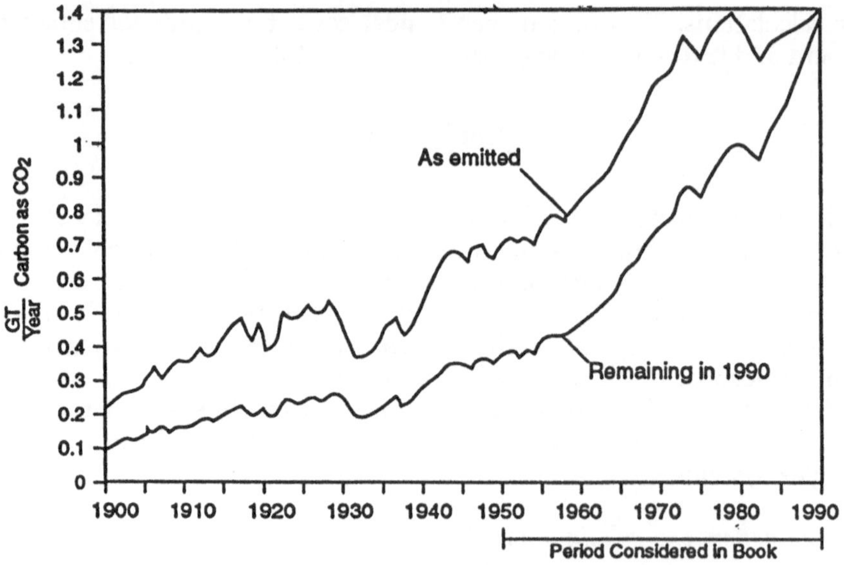

The upper curve shows the gross annual borrowing on the US natural debt in the form of carbon as CO_2 from fossil-fuel combustion (Smith, 1989b). The lower curve shows how much of the debt has been reduced from past years by natural atmospheric decay up to 1990 as determined in the model by Seigenthaler (1983). The areas under the two curves represent the gross and depleted natural debts.

Figure 2.10 *Gross and depleted natural debts of US CO_2 from fossil fuels*

large, problems are created. Just as with a financial debt, therefore, it does not seem unfair to ask nations to pay off the natural debt in the same proportion as it was borrowed.[l]

Figure 2.10 shows the relationship between current CO_2 emissions from fossil fuels and natural debt for the United States. (It leaves out other greenhouse gases and other sources of CO_2.) Current (1990) emissions are roughly 1.3 Gt (billion tonnes carbon as CO_2) per year. Cumulative emissions since 1950, however, are approximately 41 Gt. Of this, approximately 70 per cent still remains in the atmosphere, making the US natural debt to be 29 Gt, or some 116 tonnes per living US resident.[m] This natural debt might well be considered to be a reasonable measure of US responsibility (see Chapters 3 and 4). Tables 2.3 and 2.4 list the 1950–86 natural debts for the 62 largest nations in the world (with populations over ten million).

l There are other important parts of a nation's natural debt besides greenhouse gases. Excess concentrations of pollutants in lakes, rivers, and the ocean due to rates of pollutant emissions exceeding natural assimilative capacity are another part of the natural debt, for example.

m Since 1900, about 70 Gt has been released by US fossil-fuel burning, of which some 60 per cent is still in the atmosphere, giving a current per capita natural debt of 170 tonnes. Chapters 3 and 4 contain discussions about the appropriate year from which to calculate natural debt.

Table 2.3 *Current and historical carbon emissions and population data by country*

	1950–86 carbon emissions (megatonne)	*1986 total population (million)*	*1950–86 cumulative capita/year (million)*	*1986 annual emissions (t/cap)*	*1986 undepleted natural debt (t/cap)*
USA	37,284	240	7,360	5.01	155.0
Germany, United	9,123	78	2,784	3.50	117.0
Canada	2,911	26	751	4.09	112.0
Czechoslovakia	1,764	16	526	4.21	110.0
Belgium	1,081	10	349	2.68	108.0
United Kingdom	5,922	57	2,018	2.94	104.0
Australia	1,360	16	447	3.85	85.0
Poland	2,921	37	1,177	3.32	78.9
[USSR]	22,039	281	8,667	3.59	78.4
Netherlands	1,040	15	466	2.41	69.3
Bulgaria	618	9	306	3.60	68.7
France	3,646	55	1,820	1.79	66.3
Japan	6,924	121	3,929	2.11	57.2
Hungary	594	11	380	1.98	54.0
Romania	1,112	23	735	2.41	48.3
South Africa	1,456	33	835	2.79	44.1
Italy	2,294	57	1,955	1.66	40.2
Korea, Dem.	679	21	507	1.92	32.3
Spain	1,116	39	1,229	1.28	28.6
Yugoslavia	632	23	739	1.49	27.5
Greece	268	10	324	1.62	26.8
Venezuela	459	18	389	1.48	25.5
Argentina	698	31	874	0.85	22.5
Saudi Arabia	268	12	230	2.58	22.3
Portugal	156	10	339	0.79	15.6
Mexico	1,207	81	1,862	0.91	14.9
Chile	174	12	336	0.49	14.5
Korea, Rep.	584	42	1,126	1.08	13.9
Iran	534	46	1,030	0.68	11.6
Malaysia	176	16	310	0.58	11.0
Turkey	467	50	1,271	0.68	9.34
Colombia	262	29	734	0.44	9.03
Iraq	144	17	351	0.56	8.47
Syria	91	11	234	0.77	8.27
China	8,448	1,052	29,132	0.53	8.03
Algeria	160	22	518	0.69	7.27
Peru	141	20	480	0.29	7.05
Brazil	950	139	3,427	0.38	6.83
Egypt	281	48	1,194	0.41	5.90
Morocco	84	22	550	0.22	3.82
Philippines	207	56	1,343	0.16	3.70
Thailand	183	52	1,296	0.26	3.52
Vietnam	185	61	1,619	0.08	3.03
India	2,184	772	20,080	0.19	2.83
Indonesia	419	171	4,394	0.17	2.45

Table 2.3 *Continued*

	1950–86 carbon emissions (megatonne)	*1986 total population (million)*	*1950–86 cumulative capita/year (million)*	*1986 annual emissions (t/cap)*	*1986 undepleted natural debt (t/cap)*
Ivory Coast	21	10	210	0.13	2.10
Pakistan	192	103	2,544	0.13	1.86
Sri Lanka	29	17	442	0.06	1.71
Kenya	34	21	428	0.05	1.62
Cameroon	16	10	249	0.17	1.60
Mozambique	22	15	322	0.02	1.47
Ghana	19	14	317	0.05	1.36
Nigeria	132	99	2,141	0.13	1.33
Sudan	28	22	525	0.04	1.27
Zaire	30	31	733	0.03	0.97
Burma	34	38	987	0.05	0.89
Madagascar	7	10	248	0.02	0.70
Tanzania	15	23	511	0.02	0.67
Uganda	7	16	348	0.01	0.44
Bangladesh	44	103	2,632	0.03	0.42
Ethiopia	10	44	1,135	0.01	0.23
Nepal	3	16	434	0.02	0.19
				International means	
Total	123,889	4,560	124,629	1.20	28.60

Data for the 62 most populous nations (approximately 10 million or larger in 1986). The last column shows the natural debts, as defined in the text. The emissions data reflect carbon dioxide from fossil fuel combustion and cement manufacture (data mainly from Marland et al. 1988).

Two-nation, one-pollutant model

Now consider a two-nation, one-pollutant model of the world: fossil fuel carbon dioxide from India and the USA (Smith 1989a). (Before the break-up of the USSR, these were the second most populous developing nation and developed nation, respectively, each with about 22 per cent of the population of its group.) Let us ask the seemingly straightforward question, 'What will be their relative responsibilities for increasing GG emissions over the next 35 years?'

Table 2.5 lists some of the relevant parameters for 1986, as well as projections for the year 2025, given UN population estimates and three possible future energy scenarios for the USA (annual per capita fossil fuel use grows at zero per cent, at +0.5 per cent, as it has in recent years, or at –1.5 per cent, which is what would be needed to reach a 30 per cent reduction in total emissions by 2025). The one scenario for India, 4.4 per cent growth, was chosen because it is similar to recent rates and allows the average Indian in 2025 to achieve an annual emission rate of 1 tonne of carbon per capita,

Table 2.4 *Proportion of world carbon emissions and population, by country*

	1950–86 carbon emissions (% of world)	*1986 total population (% of world)*	*1950–86 cumulative capita/year (% of world)*	*1986 per capita emissions (% of mean)*	*1986 undepleted natural debt (% of mean)*
USA	30.1	5.3	5.9	503	572
Germany, United	7.4	1.7	2.2	352	431
Canada	2.3	0.6	0.6	412	412
Czechoslovakia	1.4	0.4	0.4	424	406
Belgium	0.9	0.2	0.3	269	398
United Kingdom	4.8	1.3	1.6	296	382
Australia	1.1	0.4	0.4	388	313
Poland	2.4	0.8	0.9	334	291
[USSR]	17.8	6.2	7	361	289
Netherlands	0.8	0.3	0.4	242	255
Bulgaria	0.5	0.2	0.2	362	253
France	2.9	1.2	1.5	180	244
Japan	5.6	2.7	3.2	212	211
Hungary	0.5	0.2	0.3	199	199
Romania	0.9	0.5	0.6	242	178
South Africa	1.2	0.7	0.7	280	162
Italy	1.9	1.3	1.6	166	148
Korea, Dem.	0.5	0.5	0.4	194	119
Spain	0.9	0.9	1	129	105
Yugoslavia	0.5	0.5	0.6	150	101
Greece	0.2	0.2	0.3	163	99
Venezuela	0.4	0.4	0.3	149	94
Argentina	0.6	0.7	0.7	85	83
Saudi Arabia	0.2	0.3	0.2	260	82
Portugal	0.1	0.2	0.3	80	57
Mexico	1	1.8	1.5	91	55
Chile	0.1	0.3	0.3	49	53
Korea, Rep.	0.5	0.9	0.9	109	51
Iran	0.4	1	0.8	68	43
Malaysia	0.1	0.4	0.2	58	40
Turkey	0.4	1.1	1	69	34
Colombia	0.2	0.6	0.6	44	33
Iraq	0.1	0.4	0.3	56	31
Syria	0.1	0.2	0.2	78	30
China	6.8	23.1	23.4	53	30
Algeria	0.1	0.5	0.4	69	27
Peru	0.1	0.4	0.4	29	26
Brazil	0.8	3	2.7	38	25
Egypt	0.2	1.1	1	42	22
Morocco	0.1	0.5	0.4	23	14
Philippines	0.2	1.2	1.1	16	14
Thailand	0.1	1.1	1	26	13
Vietnam	0.1	1.3	1.3	8	11
India	1.8	16.9	16.1	19	10
Indonesia	0.3	3.8	3.5	17	9

Table 2.4 *Continued*

	1950–86 carbon emissions (% of world)	*1986 total population (% of world)*	*1950–86 cumulative capita/year (% of world)*	*1986 per capita emissions (% of mean)*	*1986 undepleted natural debt (% of mean)*
Ivory Coast	0.02	0.2	0.2	13	8
Pakistan	0.15	2.3	2	13	7
Sri Lanka	0.02	0.4	0.4	6	6
Kenya	0.03	0.5	0.3	5	6
Cameroon	0.01	0.2	0.2	17	6
Mozambique	0.02	0.3	0.3	2	5
Ghana	0.02	0.3	0.3	5	5
Nigeria	0.11	2.2	1.7	13	5
Sudan	0.02	0.5	0.4	4	5
Zaire	0.02	0.7	0.6	3	4
Burma	0.03	0.8	0.8	5	3
Madagascar	0.01	0.2	0.2	2	3
Tanzania	0.01	0.5	0.4	2	2
Uganda	0.01	0.4	0.3	1	2
Bangladesh	0.04	2.3	2.1	3	2
Ethiopia	0.01	1	0.9	1	1
Nepal	0.002	0.4	0.3	2	1
				Weighted means (t/cap)	
Total	100	100	100	0.99	27.2

Note, for example, that the USA has a smaller fraction of present population (5.3%) and per capita emissions (503% of mean) than of cumulative population (5.9%) and natural debt (cumulative emissions, 572%). This is because the United States has been emitting at higher levels for a longer period and has a lower population growth rate than most of the world. The current weighted global per capita emissions are about 1 t/y, compared to the average nation's emission of 1.2 t/y shown in Table 2.3.

about the 1990 world average, or what the United States achieved before 1900. Presumably, however, advances in technology and its application by 2025 would allow much more benefit per unit of emissions than achieved by the USA at similar emission levels.

One index that could potentially answer the question of relative responsibility would be based on the expected change in national emission rates. Thus, India's annual emission rate in 1986 was 8.3 times less than that of the USA, but, at a US growth rate of zero (the middle scenario), India will catch up by 2025. Could it realistically be said that they each then have equal responsibility? Not if it is the impact on the atmosphere that is of concern, which is best indicated by natural debt.

Under the stated scenario, the USA-to-India ratio of natural debts added 1986–2025 is 2.3. In other words, this ratio might be considered their relative responsibility over the period, to be used, for example, for allocating costs of international programmes. (Higher ratios indicate that more responsibility would fall on the USA.)

Table 2.5 *Emissions indices: carbon as carbon dioxide from fossil fuels and cement, India versus the USA*

	1986 population[b] (10^6)	*1986 emissions/ capita (tonne)*	*1986 emissions/ nation (10^6 tonne)*	*1950–86 cumulative/ capita (tonne)*	*2025 population[b] (10^6)*	*2025 emissions/ capita (tonne)*	*2025 emissions/ nation (10^6 tonne)*	*1950–2025 cumulative/ capita (tonne)*	*1987–2025 incremental cumulative/ nation (10^9 tonne)*
USA (+0.5%)[a]	242	5.0	1210	155	301	6.1	1830	376	60
USA (0%)	242	5.0	1210	155	301	5.0	1500	355	54
USA (-1.5%)	242	5.0	1210	155	301	2.8	834	308	40
India (4.4%)[a]	781	0.19	150	2.8	1450	1.0	1450	22	23
Ratio, USA/India	0.31	27	8.3	55	0.21	6.1[c]	1.3	17	2.6
						5.0	1.0	16	2.3
						2.8	0.58	14	1.7

Basic data from Marland et al. 1988

a Shown in parentheses are the annual growth rates in per capita carbon emissions assumed for this scenario.
b Population inputs are medium-case estimates from the UN (1988).
c The three ratios shown in these columns correspond to the three per capita USA carbon emission growth rates (+0.5%, 0%, –1.5%).

Let us now examine this index (natural debt per nation) a bit more closely. Is it really fair to allocate responsibility on the basis of national emissions (or natural debt)? To do so would imply that responsibility is somehow a function of size, that if you are lucky enough to live in a small country you are not as responsible as someone living in a large country, even if your own personal emissions are the same. Responsibility is best judged on a per capita and not on a per nation basis. Otherwise, quite perverse results can be obtained: for example that Hong Kong should have the same allotment as China until they merge in 1997, when their total allotment would drop by half. Or that the former parts of the USSR have suddenly become less responsible now that they are smaller. Clearly we do not want an index that rewards nations for splitting up or, conversely, for taking over other nations!

On the other hand, since governments, not individuals, will be actually asked to pay the bill for greenhouse remediation projects, the population size represented by the government must be considered. If the per capita emissions (or natural debt) is simply multiplied by the population size, however, we are right back to national emissions again. Chapter 4 directly addresses this dilemma.

Returning to Table 2.5, use of per capita indices changes the ratio of present USA/India emissions from 8.3 to 27. From the standpoint of natural debt, as shown in Table 2.5, the CO_2 placed into the atmosphere since 1950 as part of the US industrialization process is equivalent to about 155 tonnes per person in 1986, about 50 times that of the average Indian, at 2.8 tonnes.

According to the scenarios in Table 2.5, the Indian per capita cumulative contribution would increase by a factor of about 6 over the period, while the American contribution would little more than double. Because of the large difference in initial values, however, during these 39 years, the Indian's contribution would increase by 19 tonnes, while the average US resident's contribution would rise more than 10 times as much, to 200 tonnes. The ratio of total cumulative amounts at the end of the period would, at 16, be lower than what it was in 1986 (50), but still high. At that point, the average Indian would still be responsible for only a minor part of the total atmospheric accumulation, less than 6 per cent of that of the average US resident. (See the box for a visual metaphor that may help put these points into perspective.)

Given these circumstances, even though their national emission rates would be the same at the end of the period, it seems presumptuous to expect the Indian to contribute much to an international effort to control greenhouse gases, at least relative to what might be expected of the American. Even with a drastic cut in emission rates, the growth of natural debt per capita is much higher in the USA.

The two-nation, two-pollutant model

To illustrate the importance of time horizon, we now upgrade the two-

Global change and the seawall metaphor

Used with care, an appropriate metaphor can sometimes assist in gaining insights about complicated relationships. In this spirit, consider the visualization in Figure 2.11, which illustrates a simple two-person model of the world.

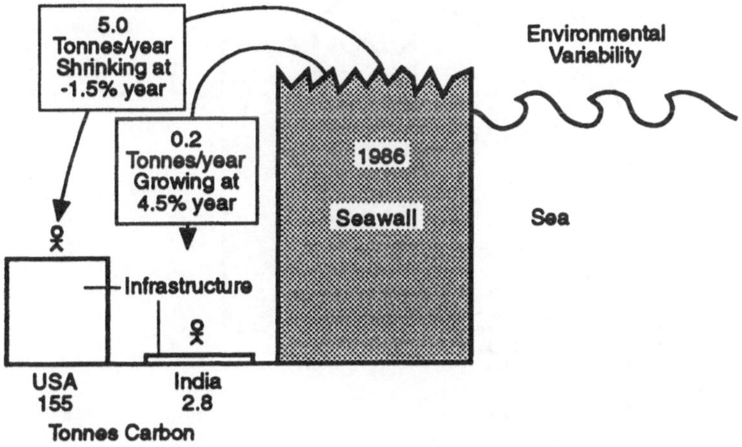

Figure 2.11 *Growth in per capita natural debt, USA and India, 1986*

The figure shows a seawall holding back a sometimes stormy and largely unpredictable sea. Occasionally, the sea overtops the wall, thus threatening to flood the average American and average Indian who live on the other side. To bolster themselves against this environmental variability, they have been building up their economic and technological infrastructure as indicated by the blocks of material on which they stand. As can be seen, the American has built a much larger base of support and thus is much less vulnerable to environmental stress than is the Indian.

One measure of the size of the support blocks is the energy that has been used to build them. In the US case, in 1986 the average American was standing on a block of 155 tonnes carbon (C in CO_2 from fossil fuels since 1950). The Indian, by contrast, was standing on a block some 55 times smaller, 2.8 tonnes.

Unfortunately, a considerable part of the material for the support blocks comes from the top of the wall holding back the sea. As long as the people take material off the top at a moderate rate, the seawall can grow back enough to counteract it naturally. Unfortunately, the rate of removal (5 t/year for the USA and 0.2 t/year for India) exceeds the natural rate and the top of the seawall is lowering. Consequently, the chance of the sea overtopping the wall seems to be increasing.

Now consider that the American, already being relatively well protected, had been able to implement a drastic energy efficiency and renewables effort and thus reduce the annual take from the top of the seawall by 1.5 per cent each year. Thus, as shown in Figure 2.12, at the end of 39 years, the American is taking just 2.8 tonnes per year, a cut of 44%. The Indian, on the other hand, seeing the increasing chance of overtoppage, feels the need to build up the support block even faster than before and thus takes 4.5 per cent more material off the top of the seawall each year. By the year 2025, therefore, the Indian is taking 1 tonne per year, which about equals the world average in the late 1980s and what the American was taking well before 1900. Presumably, however, due to better technology and planning, more benefit will be gained in the future from such a relatively modest rate of removal.

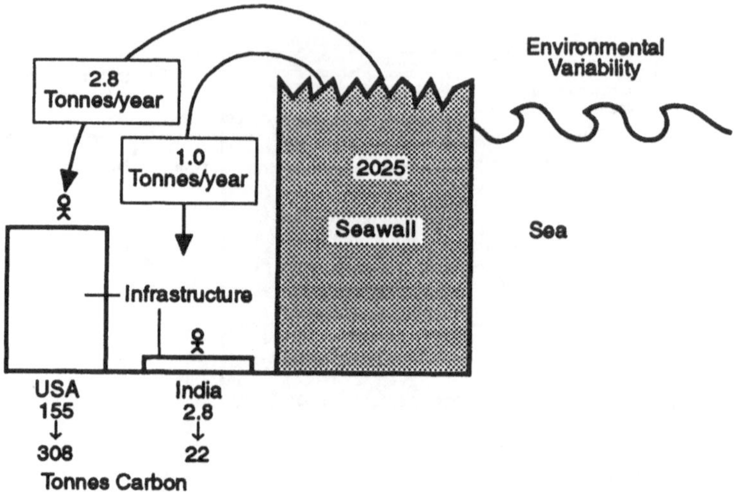

Seawall metaphor showing how greenhouse gas emissions can increase the likelihood of environmental stress from global warming, as well as decrease human vulnerability to the same stress.

Figure 2.11 shows conditions in 1986 when the average US and Indian residents were protected from stress by infrastructure built by, respectively, 155 and 2.8 tonnes of carbon (1950–86 natural debts).

In Figure 2.12 assuming the growth rates in Figure 2.11 are continued until 2025, the average Indian could reach annual greenhouse emissions of 1 tonne carbon, roughly the present world average. At this point, however, the Indian's natural debt would still be 14 times less than that of the US resident, although better technology might make it possible to achieve a smaller difference in their respective environmental vulnerabilities.

Figure 2.12 *Growth in per capita natural debt, USA and India, 2025*

Since the Indian's annual take has grown at a healthy rate and the American's has substantially fallen, the difference in their annual takes has dropped from a factor of about 30 in 1986 to 3 in 2025. From the standpoint of both benefit and risk, however, it is the amount taken from the top of the seawall that is most crucial. This is what builds up the support base that reduces vulnerability and, at the same time, increases the likelihood of environmental stress. From this standpoint, the American has taken off about eight times as much as the Indian, even though the former is undergoing a drastic annual reduction in the rate of removal and the latter's rate is increasing rapidly.

Now, consider the question posed by this book: 'If a large international programme of greenhouse remediation is to be undertaken, who should pay over the next few decades?' Looking at Figure 2.12, it does not seem quite fair to expect

the Indian to bear much of the cost during this period. Compared to the American, the Indian has not taken much material off the top of the seawall, either in total since 1950 (7 per cent of the American) or incrementally during the decades being considered (13 per cent). By starting at such a high base, even with quite rapid decline in annual rates, the American's annual and total contributions to the problem will remain much higher throughout any future period during which present policies can reasonably contend. In addition, importantly, the American will remain much more insulated from any environmental stress as well, by virtue of the infrastructure built up over many years.

In closing, it is important to remember that these conclusions refer to who is responsible for paying for climate change remediation efforts, and not to where such efforts might best be undertaken. It is quite possible, for example, that most cost-effective programmes might be in countries that have little responsibility for payment.

nation model to include the two most important greenhouse gases: CO_2 and methane. The United States and Indian emissions for 1987 are shown in Table 2.6. Note that, by the instantaneous index (time horizon = 0), the United States produces only three times as much GG as India when CH_4 is included, which is much less of a relative responsibility than the factor of 7.9 for CO_2 alone. Thus, it is not surprising that US negotiators tend to argue for including gases such as methane that are produced in relatively greater amounts by developing nations (Stewart and Weiner 1990).

On the other hand, on a 500-year basis, the ratio rises to 5.7 or 6.7 (depending on whether indirect effects are included), effectively doubling relative US responsibility. A completely integrated index (an infinite time horizon) would result in a ratio of 7.3. Note also that inclusion of indirect effects tends to obligate nations such as India with relatively high methane production, that is, the USA/India ratio decreases. Thus, US negotiators could be expected to argue for inclusion of indirect effects, for shorter rather than longer time horizons, and, perhaps, for applying a discount rate as well.

Three-nation, two-pollutant model

Figure 2.13 provides further evidence of the sensitivity to choice of time horizon. It shows the impact of time horizon on the ratio between the 1987 emissions of the United States and India compared to the ratio between the United States and the then Federal Republic of Germany (FRG). Note that with an instantaneous index, the ratios are quite different, but that with a time horizon of about 565 years, the two match. US emissions are 6.9 times either those of Germany and India, which, therefore, have equal weightings. At an infinite time horizon, however, the two are different again.

Conclusion: indices do matter

Table 2.7 on page 48 shows a summary of the various values in the one-

Table 2.6 *Greenhouse gas indices according to time horizon for India and the USA*

	1987 emissions		*Total CO_2 equivalents by time horizon (years)*				
	Million tonnes C as CO_2	*Million tonnes C and CH_4*	*0 (10^5 tonnes)*	*20 (10^9 tonne-y)*	*100 (10^9 tonne-y)*	*500 (10^9 tonne-y)*	*Infinite (10^9 tonne-y)*
USA	1200	32	1900	29 (24)	79 (73)	210 (200)	(400)
India	150	23	630	10 (6.5)	18 (13)	37 (30)	(54)
Ratio, USA/India	7.9	1.4	3	2.9 (3.7)	4.4 (5.6)	5.7 (6.9)	(7.3)

Note that including methane and using short time horizons tend to make the US contributions seem relatively smaller, i.e., the USA/India ratio goes down.

Based on coefficients in Table 2.1 and emissions in WRI (1990), Table 24.1.
CO_2 emissions from fossil fuels and cement production only.
Numbers in brackets refer to direct warming only (no indirect effects). See Table 2.1.

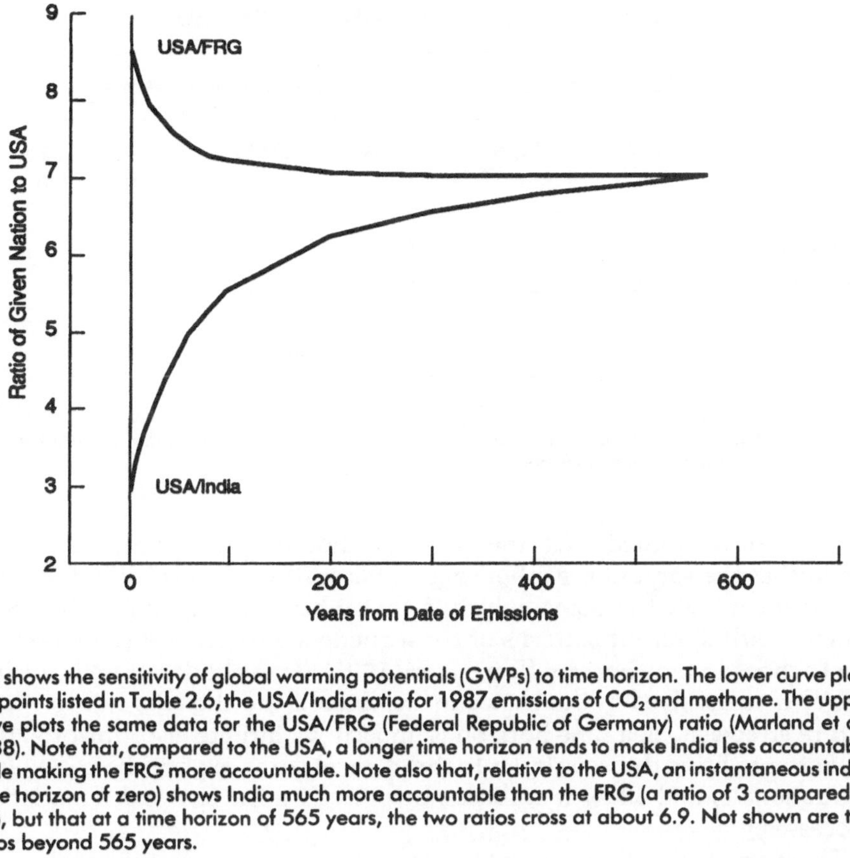

This shows the sensitivity of global warming potentials (GWPs) to time horizon. The lower curve plots the points listed in Table 2.6, the USA/India ratio for 1987 emissions of CO_2 and methane. The upper curve plots the same data for the USA/FRG (Federal Republic of Germany) ratio (Marland et al., 1988). Note that, compared to the USA, a longer time horizon tends to make India less accountable while making the FRG more accountable. Note also that, relative to the USA, an instantaneous index (time horizon of zero) shows India much more accountable than the FRG (a ratio of 3 compared to 8.5), but that at a time horizon of 565 years, the two ratios cross at about 6.9. Not shown are the ratios beyond 565 years.

Figure 2.13 *Sensitivity of global warming potentials to time horizon*

pollutant, two-nation model and illustrates the wide range of answers that could be derived from use of indices that vary only in two parameters – current versus cumulative and per nation versus per capita. It is useful to note that these values are much more sensitive to the choice of index, than to the choice of emissions scenario. Assume, for example, that the USA was able to reduce its per capita carbon emissions use by 1.5 per cent each year for the 39-year period, reaching a total emission rate more than 30 per cent below that of 1986 (and bringing the per capita emissions down by more than 40 per cent). How much difference would this rather heroic effort make in the final ratios? It would be very little for cumulative per capita indices. As shown in Table 2.5, the difference in ratios for cumulative atmospheric carbon would be less than 15 per cent (14 versus 16). Conversely, if the USA were to continue to expand its fossil fuel use by 0.5 per cent, as it has been doing recently, the ratio would only rise to 17 by 2025.

Table 2.7 *Ratio of responsibilities: USA/India*

	Current emissions		*Cumulative emissions*		
	1986	*2025*	*1986*	*2025*	*1987–2025 increment*
Per nation	8.3	1.0	16.0	3.6	2.3
Per capita	27.0	5.0	55.0	16.0	10.0

Based on data in Table 2.5, no-emissions-growth scenario for the USA.
CO_2 emissions from fossil fuels and cement production.

Any of these indices might be used to compare relative responsibilities for greenhouse gas emissions. In this book, natural debt indices are recommended, those three in the lower right quadrant, which are based on cumulative emissions per capita.

Some have argued that the need to choose a time horizon greatly diminishes the value of greenhouse-gas indices (Hammond et al. 1990). This argument misses the point, which is that when making choices between options with different patterns of consequences over time, there is no way not to choose a time horizon (*Environment* 1991). It may be done implicitly, but it is always involved in the choice made. Even if the choice is not to count future effects at all, a time horizon is implicit: an infinite discount rate. It is far better to bring the issue out into the open, make an explicit choice, explain the rationale, and allow it to become part of the review and negotiating processes.

There are many other possibilities to vary indices to reflect the real world of multiple criteria, greenhouse gases, and nations. We want to emphasize here, however, that the choice of index does indeed make a difference, sometimes a very large difference, and thus must be chosen with care to be relevant to the problem at hand and scientifically justifiable, as well as useful for policy.

References

Agarwal, A, and S Narain, 1991. *Global Warming in an Unequal World: A Case of Environmental Colonialism*, Centre for Science and Environment, New Delhi, India

Ellington, R T, and M Meo, 1990. 'Development of a Greenhouse Gas Emissions Index', *Chemical Engineering Progress*, July: 58–63

Environment, 1991. 'Solicited Commentaries on the WRI Greenhouse Gas Index,' *Environment* 22(2): 2–5, 42–43

Fujii, Y, 1990. *An Assessment of the Responsibility for the Increase in the CO_2 Concentration and Inter-generational Carbon Accounts*, WP-90-55, International Institute for Applied Systems Analysis, Laxenburg, Austria

Grubler, A, and Y Fujii, 1991. 'Inter-Generational and Spatial Equity Issues of Carbon Accounts', *Energy*, 16 (11/12): 1397–1416

Gurney, K R, 1991. 'National Greenhouse Accounting', *Nature* 353: 23

Hammond, A L, E Rodenburg, and W. Moomaw, 1990. 'Accountability in the Greenhouse', *Nature* 347: 705–706

Hansen, J, I Fung, A Lacis, et al., 1988. 'Global Climate Changes as Forecast by Goddard Institute of Space Studies Three-Dimensional Model', *J of Geophysical Research* 93: 9341–9364

IPCC (Intergovernmental Panel on Climate Change), 1990. *Climate Change: The IPCC Scientific Assessment*, Cambridge University Press, Cambridge, UK

IPCC, 1992. *Climate Change: The Supplementary Report to the IPCC Scientific Assessment*, Cambridge University Press, Cambridge, UK

Krause, F, W Bach, and J Koomey, 1989. *Energy Policy in the Greenhouse*, Vol. 1, International Project for Sustainable Energy Paths, El Cerrito, CA, USA

Lashof, D A, and D R Ahuja, 1990. 'Relative Contributions of Greenhouse Gas Emissions to Global Warming'. *Nature* 344: 529–531

Lashof, D A, and D A Tirpak, eds., 1989. *Policy Options for Stabilizing Global Climate Change*, Draft Report to Congress, US Environmental Protection Agency, Washington, DC, USA

Lashof, D A, and D A Tirpak, eds., 1990. *Policy Options for Stabilizing Global Climate Change*, Final Report to Congress, Office of Policy, Planning, and Evaluation, PM-221, US Environmental Protection Agency, Washington, DC, USA

Lelieveld, J, and P J Crutzen, 1992. 'Indirect Chemical Effects of Methane on Climate Warming', *Nature* 355: 339–342

Marland, G, T A Boden, R C Griffin, et al., 1988. *Estimates of CO_2 Emissions from Fossil Fuels Burning and Cement Manufacturing*, ORNL/CDIAC-25, Oak Ridge National Laboratory, Oak Ridge, TN, USA

McCully, P, 1991. 'Discord in the Greenhouse: How WRI is Attempting to Shift the Blame for Global Warming', *Ecologist* 21(4): 157–165

Mitchell, J K, ed., 1992. 'Greenhouse Equity: Six Commentaries on the WRI/CSE Controversy', *Global Environmental Change* 2(2): 82–100

New York Times, 1989. Sources of the Smothering Gases, November 19

Ott, W, 1978. *Environmental Indices, Theory and Practice*, Ann Arbor Science, Ann Arbor, MI, USA

Pachauri, R K, S Gupta, and M Mehra, 1992. 'A Reappraisal of WRI's Estimates of Greenhouse Gas Emissions', *Natural Resources Forum* 14: 33–38

Penner, J E, P S Connell, D J Wuebbles, and C C Covey, 1989. 'Climate Change and its Interactions with Air Chemistry: Perspectives and Research Needs,' *The Potential Effects of Global Climate Change on the United States: Appendix F – Air Quality*, May. Office of Policy, Planning and Evaluation, US Environmental Protection Agency, Washington, DC, USA

Rodhe, H, 1990. 'A Comparison of the Contribution of Various Gases to the Greenhouse Effect', *Science* 248: 1217–1219

Shine, K P, R G Derwent, D J Wuebbles, and J-J. Morcrette, 1990. 'Radiative Forcing of Climate', in Chapter 2 of *Climate Change: The IPCC Scientific Assessment*, Cambridge University Press, Cambridge, UK

Siegenthaler, U, 1983. 'Uptake of Excess CO_2 by an Outcropping Model of the Ocean', *J of Geophysical Research* 88: 3599–3608

Smith, K R, 1989a. 'Developing Countries and Climate Change: Implications for

Risk Management,' in D Street and T Siddiqi, eds., *Proceedings of the Workshop on Responding to the Threat of Global Warming: Options for the Pacific and Asia*, pp. 2–37–2–39, Argonne National Lab/East-West Center, Argonne, IL, USA

Smith, K R, 1989b. *Have You Paid Your Natural Debt?*, Environment and Policy Institute, East-West Center, Honolulu, HI, USA

Smith, K R, 1991. 'Allocating Responsibility for Global Warming: The Natural Debt Index', *Ambio* 20(2): 95–96

Smith, K R, and D R Ahuja, 1990. 'Toward a Greenhouse Equivalence Index: The Total Exposure Analogy', *Climatic Change* 17: 1–7

Smith, K R, J Swisher, R Kanter, and D R Ahuja, 1991. *Indices for a Greenhouse Gas Control Regime That Incorporates Both Efficiency and Equity Goals*, DWP–1991–22, Policy and Research Division, Environment Department, World Bank, Washington, DC, USA

Solomon, B D, and D R Ahuja, 1991. 'International Reduction of Greenhouse-Gas Emissions: An Equitable and Efficient Approach', *Global Environmental Change 1*(4): 343–350

Stewart, R B, and J B Weiner, 1990. 'A Comprehensive Approach to Climate Change', *American Enterprise*, November/December: 75–80

Subak, S, P Raskin, and D Von Hippel, 1992. 'National Greenhouse Gas Accounts: Current Anthropogenic Sources and Sinks', Stockholm Environment Institute, Boston, MA, USA

UN (United Nations), 1988. *World Population Prospects*, 1988. ST/ESA/SER.A/106, Department of International Economic and Social Affairs, New York City, USA

USNAS (United States National Academy of Sciences), 1983. *Climate Change*, Report of the Carbon Dioxide Assessment Committee, Washington, DC, USA

Victor, D G, 1990. 'Calculating Greenhouse Budgets', *Nature* 347: 431

Wang, W, M P Dudek, X Liang, and J T Kiehl, 1991. 'Inadequacy of Effective CO_2 as a Proxy in Simulating the Greenhouse Effect of Other Radiatively Active Gases', *Nature* 350: 573–577

WRI (World Resources Institute), 1990. *World Resources 1990–91*, Oxford University Press, New York City, USA

WRI, 1992. *World Resources 1992–93*, Oxford University Press, New York City, USA

3

Assessing emissions: five approaches compared

Susan Subak

Introduction

In this chapter, I present a variety of ways to assess responsibility for greenhouse gas (GHG) emissions. The parameters that could define responsibility from a polluter pays perspective include: which greenhouse gases are counted; which sources are included; and what time frame is used for estimating them. A New Zealander who lives in a country with twenty methane emitting sheep for every person may prefer to keep the gases limited to carbon dioxide only. A Swiss citizen mostly emits carbon dioxide by burning fossil fuels, and may be unhappy if only this gas is controlled. And someone from a recently industrialized country such as Singapore might feel justified in pushing for the inclusion of historic emissions in global greenhouse negotiations. The definition of GHG emissions, therefore, has great practical impact on each country's relative responsibility for emissions. The feasibility of controlling emissions sources, linking national abatement actions efficiently with global targets, and verifying emissions after targets have been set are other important considerations that policy makers must take into account when assessing responsibility for emissions.

In the following analysis, five approaches for assigning responsibility among countries for greenhouse gas emissions are examined. They comprise two historical and three current emissions assessments which vary by level of coverage of sources (Table 3.1):

1 cumulative CO_2, energy only;
2 cumulative CO_2, energy and biota (including CO_2 from both fuels and land clearance);
3 CO_2, energy only (current);
4 partial CH_4 and CO_2 (including current emissions of CO_2 from energy consumption and deforestation, and methane from energy production and landfills);
5 comprehensive (current emissions of CO_2, CO, CH_4 and N_2O from energy, industrial, biotic and agricultural sources).

Table 3.1 *Sources included in selected cumulative (1860–1986) and current (1988) emissions assessments*

	Energy (CO_2)	Biota (CO_2)	Landfills (CH_4)	Other[a]
1 Cumulative CO_2, energy only	X			
2 Cumulative CO_2, energy & biota[b]	X	X		
3 CO_2, energy (current)	X			
4 Partial CH_4 and CO_2 (current)	X[c]	X	X	
5 Comprehensive (current)	X	X	X	X

a 'Other' includes cement production, and agricultural sources, including livestock, rice cultivation, fertilizer consumption, and biomass burning apart from deforestation. The gases include CO_2, CH_4, CO, and N_2O.
b Includes estimated net CO_2 release from soil carbon and from above-ground biomass in areas converted from forests to agricultural uses only.
c CO_2 and CH_4 emissions.

All of the approaches have already entered discussions, either in a political or an academic context. Most of the approximately two dozen countries that have pledged thus far to meet specific national targets to stabilize or control greenhouse gas emissions have focussed on the control of CO_2 emissions from energy consumption. Setting targets for CH_4 (methane) from energy and industrial sources and CO_2 from biotic sources, in addition to CO_2 from fossil fuel combustion is being seriously explored by several industrialized countries. The Framework Convention on Climate Change signed at Rio de Janeiro in June 1992, which requires developed country Parties to submit plans for stabilizing emissions, can be interpreted to apply to all greenhouse gas sources with the exception of halocarbons controlled by the Montreal Protocol. Allocating future emissions based on historical release of greenhouse gases has been proposed by a number of researchers (Krause et al. 1989; Smith 1991; Gruebler and Fujii 1991).

Any of these source categories could form a broad basis for resource transfers from North to South to fund technology transfers or greenhouse gas abatement projects. But as the baseline against which national targets or the allocation of tradeable emissions permits are set, the national inventories must be accurate and verifiable. A consensus is more likely to be reached over setting targets for sources and gases that can be measured with confidence. Although in the past, regional environmental agreements have been signed before baseline national emissions estimates were completed, in the case of greenhouse gas emissions where the differences in countries' emissions rates are so great, nations are unlikely to favour setting specific targets for controlling sources for which accurate baseline inventories at the country level are not yet available and cannot yet be monitored.

In the following analysis, the relative comprehensiveness of the different source categories is briefly summarized, followed by a discussion of the problems in estimating emissions from these different sources and time frames. In addition, the implications of the five emission categories is illustrated for a selection of the major emitting countries. A brief description of the emissions totals used, and the method for calculating national inventories appear in Appendix A and Appendix B.

Comprehensiveness compared

Cumulative CO_2, energy only

As it takes many decades for CO_2 to be removed from the atmosphere, the increase in concentration of CO_2 from pre-industrial levels is largely due to CO_2 emitted in past decades. In this respect, historical CO_2 emissions are much more relevant to the level of committed atmospheric warming than are current emissions. Emissions from past energy use, however, make up a smaller portion of total CO_2 release than today's fossil fuel emissions because CO_2 from land clearing may have been roughly comparable with fossil fuel related CO_2 until the middle of this century (see Figure 3.1). Emissions from fossil fuel combustion since the start of the industrial revolution are estimated to be 175 to 215 gigatonnes (GT) of carbon (C), representing between about 55 and 70 per cent of total anthropogenic CO_2 release (IPCC 1990). Contributions to warming, however, are considerably lower because CO_2 is but one of the gases contributing to the heating effect. Considering both the change in fossil fuel emissions over time (Keeling 1973; Marland et al. 1990) and the estimated contribution of CO_2 to total warming (IPCC 1990), it is calculated that cumulative CO_2 emissions from

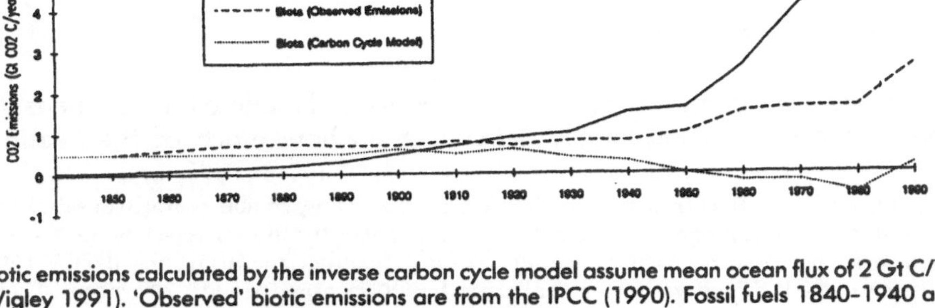

Biotic emissions calculated by the inverse carbon cycle model assume mean ocean flux of 2 Gt C/Yr (Wigley 1991). 'Observed' biotic emissions are from the IPCC (1990). Fossil fuels 1840–1940 are from Keeling (1973), 1950–1990 (Marland et al. 1990).

Figure 3.1 *Historic CO_2 emissions compared*

energy may contribute about 40 per cent of the warming effect of the trace gases now in the atmosphere.

Cumulative CO_2, energy and biota

About 60 per cent of the warming effect of anthropogenic greenhouse gas emissions in the atmosphere is thought to be from CO_2 or cumulative CO_2 emissions from energy and biota (IPCC 1990).

CO_2, energy only (current)

Although current CO_2 emissions from energy is the least comprehensive of the five categories considered here, CO_2 from energy alone is the major warming source from current emissions. Emissions from energy consumption contribute about 65 per cent of the expected warming effect of the trace gases now being emitted, if halocarbons are excluded from the total. This CO_2 share of warming reflects the use of the IPCC's GWP for a 100 year period, published in 1992 (IPCC 1992). The estimated warming is only 60 per cent of the total if the IPCC's 1990 GWPs are employed and no doubt will change as the IPCC revises in the future.[a]

Partial CH_4 and CO_2

This category, as defined above, covers about 80 per cent of the warming effect of current greenhouse gas emissions excluding halocarbons.[b] Halocarbons have been omitted from the current emissions total because they are already being phased out under the Montreal Protocol.

Comprehensive emissions

The comprehensive approach to emission measurement theoretically represents 100 per cent of current greenhouse gas emissions.[c]

The influence of time horizon

The relative comprehensiveness of the energy and modified comprehensive approaches varies considerably depending on how much of the heating

a If the warming effect from N_2O, CH_4, and CO from energy combustion are considered as well as CO_2, this proportion ranges from 54 per cent of the heating effect by the 20-year time frame to 76 per cent under the 500-year time horizon, using the IPCC's 1992 calculated GWPs; the warming potential for the 100 year time period is 70 per cent. These estimates exclude halocarbons from the total.

b The coverage ranges from 62 per cent to 87 per cent by 20 and 500 year GWPs respectively.

c Water vapour and tropospheric (low level) ozone are not included in this analysis.

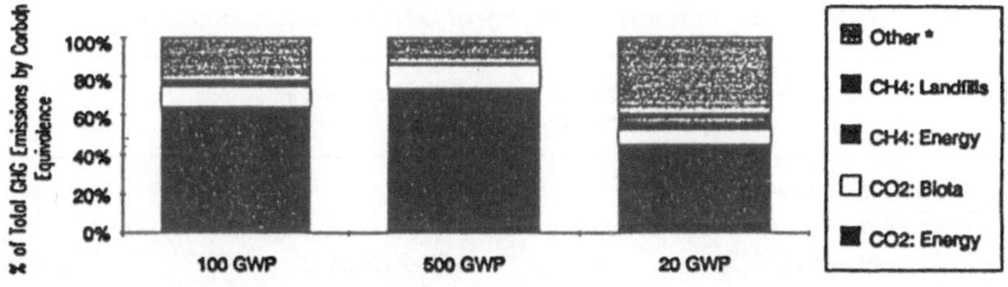

CO_2 from energy makes up about 60%, and the 'Partial CH_4 and CO_2,' which includes the additional CO_2 and CH_4 sources listed, about 80% of the heating contribution (100 GWP). Other* excludes CFCs, which make up about 12% of GHG.

Figure 3.2 *Contributions to total emissions by source*

effect of the different gases is taken into account. In Figure 3.2, current emissions are compared using three CO_2 equivalence indices including the potential heating effect of the gases over a 20, 100, and 500 year time horizon. In the 20 year time horizon, CO_2 from energy contributes only about 45 per cent of the total heating effect because the index based on the shorter time horizon does not capture the ultimate heating effect of CO_2, which continues many decades beyond the atmospheric residence time of CH_4, the next most important greenhouse gas. Accordingly, the proportion of the total heating contribution due to CO_2 from energy is much higher - about 70 per cent - over the longer time horizon.

Accuracy by category

Table 3.2 summarizes the difficulties in estimating emissions from each of the source groupings; it includes the IPCC's ranges of uncertainty in estimating emissions by source and gas globally for each of the five emissions groups.

Cumulative CO_2, energy only

CO_2 emissions from energy use have been estimated at the country level between 1950 and 1988 (Marland et al. 1989) and between 1860 and 1950 (Subak and Clark 1990). Marland et al. estimate that the uncertainty of their inventory is 6–10 per cent at the country level (Marland et al. 1988). The accuracy of the pre-1950 data set is limited because of changes in geographical borders and sovereignty, lack of information on the type of coal used in the past, and because data on fossil fuels traded in certain regions are incomplete or unavailable. To relate historical emissions to current concentrations, a coefficient or 'discount rate' must be applied to adjust for the CO_2 that has been removed from the atmosphere over time.

Table 3.2 *Estimated accuracy of GHG emissions accounts*

	Estimated accuracy	*Estimated range (IPCC 1990 unless noted)*	
1 *Cumulative CO_2 Energy* (1860–1986)			
Fossil Fuel Consumption (CO_2)	Medium	175–215 GT C	(±10%)
2 *Cumulative CO_2* (1860–1986)			
Fossil Fuel Consumption (CO_2)	Medium	175–215 GT C	(±10%)
Land Use Changes (CO_2)	Low[a]	82–152 GT C	(±30%)
3 *CO_2 Energy* (Current)			
Fossil Fuel Consumption (CO_2)	High	5.4 GT C (Marland et al. 1990)	(±5%)
4 *Partial CH_4*, CO_2 (Current)			
Above Plus:			
Landfills (CH_4)	Medium	20–70 MT CH_4	(±50%)
Land Use (CO_2)	Medium	1.1–3.6 GT C (Houghton 1991)	(±50%)
Energy Prod. and Distribution (CH_4)	Medium	44–100 MT CH_4	(±40%)
5 *Comprehensive*[b] (Current)			
Above Plus:			
Fossil Fuel Combustion (N_2O, CO)	Medium	0.5–1.4 MT N_2O (IPCC 1992)	(±50%)
Cement Production (CO_2)	High		
Biomass Burning (CH_4, CO, N_2O)	Low	20–180 MT CH_4 0.3–1.6 MT N_2O (IPCC 1991)	(±80%)
Enteric Fermentation (CH_4)	Medium	65–100 MT CH_4	(±20%)
Animal and Human Wastes (CH_4)	Low		
Rice Cultivation (CH_4)	Low	25–170 MT CH_4	(±80%)
Fertilizer Consumption (N_2O)	Low	0.01–2.20 MT N_2O	(±100%)
Halocarbons (CFCs, Halons, HCFCs)	High		
Nylon Production (N_2O)	Medium	0.6–0.9 MT N_2O (IPCC 1991)	(±20%)

a In this case, the uncertainty at the country level is far greater than the estimated global range.
b This inventory does not include stratospheric water vapour, which is thought to contribute about 4 per cent of the climate forcing (IPCC 1990), and O_3 precursors-NO_x and volatile organic compounds.

Despite these accounting difficulties, estimates of CO_2 from energy use since 1860 are more accurate than those of current emissions from land use changes and agriculture. Keeling (1973) has estimated the uncertainty of historical global estimates for CO_2 emissions from fossil fuels at 13 per cent. This range of uncertainty compares with 100 per cent or more for CO_2 emissions from biota in the current period and equivalent or greater uncertainties in estimating CH_4 from rice cultivation and biomass burning, and N_2O, NO_x, and CO from all sources (IPCC 1990; Logan et al. 1981).

Cumulative CO_2, energy and biota

Estimating national historical emissions from energy and biota involves all of the technical difficulties of estimating cumulative emissions from fossil fuels outlined above, plus the challenge of estimating biotic contributions. While data are available to calculate emissions from fossil fuels on an annual basis, comprehensive international forest surveys are conducted less frequently, generally every decade since 1949. Much of the pre-1950 data are for changes in area devoted to agricultural uses only and therefore omit forest conversion to other uses such as settlements, etc. For all periods, many of the forest surveys are considered unreliable. It is unlikely that additional scientific research will significantly improve the accuracy of these estimates on the national level as investigations of historical trace gas concentrations such as ice core and tree ring analyses shed light on global historic concentrations rather than on nation-specific emissions.

Energy, CO_2 only (current)

This is the most practical, that is, measurable and verifiable approach of the five. Carbon dioxide emissions from current energy consumption are estimated to be accurate at the country level within an error range of only about 6–10 per cent (Marland et al. 1989). A comparison of the (ORNL) Marland and Rotty inventory (Marland et al. 1988), which is based on United Nations energy statistics with a new inventory of CO_2 release (Von Hippel et al. 1992) from energy consumption that was derived from OECD/IEA statistics (OECD/IEA 1990a) suggests that the error range may be higher for some countries.[d] Regardless, the level of uncertainty in estimating emissions from this source is far lower than the uncertainty associated with inventories of the other gases and sources and should improve in the near future as a number of agencies are refining emission factor estimates and end-use data at the country level.

Partial CH_4 and CO_2

This approach is midway in practicality between the CO_2/energy only approach and the comprehensive approach. The additional sources – land use changes, landfills, and fossil fuel extraction – cannot be estimated as accurately as energy consumption. The error range for estimating CO_2 from land use changes and CH_4 from landfills and fossil fuel extraction is ± 40 to ± 50 per cent at the global level (IPCC 1990), with developing countries generally at the higher end. Nevertheless, these sources of CO_2 and CH_4 should be easier to monitor than the agricultural sources and remaining gases. The landfill and coal mine sources of CH_4 are also potentially important sources of natural gas (US/Japan Working Group on Methane 1992). Employing technology to recover and utilize natural gas from these sources should eventually enhance our capacity to control and monitor CH_4 release.

d The OECD/IEA-derived CO_2 emissions estimates differed from the ORNL Study by more than 10 per cent for about a third of all countries compared (Von Hippel et al. 1992).

For a number of countries in the tropics where CO_2 emissions from deforestation far outweigh emissions from energy consumption, per capita estimates change a great deal depending on the assumptions used to estimate land clearing and biomass levels. As the FAO's once-a-decade study of tropical deforestation and tree plantation establishment and the Brazilian Space Institutes (INPE) detailed remote sensing survey of the Amazon Basin are due to be published in the next few years, estimates of emissions from land use changes should improve significantly. In addition, new international statistics on forest growth in temperate countries recently completed by the FAO/ECE, as well as new country studies for Northern and Central Europe, provide further information on the magnitude of CO_2 uptake in northern forests.

Comprehensive emissions

The additional sources and gases not included in the above list are far more difficult to inventory. Generally, emissions from the minor greenhouse gas N_2O, and CO – which oxidizes to become CO_2 and affects the atmospheric residence time of CH_4 – are highly uncertain. All of the agricultural sources are included in this approach. Of these, the factors that determine the release of CH_4 from livestock enteric fermentation may be the best understood. But even in this case, the accuracy of national estimates for many countries is doubtful at present, because the controlling factors, which include livestock diet, breeding, and management practices, vary from country to country and accurate data are not available for many countries, particularly in the developing world. Measured CH_4 release from rice cultivation varies widely according to soil type, fertilizer application, climate, and irrigation regime, but the net effect of all these conditions on emissions is not yet understood. Calculation of CH_4 release from animal and human wastes has only started to be undertaken in the last two years, and estimates are rough, reflecting extrapolations based on only a few site-specific studies. Emissions of CH_4, CO, and N_2O from biomass burning vary with the extent of crop or forest burning, and the moisture and carbon and nitrogen content of the biota. Emissions of N_2O and CO from the remaining sources are all highly uncertain.

Unlike the sources covered only in the partial CH_4 and CO_2 approach (energy, deforestation, and landfills), the additional sources covered here (livestock, rice cultivation, cement production, and fertilizer consumption) pose greater problems as abatement targets because their control would likely entail directly curtailing economic activities rather than reducing the residuals stemming from these activities. The agricultural and industrial activities they represent may be considered essential subsistence activities by many countries (Parikh et al. 1991), although in the case of livestock management for some animals, reducing CH_4 emissions through changes in diet and breeding may be compatible with development goals (Leng 1991).

Unlike the three CO_2 approaches, the partial and fully comprehensive approaches require an index to compare the heating effect of CH_4 and CO_2 emissions. The problems involved in evaluating the relative warming contribution of the gases include the difference in estimating the atmospheric lifetime of gases (particularly CO_2), calculating indirect effects of the emitted gases, and specifying the most appropriate time period for which to calculate the warming effect (IPCC 1990). In practice, however, the choice of CO_2-equivalent applied to these sources may have little effect on most countries' relative ranking by warming contribution.[e]

Regional and national emissions by source

The difference in relative emissions contribution from industrialized and developing countries is summarized in Table 3.3.

It is clear from Table 3.3 that the more current time frame and the addition of the non-energy sources increases the emphasis of emissions from developing countries. This overall pattern holds true for emissions from selected countries. Relative per capita emissions from ten countries that account for 60 per cent of current CO_2 emissions from fossil fuels appear in

Table 3.3 *Emissions from industrialized and developing countries (% of world total CO_2 equivalent)*

Emissions category	*Industrialized*	*Developing*
1 Cumulative CO_2, energy	86	14
2 Cumulative CO_2, energy and biota	68–80[a]	32–20
3 CO_2, energy (current)	72	28
4 Partial CH_4 and CO_2 (current)	57	43
5 Comprehensive (current)	52–57[b]	48–43

a This range is based on alternate assumptions of historical land clearing rates.
b This range is calculated based only on differences between the short and longer GWPs. If CFCs are excluded from the totals, industrialized countries' emissions comprise 52 per cent of the total assuming the 100 year GWP.

e A statistical analysis of 142 countries ranked by emissions in the partial CH_4 and CO_2 source category revealed that the difference in countries' relative contributions by 20, 100, or 500 year GWP (IPCC 1990 GWPs) was negligible (Standard Pearson 0.991–0.999; Spearman Rank 0.994–0.999). The correlation for countries ranked by total greenhouse gas emissions in the comprehensive approach under the alternative GWPs was lower (Spearman Rank: 0.947–0.979), but strong enough to suggest that in practice the choice of GWPs may not be significant for most countries for any given selection of source categories requiring a gas-equivalent comparison.

Figure 3.3. The group includes the eight greatest emitters, in addition to Mexico and Nigeria.

The bars in Figure 3.3 show countries' emissions levels relative to the global mean. For example, the white bar for Germany indicates that per capita CO_2 emissions from fossil fuels are three times the global per capita mean. The per capita emissions patterns illustrated fall into three general patterns:

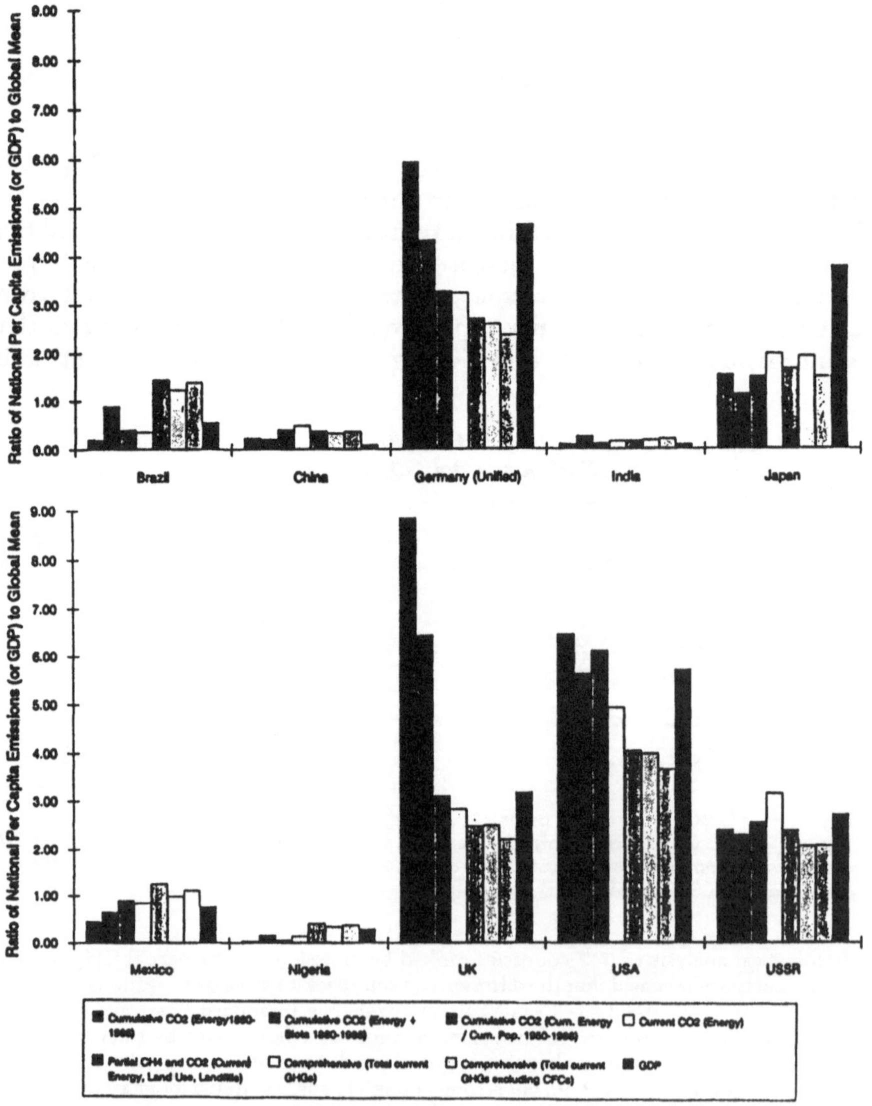

Figure 3.3 *Per capita emissions from selected countries*

1 An upward slope from the bars sequenced first as cumulative emissions followed by current emissions and fossil-related CO_2 followed by emissions from all greenhouse gases and sources. While the shape of curve is tentative for some countries, it is clear that emissions in the current period and emissions from biotic and agricultural sources emphasizes the contributions of these developing countries (Brazil, China, India, and Mexico).
2 The corresponding downward sloping pattern for industrialized countries is more dramatic. The scale of per capita emissions in the historic period and from fossil fuel emissions is significantly greater for these countries (Germany, UK, and USA).[f]
3 A horizontal pattern emerges for the more recently developed countries (Japan and USSR) where per capita emissions by time period is relatively constant and the biotic component minor.

Population-weighted emissions were selected as the most compelling form of comparison. A per capita emissions criterion is intuitively equitable in a spatial sense because it assesses individual responsibility regardless of political borders, although in practice the evenhandedness may be diminished because of individuals' disparate emissions release. Alternative allocators have major flaws. Per land area introduces undeserved entitlements to countries with large uncultivatable or uninhabitable regions. Per GDP is regressive in that late-developing countries tend to have high emissions levels per unit of output, as many developing economies are especially energy intensive. Several regions have advanced a population-based approach. Japan recently pledged to cap future CO_2 emissions at current per capita levels. In preliminary discussions on approaches to meeting its overall CO_2 stabilization goal, the European Community indicated a preference for national targets based on per capita CO_2 emission levels. Analysts and scholars have also favoured the per capita approach, although with added variations and qualifications, e.g. considering cumulative population (Smith et al. 1990), weighting by adult population (Grubb et al. 1992), crediting carbon sinks on a per capita basis (Agarwal and Narain 1991) and designating an intergenerational per capita emissions allotment (Gruebler and Fujii 1991).

In Figure 3.4, I compare three sets of data – greenhouse gas emissions, population, and GDP – for the selected countries. Two general patterns emerge. For the developed countries, the share of the world's total for each of the indicators forms an inverse U-shape, with the share of GDP (and UN

f Note that the cumulative emissions begin to approach current fossil fuel emissions for some countries if CO_2 is weighted by cumulative population rather than by current population only.

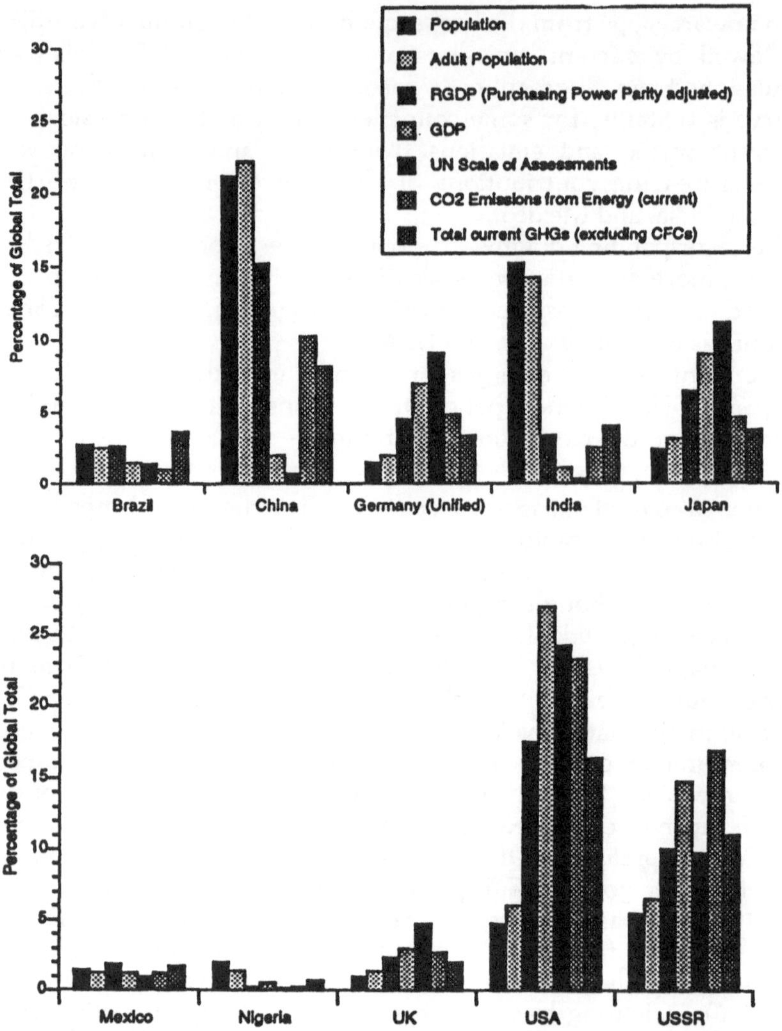

Figure 3.4 *Socioeconomic assessments compared with GHG emissions*

contributions[g]) greater than the countries' contribution to world population and emissions. For instance, Japan makes up 2 per cent of the world's population, produces 9 per cent of the global GDP, and releases only 5 per cent of annual fossil-fuel related CO_2. In contrast, the pattern for developing countries, although not as pronounced, tends to be a U-shape, with relative GDP less than the other indicators. One implication of this pattern found

g The United Nations Scale of Assessments, which is based on a number of economic, geographic and demographic factors, was used as a basis for collecting contributions to the costs associated with the Montreal Protocol.

among the ten countries examined here is that greenhouse gas intensity (GHG/GDP) is greater in developing countries relative to the corresponding ratios in developed countries than is the GHG/population ratio. Developing regions often release more CO_2 (energy related) per unit of output than do developed countries (Grubb et al. 1992). In Figure 3.4, a comparison is also made with relative GDP, or purchasing power parity adjusted GDP, a newer index that attempts to adjust for differences in purchasing power in different parts of the world and for fluctuating exchange rates (Summers and Heston 1988).

While analysts are already discussing ways of adjusting assessments in order to find an equitable solution and the right incentives, in practice the differences between alternate indicators such as population and adult population are often not as great as the differences between the levels of responsibility implied in the alternative emissions source categories discussed above. By way of comparison, Figure 3.4 shows that the difference between developing countries' share of the world's population versus adult population is not as great as the difference between countries' share of emissions from all greenhouse gases versus CO_2 from energy only.

Conclusions

The choice of emissions source category can bear upon countries' implied responsibility for emissions if an agreement is based on some form of the polluter-pays principle.[h] The emissions categories involving the greatest uncertainties in measurement – the more comprehensive approaches – also place relatively greater emphasis on emissions from developing regions, where a high proportion of emissions from agricultural and biotic sources originate. Per capita emissions of CH_4 and N_2O are frequently greater in developing countries.

Cumulative CO_2 emissions from fossil fuels can now be estimated with more precision than can most of the sources in the comprehensive source category. However, the uncertainty associated with estimating current emissions from biotic and agricultural sources should diminish with time whereas flaws in historical data, in particular those relevant to land use, may be immutable. CO_2 from fuel combustion is the most verifiable and

h It should be noted that certain approaches to controlling emissions, such as international tradeable permits schemes and national emissions targets, require highly accurate emission baseline inventories of activities that can be monitored (Victor 1991). There may be a place, however, for simultaneously providing incentives for controlling the less verifiable emissions sources. For example, a permits allocation scheme could be based on per capita fossil-fuel-derived CO_2 only, but provide less precisely defined 'credits' for afforestation or CH_4 recapture facilities in landfills and coal mines. Such controls, as suggested by Grubb et al. (1991) could be matched to measurable criteria such as area of forest cleared.

comprises about 65 per cent of current contribution to global warming.

The partial CH_4 and CO_2 category, which includes CO_2 and CH_4 from energy, industrial, and biotic sectors (and excludes the more difficult-to-measure agricultural activities and the minor trace gases) makes up about 80 per cent of the total warming effect from current emissions (excluding halocarbons). It would be significantly more difficult to monitor emissions from these sources than to monitor CO_2 emissions from fossil fuel combustion, but for a variety of reasons, economic as well as environmental, it may be time to develop emissions assessments applied to this more inclusive, but not comprehensive, approach.

References

Agarwal, Anil, and Sunita Narain, 1991. *Global Warming in an Unequal World: A Case of Environmental Colonialism*. New Delhi: Centre for Science and Environment

Bingemer, H G, and P J Crutzen, 1987. 'The Production of Methane from Solid Wastes.' *Journal of Geophysical Research* 90(D2): 2181–2187

Bolin, Bert, 1986. 'How Much CO_2 Will Remain in the Atmosphere?' in *The Greenhouse Effect, Climatic Change, and Ecosystems*. Chichester, West Sussex: John Wiley & Sons

Brown, S, A J R Gillespie and A E Lugo, 1989. 'Biomass Estimation Methods for Tropical Forests with Applications to Forestry Inventory Data.' *Forest Science* 35: 881–902

Casada, M E, and L M Safely, Jr, 1990. *Global Methane Emissions from Livestock and Poultry Manure*. A report submitted to the US Environmental Protection Agency by the Biological and Agricultural Engineering Department, North Carolina State University

Crutzen, P J, and M O Andreae, 1990. 'Biomass Burning in the Tropics: Impact on Atmospheric Chemistry and Biogeochemical Cycles.' *Science* 250: 1669–1677

Crutzen, P J, I Aselmann and W Seiler, 1986. 'Methane Production by Domestic Animals, Wild Ruminants, Other Herbivorous Fauna, and Humans.' *Tellus* 38B: 271–284

ECE/FAO, 1985. *The Forest Resources of the ECE Region* (Europe, USSR, North America). Geneva: The United Nations

Eichner, M J, 1990. 'Nitrous Oxide Emissions from Fertilized Soils: A Summary of Available Data.' *Journal of Environmental Quality* 19: 272–280

FAO, 1987. *1986 Production Yearbook*. Rome: FAO

FAO, 1988a. *FAO Fertilizer Yearbook*. Rome: FAO

FAO, 1988b. *An Interim Report on the State of the Forest Resources in the Developing Countries*. Rome: FAO

FAO, 1990a. *Interim Report on Forest Resources Assessment 1990 Project*. Committee on Forestry, Tenth Session, Food and Agricultural Organization of the United Nations, Rome.

FAO, 1990b. *1988 Production Yearbook*. Rome: FAO

Fearnside, P M, A T Tardin and L G M Filho, 1990. *Deforestation Rate in Brazilian Amazon*. National Secretariat of Science and Technology, Brazil

Grubb, M, J Sebenius, A Magalhaes and S Subak, 1992. 'Sharing the Burden' in *Confronting Climate Change*. I M Mintzer (ed). Cambridge: Cambridge University Press

Grubb, M J, D G Victor and C W Hope, 1991. 'Pragmatics in the Greenhouse.' *Nature* 354(6352): 348–350

Gruebler, A, and Y Fujii, 1991. 'Inter-Generational and Spatial Equity Issues of Carbon Accounts.' *Energy* 16(11/12): 1397–1416

Houghton, R A, 1991. 'Tropical Deforestation and Atmospheric Carbon Dioxide.' *Climatic Change* 19: 99–118

ICF, 1990. 'Emissions Estimates by Country.' Memo from Craig Ebert and Amy Kim of ICF to Paul Schwengels and Dillip Ahuja of US EPA Office of Global Change (September 25)

IPCC, 1990. *Climate Change – The IPCC Scientific Assessment*. Cambridge, UK: Cambridge University Press

IPCC, 1991. IPCC Working Group One. *Greenhouse Gas Sources and Sinks: Update*

IPCC, 1992. *Climate Change 1992: The Supplementary Report of the IPCC Scientific Assessment*, Cambridge, Cambridge University Press

Keeling, Charles, 1973. 'Industrial Production of Carbon Dioxide from Fossil Fuels and Limestone.' *Tellus* XXV(2)

Krause, F, W Bach and J Koomey, 1989. *Energy Policy in the Greenhouse*. El Cerrito, CA: International Project for Sustainable Energy Paths

Leng, R A, 1991. *Improving Ruminant Production and Reducing Methane Emissions from Ruminants by Strategic Supplementation*. Office of Air and Radiation, US EPA, EPA/400/1-91/004. June

Logan, J A, M J Prather, S C Wofsy and M B McElroy, 1981. 'Tropospheric Chemistry: A Global Perspective' *Journal of Geophysical Research* 86: 7210–7254

Marland, Greg, T A Boden, R C Griffin, S F Huang, P Kanciruk and T R Nelson, 1988, 1989, 1990. *Estimates of CO_2 Emissions from Fossil Fuel Burning and Cement Manufacturing Using the United Nations Energy Statistics and the U.S. Bureau of Mines Cement Manufacturing Data*. NDP030. Oak Ridge, Tennessee: Carbon Dioxide Information Analysis Center, Oak Ridge National Laboratory

Mitchell, B R, 1981. *European Historical Statistics: 1750–1975*. New York: Facts on File

Mitchell, B R, 1982. *International Historical Statistics: Africa and Asia*. New York: New York University Press

Mitchell, B R, 1983. *International Historical Statistics: The Americas and Australia*. Detroit: Gale Research Co

Myers, N, 1989. *Deforestation Rates in Tropical Forests and Their Climatic Implications*. A Friends of the Earth Report (December)

OECD, 1991. *Country Positions on Climate Change*. Group on Energy and Environment. ENV/EC/EN (91)4

OECD/IEA, 1990a. *Energy Balances of OECD Countries 1987–1988*. OECD, Paris

OECD/IEA, 1990b. *World Energy Statistics and Balances 1985–1988*. OECD, Paris

Parikh, J, K Parikh, S Gokarn, J P Painuly, B Saha and V Shukla, 1991. *Consumption Patterns: The Driving Force of Environmental Stress*. Bombay, India: Indira Gandhi Institute of Development Research

Richards, John, Jerry Olson and Ralph Rotty, 1983. *Development of a Data Base for Carbon Dioxide Releases Resulting from Conversion of Land to Agricultural Uses*. Publication No. 2181. Oak Ridge, Tennessee: Institute for Energy Analysis, Oak Ridge Associated Universities

Schuetz, H, A Holzapfel-Pschorn, R Conrad, H Rennenberg and W Seiler, 1989. 'A Three-Year Continuous Record on the Influence of Daytime, Season and Fertilizer Treatment on Methane Emission Rates from an Italian Rice Paddy.' *Journal of Geophysical Research* 94(D13): 16405–16416

Siegenthaler, U, and H Oeschger, 1987. 'Biospheric CO_2 Emissions During the Past 200 Years Reconstructed by Deconvolution of Ice Core Data.' *Tellus* 39B: 140–154

Smith, K, 1991. 'Allocating Responsibility for Global Warming: The Natural Debt Index.' *Ambio* 20(2)

Smith, K R, J Swisher, R Kanter and D R Ahuja, 1990. *Indices for a Greenhouse Gas Control Regime: Incorporating Both Efficiency and Equity Goals.* Prepared for the Environmental Policy and Research Division of the World Bank (December); see Chapter 4 of this book

Subak, Susan, and William Clark, 1990. 'Accounts for Greenhouse Gases: Towards the Design of Fair Assessments,' in *Usable Knowledge for Managing Global Climatic Change.* William C Clark (ed.). Stockholm, Sweden: Stockholm Environment Institute

Subak, Susan, Paul Raskin and David Von Hippel, 1993. National Greenhouse Gas Accounts: Current Anthropogenic Sources and Sinks. *Climatic Change.* November

Summers, R, and A Heston, 1988. 'A New Set of International Comparisons of Real Product and Prices: Estimates for 130 Countries, 1950–1985.' *Journal of International Association for Research Income and Wealth* 34(1): 1–26

United Nations, 1988. *1986 Energy Statistics Yearbook.* New York: United Nations Department of Economic and Social Affairs

United Nations, 1990. *1988 Demographic Yearbook.* New York: United Nations

UNEP, 1990. *Report of the Secretariat on the Reporting of Data by the Parties in Accordance with Article 7 of the Montreal Protocol* (November 14) Nairobi, Kenya

US/Japan Working Group on Methane, 1992. *Technological Options for Reducing Methane Emissions: Background Document of the Response Strategies Working Group,* Washington DC Environment Agency of Japan, US AID and US EPA

Victor, D, 1991. 'Limits of Market-based Strategies for Slowing Global Warming: The Case of Tradeable Permits.' *Policy Sciences* 24: 199–222

Von Hippel, D, P D Raskin and S Subak, 1993. 'Estimating Greenhouse Gas Emissions from Energy: Two Approaches.' *Energy Policy,* June

Wigley, T M, 1991. 'A Simple Inverse Carbon Cycle Model' *Global Biogeochemical Cycles* 5: 373-382

World Bank, 1990. *World Development Report 1988.* Oxford: Oxford University Press

Appendix A: Estimates of greenhouse gas emissions

This list gives estimates of the emissions used in this chapter.

1 Cumulative CO_2 energy (1860–1986):

Fossil fuel combustion:	178 GT carbon as CO_2 (CO_2 C)

2 Cumulative CO_2 (1860–1986):

Fossil fuel combustion:	178 GT CO_2 C
Land use changes:	66 GT CO_2 C
Total:	244 GT CO_2 C

3 Current CO_2 energy (1988):	
Fossil fuel combustion:	5.4 GT CO_2 C/year
4 Partial CH_4 and CO_2 (1988):	
Fossil fuel combustion:	5.4 GT CO_2 C
Land use changes:	0.9 GT CO_2 C
Landfills:	36 MT CH_4 (170 MT Carbon Equivalence (CE))
Fossil fuel production:	74 MT CH_4 (424 MT CE)
Total:	6.9 GT CE
5 Comprehensive emissions (1988)	
In addition to the above:	
Fossil and wood combustion:	201 GT CO C (341 MT CE) 1.3 MT N_2O (103 MT CE)
Cement production:	150 MT CO_2 C
Halocarbons:	1.4 MT CFC-11 equivalent (1,337 MT CE)
Biomass burning:	36 MT CH_4 (170 MT CE), 276 MT CO C (251 MT CE), 1.6 MT N_2O (126 MT CE)
Soil release, tropical pasture:	0.1 MT N_2O (8 MT CE)
Enteric fermentation:	75 MT CH_4 (354 MT CE)
Animal wastes:	28 MT CH_4 (132 MT CE)
Rice cultivation:	98 MT CH_4 (463 MT CE)
Fertilizer consumption:	0.8 MT N_2O (63 MT CE)
Total:	10.3 GT CE

Appendix B: Calculating cumulative and current emissions

This Appendix summarizes the sources and methods used for calculating emissions listed above.

Cumulative CO_2, energy

Carbon dioxide emissions from fossil fuel combustion between 1860 and 1986 rely on Marland et al. (1988) for the 1950–1986 period, and Subak and Clark (1990) for emissions between 1860 and 1949. The cumulative estimates do not take into account the proportion of trace gas removed from the atmosphere. Energy consumption data used in Subak and Clark (1990) are based on Mitchell's (1981, 1982, and 1983) International Historical Statistics series. Global emissions factors were derived from Marland et al. (1988) and weighted by carbon density estimates by nation published in the United Nations *1986 Energy Statistics Yearbook* (1988). In cases where political borders have changed since 1860, emissions were assigned to countries based on estimated energy use share. For example, fossil fuel consumption in the Indian States was assigned as follows: India, 80 per cent; Pakistan, 15 per cent; and Bangladesh, 5 per cent.

Cumulative CO_2, energy and biota

Emissions of CO_2 between 1860 and 1986 are based on the fossil fuel data set described above and the Richards et al. (1983) database on CO_2 release from forest conversion to agricultural purposes. The Richards et al. data set for the 1860–1978 period is based on historical agricultural censuses and FAO land use surveys completed in 1950. To update the database to 1986, we used the FAO *1986 Production Yearbook* (FAO 1987). As forest conversion to non-agricultural uses was not included, this database is not intended to be a comprehensive survey of CO_2 emissions from land use changes.

Current CO_2, energy

Current CO_2 emissions from fossil fuel combustion are calculated at the Stockholm Environment Institute (Von Hippel et al. 1992) based on 1988 energy consumption data published by the OECD/IEA (1990a, 1990b). Carbon dioxide emissions from oil flaring were taken from the Marland et al (1990) compendium. As in the cumulative CO_2 inventories, emissions from renewables, that is, fuelwood, are assumed to be in a steady-state, with no net CO_2 emissions.

Partial CH_4 and CO_2

Methane emissions from coal mining are derived from ICF (1990b) and natural gas transportation and distribution from OECD/IEA (1990a, 1990b). CO_2 release from deforestation is based on land clearing estimates from FAO (1990), Fearnside et al. (1990), FAO (1988b) and Myers (1989), biomass levels by Brown et al. (1989) and carbon soil emission rates by Houghton (1991). Afforestation rates are primarily from ECE/FAO (1985) and FAO (1988b). The landfill CH_4 source is based on a methodology outlined by Bingemer and Crutzen (1987) and waste generation, landfilling and waste composition information is compiled from disparate sources.

The current emissions are expressed in CO_2 equivalent units, which compare the relative warming contribution of the trace gases. The CO_2 equivalents are based on each trace gas Global Warming Potential (GWP), an index that includes the immediate radiative effect of the gases and the potential warming effect over the time the trace gas resides in the atmosphere. The GWP used in this study is calculated by the Intergovernmental Panel on Climate Change (IPCC) and corresponds to a 100 year time horizon.

Comprehensive emissions

To estimate methane emissions from livestock production, emission factors (Crutzen et al. 1986) were applied to FAO livestock population estimates (1990b), and Casada and Safely's (1990) study of CH_4 release from animals wastes was used. Methane emissions from rice cultivation are derived from emission factors (Schuetz et al. 1989) and rice cultivation area (FAO 1990b). Emissions of N_2O from fertilizer consumption was calculated using the mid-range of Eichner's (1990) emission factors and data from the *FAO Fertilizer Yearbook* (1988a). Halocarbon emissions were calculated using ICF's (1990) methodology for converting from UNEP's (1990) production figures to emissions. Release of CO_2 from cement manufacturing was derived using emission factors from Marland et al. 1988. Biomass burning estimates were taken from Crutzen and Andreae (1990), and adjusted to avoid double counting with the fuelwood and deforestation emissions.

4

Who pays (to solve the problem and how much)?

Kirk R Smith, Joel Swisher and Dilip R Ahuja

Those advocating creation of an international programme to address global warming from greenhouse gas (GG) emissions are required to face, among other tasks, five categories of questions (Smith et al. 1991):

1 Is there adequate theoretical and observational evidence of significant potential harm if nothing is done?
2 If so, could a feasible programme of greenhouse remediation accomplish sufficient benefits to be justified?[a]
   - 2a What part of this programme is best devoted to reduction of GG emissions or to increases in GG sinks (natural or anthropogenic processes that absorb GGs from the atmosphere)?
   - 2b What part of this programme is best devoted to reduction of human vulnerability to global warming through, for example, accelerated economic growth of certain kinds in countries with large poor populations?
3 If so, is there a rational and politically acceptable way of establishing priorities among potential remediation projects?
4 If so, is there a rational and politically acceptable way of allocating the costs for these projects?
5 If so, what kind of international institutional mechanisms are needed to facilitate the financing and implementation of such projects?

Although there is by no means universal agreement, many observers believe that the answers to questions 1 and 2 are likely to be in the affirmative, that

a Here, by remediation, we refer to a wide range of actions that result in the reduction of net emissions from GG sources, an increase in net GG sequestering by GG sinks, or decreases in the vulnerability of human populations to warming. Potentially there are also other ways to reduce the threat of climate change from GG emissions, such as artificially blocking sufficient incoming sunlight from reaching Earth to balance increased warming (USNAS 1991). At present, these options do not seem likely to be feasible, but should their prospects improve, they could be judged in the same type of indexing framework considered here.

is, there could be significant risk without action and significant reduction of risk with action. In any case, it is not our purpose to address these issues directly. Rather, we focus on the last three questions, with particular emphasis on 3 and 4, the means to decide both what needs to be done and who will pay.

The most common approach to question 3 (what should be done) in both international negotiations and unilateral declarations has been uniform cuts. Several European nations, for example, have proposed to unilaterally cut their own emissions by 5–25 per cent. Alternatively, with nearly the same result, it may be proposed to limit emissions to those of a particular year, 1990, for example, in some of the UNCED discussions. These approaches are similar to that followed in the original Montreal Protocol where the signatory nations agreed to cut production of selected compounds to 50 per cent by a specified time. A uniform cut in greenhouse gases was proposed by a number of European countries at the UNCED meeting, but not accepted by the USA and Japan. There are major problems with this (two political and one economic):

- By grandfathering currently inefficient emissions, uniform emissions reductions may seem to penalize those countries, like Japan, that have been able to develop economies that already emit less per unit of economic output.
- Equal reductions based on current emissions would be clearly unacceptable to developing countries as it would not allow the growth required to meet their development needs.
- Uniform cuts, by ignoring that the marginal costs of reductions may be quite different among countries, are likely to lead to substantial economic inefficiencies, that is, to be unnecessarily expensive.

Alternatives to uniform cuts that consider both equity and efficiency are described in the next two chapters. Here, our focus is on question 4: who should pay?

Indices of allocation: a brief review

Several investigators have attempted to allocate the global carbon budget based on exogenous considerations of the maximum acceptable warming or its rate of increase (for example, Krause et al. 1992), world averages (Mukherjee 1992), economic optimization models (Michaelis 1992), or other factors (Gurney 1991).[b]

Dividing emissions rights equally among countries, coupled with the

b See also the discussion and references in Chapter 2. Solomon and Ahuja (1991) have reviewed this subject.

ability to sell or lease those rights, is the simplest scheme, yet fraught with inequities because it does not link emissions to human beings or activities. Thus it has few, if any, proponents. Another straightforward basis for allocating rights is land area (Westing 1989). Since 1950, national boundaries have not changed much (leaving aside the national break-ups of the early 1990s). Its stability as a measure, the ease of measurement, the avoidance of monitoring and verification difficulties are what recommend it. (Cheating is difficult.) There was a time, according to Grubb (1989), when the United States was arguing informally in international fora that its continental land mass necessitated enormous energy expenditures in having to move goods and people. Ultimately, with the possible exceptions of those countries with large wastelands (for example, Mongolia), land area is a measure of natural resources. Using it as an index to allocate emissions rights, however, favours large but sparsely populated nations (for example, Australia) and discriminates against small densely populated nations (for example, Japan).

If it is accepted that every person has an equal right to atmospheric resources – the ultimate global commons – then the most obvious and equitable basis is to distribute emissions permits in proportion to national populations (Feiveson et al. 1988; Agarwal and Narain 1991). If rights in subsequent years continue to be proportional to contemporaneous populations, however, a perverse incentive for population growth may be created. For this reason, and to make his scheme more palatable to industrialized countries, Grubb (1989) has suggested that allocations be based on adult populations. This would have the effect of reducing net transfers from countries with rectangular age distributions to developing countries with pyramidal age structures, but could be seen as discrimination against children. Depending on the definition of 'adult,' it would provide a 15–21 year delay between births and receiving the allotment, and thus reduce the pro-natalist incentive.

An alternate incentive for population stabilization could be built into the scheme by pegging the allotment to the entire population in a recent year and not increase future allotments. Compared to an index based on adult population, this would seem to represent less discrimination against children in the first years of an international protocol and no more discrimination in later years.

Arguing that any index based on per capita emissions alone would require unacceptably huge reductions in industrial countries (up to 75 per cent) or entail massive transfer payments to developing countries, Wirth and Lashof (1990) have proposed apportionment based half on per capita and half on per GDP, all the quantities being for the current or a recent year.[c]

Similarly a multiplicative index could be structured that is directly proportional to emissions and inversely proportional to both GDP and

c Allocation of Emissions (AE) = (0.5 (emissions/population) + 0.5 (emissions/GDP)).

population, the ratio being integrated over time. It is not clear, however, if GDP should find a place in an index for allocation, since countries would have already benefited from that economic activity.

Accountability

In this book, we are coming to these issues from a somewhat different direction. Rather than decide on what the ideal allocation of emissions ought to be, we first seek ways that the present and historical patterns of emissions can be used in international negotiations to determine who should pay for any needed mitigation efforts and then, in later chapters, ways that the best mitigation efforts can be chosen. Thus, rather than concerning ourselves directly with *allocation*, we address *accountability*. In the long run, of course, consistent application of accountability should lead to a desired allocation by the simple process of nations attempting to reduce their accountability, a sort of 'invisible hand'. In the interim, however, rather than putting an onus on those countries that have exceeded their allocations, a focus on accountability simply asks that nations should accept responsibility for the emissions they have made, no matter how small or large. The result can be the same, but the moral implications are different.

To make practical the concept of individual rights over time, in this book, we link accountability at any one time to the amount of atmospheric assimilative capacity that has been 'borrowed' from the natural environment, individuals' natural debt as presented in Chapter 2 (Smith 1989b, 1991). The borrowed capacity at any one time is the greenhouse gases remaining in the atmosphere from past emissions (above natural levels). This is less than what was actually emitted, since various natural and human-influenced sinks have absorbed the different gases in amounts depending on the time since emissions. The longer ago the gases were released, the less remains today. We argue that an appropriate indicator of international accountability is the amount of assimilative capacity borrowed to date, the natural debt.[d]

Equity and efficiency

It may be easier to find a point of international agreement on mitigation costs by separating the negotiating criteria, and the indices to measure them, according to the two general questions that we have set out to address (numbers 3 and 4 above). Using this approach, we can consider the

d Grubb (1989) objects to this scheme on grounds of impracticality - industrial countries would not agree - and ignorance - countries in the past were unware that they were depleting a finite resource. See counter-arguments below.

competing goals of equity and efficiency, while maintaining a rational (but still negotiable) basis for assigning obligations at the national level.

We categorize the basic negotiating criteria in terms of the following questions:

- To determine the best projects (question 3) these questions must be addressed:
  - What are the goals (globally, and who should do what)?
  - What are the best opportunities (who can do what)?
- To determine who will pay (question 4) these questions must be addressed:
  - Where are the resources available (who can pay)?
  - Who has responsibility for the problem (who should pay)?

These four questions are organized in Figure 4.1. The 'Who can . . .' questions in the left column are addressed by criteria that can be measured according to physical and financial quantities, and the 'Who should . . .' questions in the right column are addressed by criteria that, while they can be quantified, must involve a large degree of value judgment. The 'Pay for . . .' questions in the top row involve equity criteria, based on past and

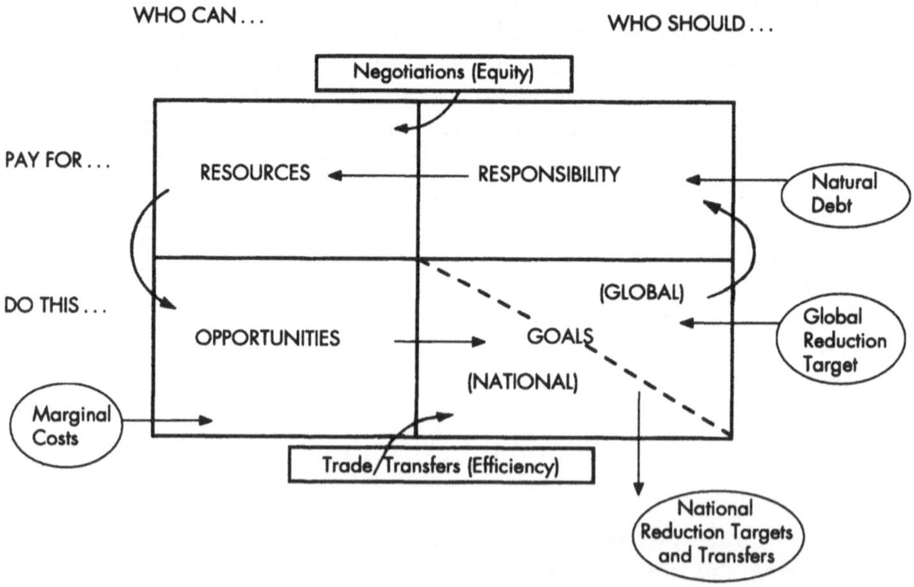

This matrix divides the two questions (#3, which mitigation projects should be chosen? and #4, who should pay for them?) into practical and normative (ethical) components based on available resources and opportunities and historical responsibility and future goals (Smith et al. 1991).

Figure 4.1 *Typology and indices for allocating greenhouse gas emission reductions*

present activities, while the 'Do this . . .' questions in the bottom row involve efficiency criteria, based on present and proposed future action.

In Figure 4.1, the intersection of the rows and columns form four cells that contain the four criteria just defined. The ovals represent information flows that measure the corresponding criterion. For example, marginal cost is a measure of emissions reductions opportunities. The rectangles indicate allocation processes that would be needed to reconcile different criteria: resources and responsibility through international negotiation; and opportunities and goals through international trade or transfers.

The starting point of a negotiating process is the global goal for emissions reductions, based on a perceived common vulnerability to climate change. This goal may be the result of any combination of scientific, economic, and political considerations (questions 1 and 2, above). Once the general goal is set, each country's share of the responsibility for causing the problem can be determined, based on its past and present contribution to the source of the problem, namely carbon dioxide and other greenhouse gas emissions.

Responsibility is a useful but incomplete measure of a country's accountability for financing emissions reductions, however. A negotiated solution must also consider a country's available resources with which to pay. The overall obligation to pay (OTP), thus, addresses two issues, one ethical and one practical. The ethical issue is that those countries that have contributed most to the problem (and benefited thereby) should have some obligation to pay for its amelioration. The practical consideration is simply that a solution to the problem is more likely if those countries that have greater resources are willing to pay relatively more of the total cost.[e]

By having a separate indicator for each of the top two boxes of Figure 4.1 (resources and responsibility), therefore, international negotiations can proceed in an orderly way to trade one against the other to obtain the politically optimum mixture that becomes the obligation to pay.

Once the obligations to pay for each nation are determined, they can be compared with the international distribution of opportunities for emissions reductions. Clearly, it should not be expected that the opportunities will be distributed among countries in the same way as obligations. Some countries will have a relatively high concentration of opportunities, while others will have relatively large obligations to pay for emissions reductions. The resolution of these differences will be taken up in later chapters.

As there are several concerns that need to be addressed in any scheme to determine obligations, the primacy given to such criteria as simplicity, equity, efficiency, the perceived ease of reaching agreements, etc., leads to different indices for obligation. Yet, it seems to us that it is preferable, and perhaps easier to obtain agreements as well, if the indices for different

e In addition, it can be argued that those nations with less resources have a relatively greater need to apply resources to reduce their future vulnerability to climate change rather than to international schemes for greenhouse gas reduction (Smith 1989a).

objectives are kept separate (and reconciled later in the negotiating process) than to assert that they are indicators of something that they do not measure.

We thus divide the question of 'Who pays?' in Figure 4.1 into two parts: (1) the 'ability to pay', which is indicated by present wealth and is completely separate from greenhouse gas emissions, and (2) a 'responsibility' index, based on cumulative per capita emissions. These address the respective sub-questions 'Who can pay?' and 'Who should pay?' of Figure 4.1.

Ability to pay (ATP)

In spite of its well-known difficulties, gross national or domestic product (GNP or GDP) is accepted widely as an index of national economic resources. GDP is the basic determinant of a country's contribution to the UN system, for example.[f] With certain modifications, the international fund set up under the Montreal Protocol (Table 4.1) also relies on the UN scale to calculate a country's contribution to be used for technology transfer and financial assistance to signatory developing countries to use safer substitutes for CFCs. As seen in Table 4.1, however, by comparison, the contributions to the Global Environment Facility have come more from Western Europe and a few of the larger developing countries.

There are a few methodological problems with international comparisons of GNP statistics, arising from the presence of large informal sectors in some economies, the vagaries of fluctuating exchange rates, and from the differences in purchasing power. To some degree, adjustments can be made to correct these problems, for example in the development of purchasing power parity statistics (Summers and Heston 1988).

There is one drawback to using GNP, even if corrected for purchasing power, as an index of payments into a fund for the mitigation of the climate-change problem. This is that no allowance is made for 'disposable national income', so that poor countries like Indonesia and rich ones like Sweden would be expected to make similar contributions because their population ratio happens to be approximately equal to the inverse of their ratio of per capita GNP. To determine a country's ability to pay (ATP), it seems obvious that some measure of wealth based on per capita income is required.

One way to define an ability to pay would be to subtract from the GNP, some threshold of 'basic need.' In the case of the global fund set up under the amendments to the Montreal Protocol, countries that emit less than 300 grams of CFC-equivalent per capita per year are exempted from contributing. In a similar fashion, the Global Environment Facility uses a cut-off of $4,000 per capita for determining certain categories of contributions and

f The UN scale is not by any means static, but subject to continuous re-evaluation and negotiation as well as periodic updating (UN 1989, 1992), processes that would need to be part of any international greenhouse regime.

Table 4.1 *Relative and total contributions to the Global Environment Facility, the Montreal Protocol and the two combined*

	Global Environment Facility[a]	*Montreal Protocol*[b]	*Total (weighted)*[a,b]	*United Nations scale*[a]
High Income				
USA	12.4	25.0	13.5	25.0
Germany, United	13.0	10.8	12.8	9.31
France	13.3	7.3	12.7	6.25
Japan	12.6	13.2	12.7	11.38
Italy	7.5	4.7	7.2	3.99
UK	6.2	5.7	6.2	4.86
Switzerland[b]	4.7	1.3	4.4	1.08
Netherlands	4.4	1.9	4.2	1.65
Austria	3.1	0.9	2.9	0.74
Sweden	2.9	1.4	2.8	1.21
Norway	2.3	0.6	2.1	0.55
Canada	1.8	3.6	2.0	3.09
Finland	2.1	0.6	1.9	0.51
Australia	1.9	1.8	1.9	1.57
Denmark	1.9	0.8	1.8	0.69
Spain	1.2	2.2	1.3	1.95
(USSR)[c]	0	13.5	1.3	11.57
Belgium	1.2	1.3	1.2	1.17
New Zealand	0	0.3	0.03	0.24
Ireland	0	0.2	0.02	0.18
UAE	0	0.2	0.02	0.19
Singapore	0	0.2	0.01	0.11
Luxembourg	0	0.1	0.01	0.06
Iceland	0	0.0	0.003	0.03
Bahrain	0	0.0	0.002	0.02
Liechtenstein[b]	0	0.0	0.001	0.01
Middle Income				
Mexico	0.47	0	0.4	0.94
Brazil	0.47	0	0.4	1.45
South Africa	0	0.6	0.1	0.45
Czechoslovakia	0	0.5	0.04	0.66
Greece	0	0.4	0.04	0.4
Poland	0	0.4	0.04	0.56
Portugal	0	0.2	0.02	0.18
Hungary	0	0.2	0.02	0.21
Bulgaria	0	0.1	0.01	0.15
Malta	0	0.0	0.0007	0.01
Low Income				
China	0.47	0	0.4	0.79
Egypt	0.47	0	0.4	0.07
India	0.47	0	0.4	0.37
Indonesia	0.47	0	0.4	0.15

Table 4.1 *Continued*

	Global Environment Facility[a]	*Montreal Protocol*[b]	*Total (weighted)*[a,b]	*United Nations scale*[a]
Morocco	0.47	0	0.4	0.04
Pakistan	0.47	0	0.4	0.06
Turkey	0.47	0	0.4	0.32
World Bank	3.4	0	3.0	-
TOTAL (%)	100	100	100	94.2
Total (million US$)	1212	127	1338	2147

Listed for each nation are the percentages of the global totals as of mid-1992. The nations are divided into income classes, and their total contributions are ranked within each class. Shown in the last column are 1989-91 assessments for dues to the United Nations. Data from UN (1989), NZMERT (1990), UNEP (1992), and World Bank (1992). 0 = zero; 0.0 = very small.

a The present maximum UN contribution is set at 25 per cent, although it started at 40 per cent in 1946. Otherwise the US contribution would be higher than the 25 per cent shown here. The original floor was set at 0.04 per cent, but in 1973 was set to 0.01 per cent (UN 1989).
b Not a full UN member.
c This total is divided as follows: Russian Republic = 86 per cent; Ukraine = 11 per cent; Belarus = 3 per cent.

recipients. Another such cut-off that could be used, for example, would be one of the 'poverty line' estimates discussed in the *World Development Report* (World Bank 1990). The original precedent is the UN scale of assessments set up in 1946, which subtracts a threshold income (originally $1,000 per capita, more recently set to $2,200), before calculating dues (UN 1989).[g]

One argument for using such a cut-off is that countries below this level require all the resources they have to bring their populations up to a minimum acceptable income. A rationale related directly to climate change is that countries below this limit are more vulnerable to the effects of adverse climate change and should devote most of their resources to reduce this vulnerability, which will have great advantages even if global warming does not occur in the period of concern.

Many would argue, however, that the poverty line is not a sufficiently high goal, and yet it is difficult to define some other income level as an acceptable minimum. Here, we take the somewhat different approach of choosing an income that seems to be capable of achieving a minimum 'quality of life' based on the past ability of countries to achieve adequate levels of

g The UN scale also gives special consideration to those nations needing debt relief and is constrained by limits as to maximum assessments. See Table 4.1 for a comparison of the present international assessments.

infant mortality, life expectancy, and literacy. These are combined in the Physical Quality of Life Index (PQLI) developed by Morris (1979).[h]

Thus, we choose as a cut-off, income which is the average income of nations with a Physical Quality of Life Index of 80–90.[i] This represents the approximate inflection point in the relationship of PQLI and GNP, or purchasing power parity (PPP) (see Figure 4.2). Below this point, small increases in income often lead to large increases in PQLI, while above it, large increases in GNP only produce modest increases in PQLI.[j]

The second column of Table 4.2 shows this index calculated with two indicators of income: GNP and GDP corrected for PPP. The table shows that the US ATP changes by only 10 per cent (from 37 to 40) if income is corrected for PPP, but that other countries' ATP can change dramatically. The USSR ATP, for example, changes from 18 to 12 per cent (of the world total),[k] while that of Mexico increases by a factor of 18 and Romania goes down by almost a factor of five.[l]

Responsibility to pay

Indices of ability to pay do not take into account the responsibility of a country for having caused the problem (the polluter-pays-principle). Responsibility itself can be measured in different ways. The two main ways are: direct historic (including current) contributions to causing the problem; and how efficiently a country has been using its resources. As with ATP, we present a formulation that allows us to address both separately.

As discussed in Chapter 2, the contribution of a gas to global warming is a result of Earth's exposure to the gas, which in turn is a function both of atmospheric concentration and residence times (Smith and Ahuja 1990).

---

h PQLI is composed of three equally weighted indicators: life expectancy, infant mortality, and literacy (Morris 1979). More recently, the UNDP (1991) has proposed the Human Development Index (HDI), which is composed of three indicators: life expectancy, educational attainment, and adjusted income per capita.

i This figure was about $1,800 GNP per capita in 1986 or $3,400 per capita PPP in 1980.

j Thus, a country's ability to pay can be defined as being:

$$ATP = GNP - [POP \times (GNP/cap_{PQLI})]$$

where GNP and POP total national income and population, respectively, and GNP/cap_{PQLI} is the threshold income per capita. PPP can be substituted for GNP.

k These calculations were done before more recent PPPs were available. Using 1988 data (Summers and Heston 1991), the average PPP per capita of those countries with PQLIs of 80–90 was about $4,500, while the shift from GNP to PPP would increase the US ATP by about 25 per cent.

l These calculations need to be redone for the independent states that formerly made up the USSR. The relative ATPs would reduce substantially because the economies of these states have declined dramatically as political and economic transitions proceed. Note that ATP is a current, not a cumulative, measure. If income goes down, so does ATP, no matter what the income was in the past. This gives a reduction in international obligation for nations having current economic problems that might threaten the quality of life of their citizens.

Table 4.2 *Ability to pay, responsibility and obligation to pay for reduction of greenhouse gas emissions by country*

	Ability to pay		*Responsibility*			*Obligation to pay (GNP)*		*Obligation to pay (PPP)*	
	GNP	*PPP*	*5t*	*10t*	*20t*	*5t*	*10t*	*5t*	*10t*
USA	37	40	34	37	40	35	37	37	39
Germany, United	7.6	7.9	8.3	8.8	9.3	7.9	8.2	8.1	8.3
Canada	3.1	3.2	2.6	2.8	2.9	2.8	2.9	2.9	3.0
Czechoslovakia	1.1	0.92	1.6	1.7	1.8	1.4	1.4	1.3	1.3
Belgium	0.71	0.96	1.0	1.0	1.1	0.84	0.9	1.0	1.0
United Kingdom	3.9	6.8	5.3	5.7	5.9	4.6	4.8	6.1	6.2
Australia	1.6	1.2	1.2	1.3	1.3	1.4	1.4	1.2	1.2
Poland	0.11	0.96	2.6	2.7	2.7	1.3	1.4	1.8	1.8
USSR	18	12	20	20	20	19	19	16	16
Netherlands	1.2	1.3	0.9	0.9	0.9	1.0	1.1	1.1	1.1
Bulgaria	0.44	0.22	0.5	0.6	0.5	0.49	0.50	0.38	0.39
France	4.8	5.6	3.2	3.3	3.1	4.0	4.0	4.4	4.5
Japan	13	9.2	6.0	6.0	5.5	9.5	10	7.6	7.6
Hungary	0.02	0.38	0.5	0.5	0.5	0.26	0.27	0.44	0.45
Romania	0.94	0.20	0.9	0.9	0.8	0.94	0.94	0.57	0.57
South Africa	0	0	1.2	1.2	1.0	0.61	0.60	0.82	0.81
Italy	3.8	3.6	1.9	1.8	1.4	2.8	2.8	2.7	2.7
Korea, Dem.	0	0	0.5	0.5	0.3	0.27	0.25	0.27	0.25
Spain	1.1	1.7	0.9	0.8	0.4	1.01	0.95	1.29	1.2
Yugoslavia	0.12	0	0.5	0.4	0.2	0.30	0.27	0.24	0.21
Greece	0.18	0.16	0.2	0.2	0.1	0.19	0.18	0.18	0.17
Venezuela	0.30	0	0.3	0.3	0.1	0.32	0.30	0.17	0.15
Argentina	0.17	0.45	0.5	0.4	0.1	0.34	0.29	0.48	0.43
Saudi Arabia	0.60	1.2	0.2	0.2	0.03	0.40	0.38	0.69	0.67
Portugal	0.05	0.06	0.1	0.1	0	0.08	0.05	0.08	0.06
Mexico	0.06	1.1	0.8	0.4	0	0.41	0.24	0.92	0.75
Chile	0	0.16	0.1	0.1	0	0.05	0.03	0.13	0.11
Korea, Rep.	0.22	0	0.4	0.2	0	0.29	0.20	0.18	0.09
Iran	0.02	0	0.3	0.1	0	0.15	0.05	0.14	0.04
Malaysia	0.01	0	0.1	0.02	0	0.05	0.01	0.05	0.01
Turkey	0	0	0.2	0	0	0.10	0	0.10	0
Colombia	0	0	0.1	0	0	0.06	0	0.06	0
Iraq	0.16	0	0.1	0	0	0.11	0.08	0.03	0
Syria	0	0	0.03	0	0	0.02	0	0.02	0
China	0	0	3.0	0	0	1.5	0	1.5	0
Algeria	0.18	0	0.05	0	0	0.11	0.09	0.02	0
Peru	0	0	0.04	0	0	0.02	0	0.02	0
Brazil	0.04	0	0.2	0	0	0.14	0.02	0.12	0
Egypt	0	0	0.04	0	0	0.02	0	0.02	0
Morocco	0	0	0	0	0	0	0	0	0
Philippines	0	0	0	0	0	0	0	0	0
Thailand	0	0	0	0	0	0	0	0	0
Vietnam	0	0	0	0	0	0	0	0	0
India	0	0	0	0	0	0	0	0	0
Indonesia	0	0	0	0	0	0	0	0	0
Ivory Coast	0	0	0	0	0	0	0	0	0
Pakistan	0	0	0	0	0	0	0	0	0
Sri Lanka	0	0	0	0	0	0	0	0	0
Kenya	0	0	0	0	0	0	0	0	0

Table 4.2 *Continued*

	Ability to pay		*Responsibility*			*Obligation to pay (GNP)*		*Obligation to pay (PPP)*	
	GNP	*PPP*	*5t*	*10t*	*20t*	*5t*	*10t*	*5t*	*10t*
Cameroon	0	0	0	0	0	0	0	0	0
Mozambique	0	0	0	0	0	0	0	0	0
Ghana	0	0	0	0	0	0	0	0	0
Nigeria	0	0	0	0	0	0	0	0	0
Sudan	0	0	0	0	0	0	0	0	0
Zaire	0	0	0	0	0	0	0	0	0
Burma	0	0	0	0	0	0	0	0	0
Madagascar	0	0	0	0	0	0	0	0	0
Tanzania	0	0	0	0	0	0	0	0	0
Uganda	0	0	0	0	0	0	0	0	0
Bangladesh	0	0	0	0	0	0	0	0	0
Ethiopia	0	0	0	0	0	0	0	0	0
Nepal	0	0	0	0	0	0	0	0	0
TOTAL (%)	100	100	100	100	100	100	100	100	100

Resources (capacity), accountability (ethics), and combined indices are normalized to 100 per cent, thus indicating each nation's percentage of the world total. Shown are the same 62 nations (10 million or more) in the same order (descending natural debt) as in Table 2.3. Different thresholds of tonnes per capita are shown for RESP and OTP.
Resources: Ability to pay (ATP) is a function of income, measured either by GNP or PPP, and the threshold chosen as the minimum required to attain a reasonable physical quality of life (PQLI).
Accountability: The responsibility index (RESP) is a function of the natural debt (cumulative emissions per capita since 1950) and the threshold natural debt needed to achieve a reasonable PQLI.
Combined: Obligation to pay (OTP) is here defined as an equal weighting of ATP and RESP.

This argues that the responsibility of countries for the present situation is best indicated by total historical emissions integrated over time – natural debt. From the standpoint of physical reality, this is a better measure of responsibility than current emissions or growth rates, because integrated emissions directly drive climate warming.

Although the polluter-pays-principle is conceptually attractive, we recognize that there may be discomfort in applying it historically, for example, back to the beginnings of the industrial revolution. There are basically two arguments: one political and one practical.

Although the first greenhouse warming paper was written in the last century (Arrhenius 1896), it can be argued that past generations acted out of ignorance and thus their descendants should not be penalized. There is a strong counter-argument, however. If we are asking the present generation to take responsibility for the future, it must be given a feeling of control over the future. Without any control, there can be no true responsibility, because there is no reason to think the values and consequent sacrifices of today will be honoured in the future. Paradoxically, however, in order to impart a perception of control over the future, the present generation must feel

Figure 4.2a

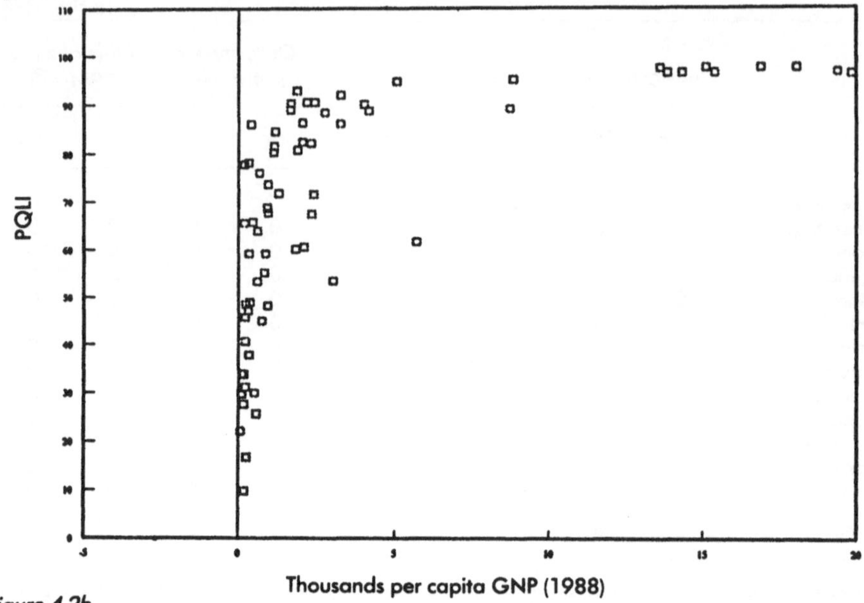

Figure 4.2b

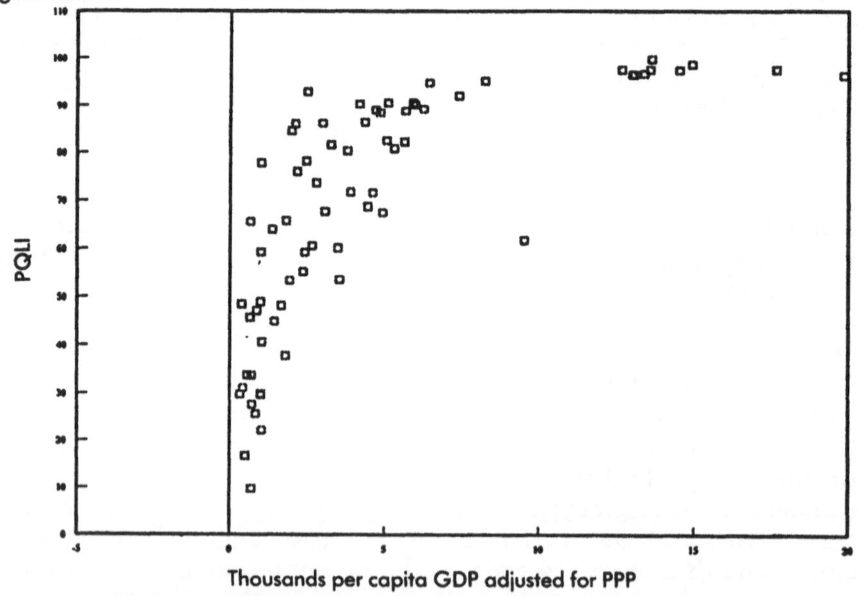

The relationship between PQLI and two measures of per capita income: (a) Gross National Product (GNP), and (b) Gross Domestic Product adjusted for purchasing power parity (PPP). The text describes the use of these figures to determine appropriate thresholds for adjusting incomes as part of schemes to allocate costs of such international efforts as greenhouse gas mitigation. GNP and PPP data from Summers and Heston (1991). PQLI calculated by the method of Morris (1979) from UNDP (1991) data.

Figure 4.2 *Relationship between PQLI and two measures of per capita income* somewhat constrained by the past. Only then will it believe that its efforts will not be for nothing. If we dismiss historical responsibility, what is to keep the next generation from doing so (Smith 1977)?

Put another way, one of the best ways to encourage this and future generations to take more account of the longer term impacts of technology and other human innovations is to make it responsible for problems that arise. Only then will it take a really serious look and apply the appropriate caution in its choices. Not to do so is to provide great incentives to stay ignorant.

As discussed in Chapter 3, given complete information on historic emissions of all greenhouse gases, their sinks, and their transformations in the atmosphere, one could calculate the contribution of each country to increases in the current concentrations from emissions of greenhouse gases since pre-industrial times, and require that remedial action by each country be proportional to that contribution. In practice, this approach to solving the greenhouse issue may be unworkable at present for several reasons:

- Generally, data on emissions of most gases become unreliable the farther back in time one goes. Country-specific data on many non-CO_2, non-CFC greenhouse gases are not known even for recent years.
- Shifting political boundaries and dominion of one country over another causes assignments before 1950 for many parts of the world to be problematic, even for gases such as CO_2 for which data may be available. (A problem exacerbated again in the late 1980s.)
- Historical information about sinks and rates of atmospheric transformation is even less well known than for emissions (see box on page 84).

Thus, to take this discomfort about historical responsibility into account, as well as the practical difficulties in actually determining remaining historical emissions, we use here historical records going back only to 1950 and only for fossil fuel CO_2. (See the box on project evaluations, page 89, however, for an illustration of how other gases can be considered when evaluating mediation options.)

International agreements to limit climate change will be easier to negotiate if they are perceived to be equitable. Hence, they must begin with the premise that every human being has the same equal right to atmospheric resources (Grubb 1989). Thus, comparisons based solely on present national emissions are not generally applicable, because no allowance is made for population size (Smith 1991). Similarly, as shown in Chapter 2, comparisons based on growth rates alone are misleading because absolute additions are ignored. (Such analyses, however, are essential for indicating where potentials for reduction lie.)

On the other hand, since it is national governments and not individuals that eventually will be charged for global remediation efforts, per capita emissions by themselves are inadequate. If, however, per capita emissions

Carbon sinks and responsibility indices

The amount of gases remaining in (and thus warming) the atmosphere is determined by the balance between emissions from natural and anthropogenic sources and the amount taken up in natural and human-engendered processes, called sinks. The ocean is a major carbon dioxide sink, mainly through physical and chemical absorption at the surface but also through phytoplankton that may fall from the surface to carry carbon into ocean sediments. Indeed, any place photosynthesis results in a net increase in biomass will be classified as a sink until the increase stops. In addition, bogs where carbon is being buried, forests where fire causes carbon to be sequestered in the soil as charcoal, places where soil carbon is growing, and many other widespread phenomena act as sinks.

There have been thoughtful suggestions that the distribution of natural sinks for atmospheric carbon be considered in allocating emissions rights (for example, Epstein and Gupta 1990; Agarwal and Narain 1991; Mitchell 1992). Since the operation of sinks is just as important as the magnitude of emissions in determining atmospheric carbon levels, this approach has some appeal.

Unfortunately, unlike important emissions sources, the sizes of important sinks change as the concentrations of the two most important carbon-based greenhouse gases, carbon dioxide and methane, change. When atmospheric concentrations rise, some sinks increase because the processes that drive most of them are functions of concentrations. On the other hand, some other sinks may decline. Unfortunately, therefore, each sink type reacts differently to changes in the atmospheric concentrations. Sinks, thus, are moving targets as well as being substantially different for the different gases.

To avoid this problem, one might take 'pre-human' CO_2 concentrations as the baseline from which to determine the 'natural' size of sinks. The 'natural' sinks could then be allocated on a per capita basis or whatever. By definition, however, 'natural' sinks and sources were essentially in balance. More precisely, the pre-human carbon balance was maintained by a net terrestrial flow that was slightly (some tenths of a GT out of about a 200 GT annual flow) positive (a source), but countered by an equivalent slightly negative flow for the oceans (a sink) (Schmidbauer et al. 1991). Since most people agree that the ocean sink ought to be part of the common human heritage, that is, allocated equally to each person, it is not clear how useful (or practical) it is to attempt to allocate the pre-human natural terrestrial net source. As a whole, both are extremely small in any case.

Although small as a whole, it might be countered that there are large net terrestrial sinks and sources in individual localities that ought to be considered in allocations. This approach is frustrated, however, by our inability to define the 'natural' state of the landscape in any one place. A common assumption is that the landscape/atmosphere relationship was in a 'natural' condition before humanity began to burn significant amounts of fossil fuel, that is, before the early 1800s. Since this seems also to correspond roughly to the date at which atmospheric levels of the major carbon greenhouse gases (CH_4 and CO_2) began to rise, the changes are attributed to the combination of fossil carbon input and land-use changes that have occurred since. The IPCC, for example, subsequent to stating that 'concentrations of CO_2 and CH_4, after remaining relatively constant up to the 18th century, have risen sharply since then due to man's activities' (IPCC 1990: xvi), confines its examination of the impact of land use changes to the period after 1850 (Section 1.2.2.2.ff.)

There is a questionable assumption hidden here, however, that apparently relatively stable atmospheric concentrations previous to 1800 meant that there were no major land-use changes. In fact, humanity has caused rather dramatic land-use changes ever since learning how to use fire. Much of the landscape that may seem to us 'natural' is actually the result of pre-industrial human management by fire and other means.[a] No continent has been spared this management. Starting 12,000 years back, for example, an estimated 25 per cent of subsequent human deforestation had actually occurred by 1700, 50 per cent by 1850, and 75 per cent by 1915 (Turner et al. 1990). Furthermore, human land management had started long before 10,000 B.C. The deforestation going on now, while vitally important as a fraction of the remaining forest and perhaps occurring at a greater rate than in the past, is not excessively large compared to what our ancestors accomplished.

Most of this land-use change, then as now, was undoubtedly related to food production and security, but some must have been due to the harvesting of biomass as fuel, which remains even today an important component of total biomass combustion. Reliance on biomass fuel was, of course, even more widespread before use of fossil fuels and may have resulted in significant net emissions of carbon (Kammen and Marino 1993).

Unfortunately, our knowledge of historical population distribution and its impact on the biosphere is not sufficient to allow adequate estimates of land-use changes and the corresponding gross emissions of carbon, let alone the net result for the atmosphere taking into account changing carbon sinks (Andreae 1991). That there do not seem to be any major historical (human history) changes in atmospheric carbon levels previous to 1800[b] does not prove, however, that there were no major land-use changes and corresponding gross carbon emissions, only that over time they must have been balanced by appropriate sinks.

It does seem clear, however, that the idea of ever being able to define a 'natural' baseline in every locality is questionable. Indeed, it may well be that the same climate, soil, and other physical conditions can support entirely distinct 'climax' landscapes depending on the pathway taken. Even short-term temporary intervention by human management can completely change all subsequent history in an area by instituting positive feedback mechanisms. Alternatively, removal of human management may not cause reversion to a unique 'natural' landscape (Woodcock 1992).

We are greatly hampered by lack of data and knowledge not only about past sinks, but also about those in operation today. As shown in Figure 4.3, out of the 7 GT of carbon thought to be released annually by human activities (5.4 from fossil fuels, 1.6 from land-use changes), less than 50 per cent (3.4 GT) stays in the atmosphere and only 2 GT goes into the ocean (IPCC 1990, 1992). The remaining 1.6 GT goes to the infamous 'missing carbon' sink. As indicated on the figure, there is some evidence that the carbon is going into terrestrial sinks in the northern hemisphere, but much uncertainty remains. This uncertainty, alone, handicaps any effort to incorporate sinks into allocation schemes.

Assume, for the moment, that the missing carbon is shown conclusively to be absorbed by sinks in temperate and boreal forests, as many today believe. What would be the implications of this for global negotiations? Should the tropical, mostly developing, countries now undergoing deforestation be charged with emissions from land-use changes on top of those from fossil fuel, as is the case in many proposed allocation schemes (such as in WRI 1990)? If so, should not the

a See, for example, Turner and Butzer (1992) for a look at the pre-Columbian Western Hemisphere, the detailed documentation in Pyne (1982, 1991) for North America and Australia, and a contemporary look at slash-and-burn agriculture by Rambo (1984).

b A conclusion based on ice-core measurements that have come to be questioned in some quarters (Jaworski et al. 1992).

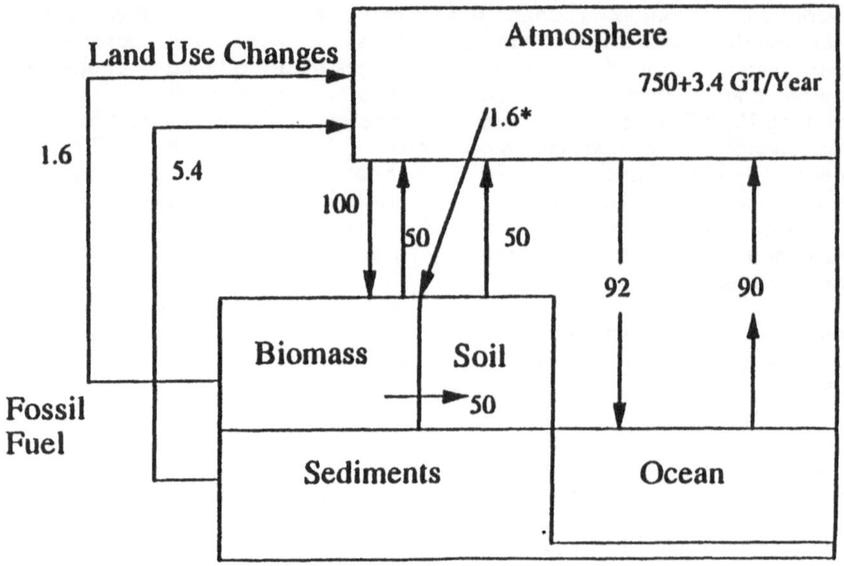

The flow marked 1.6* GT is the 'missing carbon,' the fate of which is not known. Most likely, as shown, it is being taken up by biomass growth on land. Note that is roughly equal in size to the net emissions from land-use changes, such as deforestation. Data from IPCC (1990) and Schmidbauer et al. (1991).

Figure 4.3 *The global carbon cycle: approximate current flows (GT carbon)*

high-latitude, mostly developed, countries be accorded a net decrease in their fossil-fuel totals to account for the sinks in their forests? If so, the result would be a great leveling of international accountability.

On the other hand, whereas the forests of the currently developed countries may be absorbing carbon today, they are generally still far from their 'original' (that is, pre-human) size. Should these countries then be held responsible for the carbon released in the great deforestations of the past? If so, since we do not know much about the 'natural' condition before these deforestations, how would this be done? Should some currently developing countries be handicapped because they are deforesting relatively late in human history? It is certainly not clear that they are, or will be, deforesting more than others did in the past. Other developing countries, such as China, may have completed much of their deforestation even before North America and Europe. Mesopotamia may have done so even earlier.

The current global flow of biospheric carbon from low to high latitudes is likely just the reverse of what went on previously when the high-latitude forests were being cleared. Some, and perhaps a significant amount, of the gross carbon released by high-latitude clearing was once taken up by sinks in the tropics, through CO_2 fertilization, for example. This balance is consistent with the (apparent) fact that it did not result in significant increases in atmospheric carbon. In effect, people have been using each other's carbon sinks for a long time. Today, this is called the 'missing carbon' phenomenon.

These ecological/biophysical factors affect how responsibility can be attributed in negotiations today. Someday, perhaps, we may be able to determine which local parts of the natural pre-human landscape were sources and which were sinks. We would then be able to judge what is occurring in these localities today relative to a meaningful baseline. The value of doing so for purposes of allocation is questionable, however. Just as we would not want to bias the allocation of sink rights against people who do not happen to live near the ocean, so too there is no reason to punish those who happen to live near swamps or reward the people who happen to live near forests (or wherever the natural sources and sinks are).[c] This logic implies that the natural terrestrial sources and sinks also ought to be allocated equally to all people (we certainly do not have the data at present to do otherwise).[d]

Although conceptually tricky, this analysis seems to argue that in addition to sinks, changes in the biosphere itself should be left out of responsibility indices. This conclusion follows because:

1 We cannot and may never be able to define in any meaningful way the natural baselines from which we have departed, either on a total or local basis.
2 Even if we could quantify them, departures from these baselines are much too ancient for us to link them rationally to particular human populations today.
3 We cannot define the baseline as today's biospheric situation, because we still understand it quite poorly. (Some 1.6 GT of carbon of the approximately 7 GT total emitted is going into sinks, probably biospheric, that we have not identified.)

For fossil fuels, on the other hand, the situation is quite different:

1 We do know accurately the natural baseline (near zero) and the geographic and temporal distributions of emissions (Ebert and Karmali 1992).
2 The bulk of the emissions has occurred within a period such that they can be rationally allocated to countries today.
3 Fossil fuel emissions are larger, growing more rapidly, and less intrinsically limited than net biospheric emissions.

As a result, in this book, we do not include biospheric changes as part of the responsibility index.

To optimize present management of the carbon sphere, however, humanity clearly must address biospheric sinks. Nations should be encouraged to maintain and enhance whatever sinks they control and to develop others. The techniques to estimate the marginal costs of doing so and to compare these costs with other ways of limiting carbon emissions are the subjects of Chapters 5 and beyond. Of necessity, however, is choosing a baseline, that is, the mixture of biospheric sinks and sources from which departures will be accounted henceforth. This baseline is taken to be 1990, an arbitrary, but practical, choice, one also selected in the Framework Convention on Climate Change discussed at the June 1992 UNCED meeting in Rio.

Conceptually intermediate between fossil fuel emissions and human-engendered changes in the 'natural' biosphere are the atmospheric consequences of managed agroecosystems. Of most importance for global warming is the methane from ruminants, rice paddies, and biomass burning. Although obviously 'managed,' given the scale of total human intervention in the biosphere, agroecosystems are merely at one end of a spectrum of management. We thus

c Compared to carbon dioxide, trying to allocate sinks for methane would be even more complicated, since certain atmospheric chemical species play an important role in methane degradation. How would these be allocated?

d Ahuja (1992) has argued persuasively that treating emissions on a per capita basis is equivalent to treating sinks in a like manner.

classify them with the rest of the biosphere and do not include them in the responsibility index, but do take account of them in evaluating mitigation proposals. For calculating marginal costs of emissions reduction, we again take 1990 as the base year.

References

Agarwal, A, and S Narain, 1991. *Global Warming in an Unequal World: A Case of Environmental Colonialism*, Centre for Science and Environment, New Delhi, India

Ahuja, D R, 1992. 'Estimating National Contributions to Greenhouse Gas Emissions,' in J K Mitchell, ed. *Global Environmental Change* 2(2): 83-87

Andreae, M O, 1991. 'Biomass Burning: Its History, Use, and Distribution and Its Impact on Environmental Quality and Global Climate,' in J S Levine, ed., *Global Biomass Burning*, MIT Press, Cambridge, MA, USA

Ebert, C, and A Karmali, 1992. 'Uncertainties in Estimating Greenhouse Gas Emissions,' Environment Department Working Paper No. 52, World Bank, Washington, DC, USA

Epstein, F, and R Gupta, 1990. *Controlling the Greenhouse Effect: Five Global Regimes Compared*, Brookings Institution, Washington, DC, USA

IPCC, 1990, 1992. *Scientific Assessment, Supplement to the Scientific Assessment*, Cambridge University Press, Cambridge, UK

Jaworski, Z, T V Segalstad and N Ono, 1992. 'Do Glaciers Tell a True Atmospheric CO_2 Story?' *Science of the Total Environment* 114 (April): 227-284

Kammen, D M, and B D Marino, 1993. 'On the Origin and Magnitude of Pre-industrial Anthropogenic CO_2 and CH_4 Emissions,' *Chemosphere*, 26(1-4): 69-86

Mitchell, J K, ed., 1992. 'Greenhouse Equity' (six authors comment on Agarwal and Narain 1991) *Global Environmental Change* 2(2): 82-100

Pyne, S J, 1982. *Fire in America*, Princeton University Press, NJ, USA

Pyne, S J, 1991. *Burning Bush, A Fire History of Australia*, Henry Holt, NY, USA

Rambo, A T, 1984. 'No Free Lunch: A Reexamination of the Energetic Efficiency of Swidden Agriculture,' in A T Rambo and P E Sajise, eds, *An Introduction to Human Ecology Research on Agricultural Systems in Southeast Asia*, University of the Philippines, Los Banos, pp. 154-163

Schmidbauer, B, et al., 1991. Vol. 1, *Protecting the Earth*, German Bundestag, Bonner Universitats-Buchdruckerei, Bonn

Turner, B L, and K W Butzer, 1992. 'The Columbian Encounter and Land-Use Change,' *Environment* 34(8): 16-20, 37-44

Turner, B L, W C Clark, R W Kates, et al., 1990. *The Earth as Transformed by Human Action: Global and Regional Changes in the Biosphere over the Past 300 Years*, Cambridge University Press, Cambridge, UK

Woodcock, D W, 1992. 'The Rain on the Plain: Are There Vegetation-Climate Feedbacks?' *Paleogeography, Paleoclimatology, Paleoecology (Global and Planetary Change Section)* 97: 191-201

WRI (World Resources Institute), 1990. *World Resources 1990-91*, Oxford University Press, NY, USA

are simply multiplied by national population, one obtains national emissions again, which is also unacceptable.

The way out of this paradox is to recognize the need for either a non-linear weighting of emissions in something like a progressive income-tax schedule, a threshold, or both. This is parallel to the way the ability-to-pay index was constructed, which recognized that a dollar of GNP above the threshold is to be counted differently to one below. Just so with natural debt per capita. This natural debt threshold is analogous to the already accepted 300 g/capita threshold under the Montreal Protocol (Table 4.1).

Using global warming potentials in project evaluations

Even though comprehensive indices that aggregate a range of greenhouse gases may yet be impractical for international agreements because of data uncertainties and verification difficulties, they can still find a place in project evaluation (Levander 1990; Wilson 1990; Ellington et al. 1992; Pitstick et al. 1992). When choosing among possible greenhouse gas mitigation measures, for example, there is a need to compare costs per unit greenhouse reduction. In many cases, in project evaluation it will be possible to estimate and verify the changes in emissions of the other greenhouse gases as well as of CO_2. In some cases, inclusion of these non-CO_2 greenhouse gases can be crucial to making the right choice.

To illustrate the importance of non-CO_2 gases, consider the following question: 'What are the greenhouse-gas implications of moving up the household energy ladder from wood to kerosene or LPG for cooking?' Given that something like half the world's households still use biomass fuels for cooking, this question is important.

Essentially all the products of incomplete combustion (PIC) that are produced during biomass and fossil-fuel combustion are also effectively greenhouse gases. These include methane, carbon monoxide, and non-methane hydrocarbons, the last two mainly exert their effects indirectly. Indeed, as shown in Table 2.1 of Chapter 2, considering indirect warming effects, if carbon from fuel combustion is to go into the air, carbon dioxide is actually about the least damaging form.

Unfortunately, the combustion efficiency of traditional biomass burning in simple stoves is usually much less than 100 per cent. It is not uncommon, for example, for well over 10 per cent of the carbon to be released as PIC rather than carbon dioxide, which would be the only carbon product if combustion was complete. Because PIC on average have higher global warming potentials (GWP) than carbon dioxide, the total impact can be substantially higher than would be indicated by an evaluation based on carbon dioxide alone. A recent pilot study in the Philippines, for example, has shown that the total GWP of woodstoves can, depending on circumstances, be more than double that of the carbon dioxide (i.e., the GWP of the PIC can rival that of the carbon dioxide alone) (Smith et al. 1993).

This point is illustrated in Figure 4.4, which shows the flow of carbon through one of these stoves. Note that applying the 20-year GWP to the PIC gives a total CO_2-equivalent even greater than that of the CO_2 itself, even though the latter has more than seven times more of the original fuel carbon. By comparison, a 100-year time horizon gives a much lower, but still significant GWP (40 per cent of CO_2).

For comparison, the carbon balance of one possible alternative, a kerosene stove cooking the same meal, is shown in Figure 4.5. Since kerosene has 1.5 times more energy per carbon atom than wood, and kerosene stoves are 2.5 times more energy efficient, the total fuel carbon needed is much less: 27 per cent. In addition, in a 100-year horizon, kerosene stoves seem to produce less than 10 per cent additional GWP over the CO_2 (20 per cent for a 20-year GWP).

There are several tentative but potentially important implications of these findings. First the overall greenhouse gas benefit of improved biomass cookstove programmes may be much larger than previously estimated. This outcome depends, however, on the degree to which the stoves actually improve combustion efficiency, rather than just overall efficiency, which is also affected by the efficiency of heat transferring to the cooking pots. Assuming that the wood is not being harvested renewably, for example, the cost of improved stoves to reduce carbon emissions (CO_2-equivalents) might be something like $40/tonne without considering the PIC, but only $25/tonne considering the PIC at a 100-year horizon. At a 20-year horizon, it would only be $14/tonne.

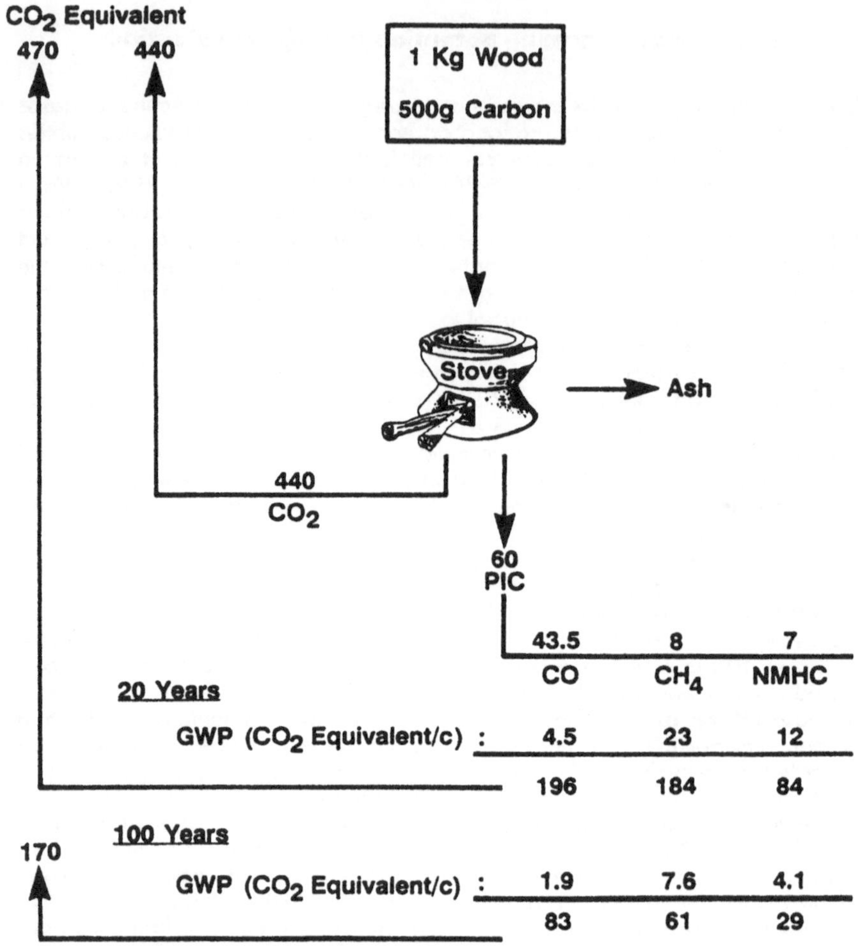

Based on measurements in Manila, this shows the carbon flows through a simple wood-fired cookstove starting with 1 kg of wood (0.5 kg carbon). Note that 60g of carbon is not completely burned (89% combustion efficiency) but is released as products of incomplete combustion (PIC). With indirect warming effects included and depending on the time horizon chosen (Table 2.1), the resulting global warming potentials for the PIC can rival that of the carbon dioxide (for example, at 20 years, 470g CO_2 equivalent compared to 440g CO_2 itself). CO = carbon monoxide; CH_4 = methane; NMHC = non-methane hydrocarbons. From Smith and Thorneloe (1992).

Figure 4.4 *Flow of carbon through a wood-fuelled cooking stove*

Second, surprisingly, in some cases, there may be substantial greenhouse benefits in switching from some kinds of biomass stoves to modern fuels such as kerosene and LPG. Although these fossil fuels produce significant GWP because they are non-renewable, the overall GWP impact may be lower because of the high emissions of PIC from some biomass stoves. This result is shown in Table 4.3

(case A) where, at all time horizons, kerosene and LPG have much lower total GWP than a stove fuelled with non-renewably harvested wood. Less obvious, however, are the results shown in case B, based on a completely renewable wood harvesting system. Because the wood stove produces so much PIC, there is an advantage of moving to the fossil fuels in all but the 500-year time horizon, even though burning the modern fuels releases fossil carbon (net CO_2 emissions from the stove are considered to be zero). This conclusion may provide the extra incentive needed in some areas to shift policy toward encouraging fuel substitution rather than improved woodstoves.

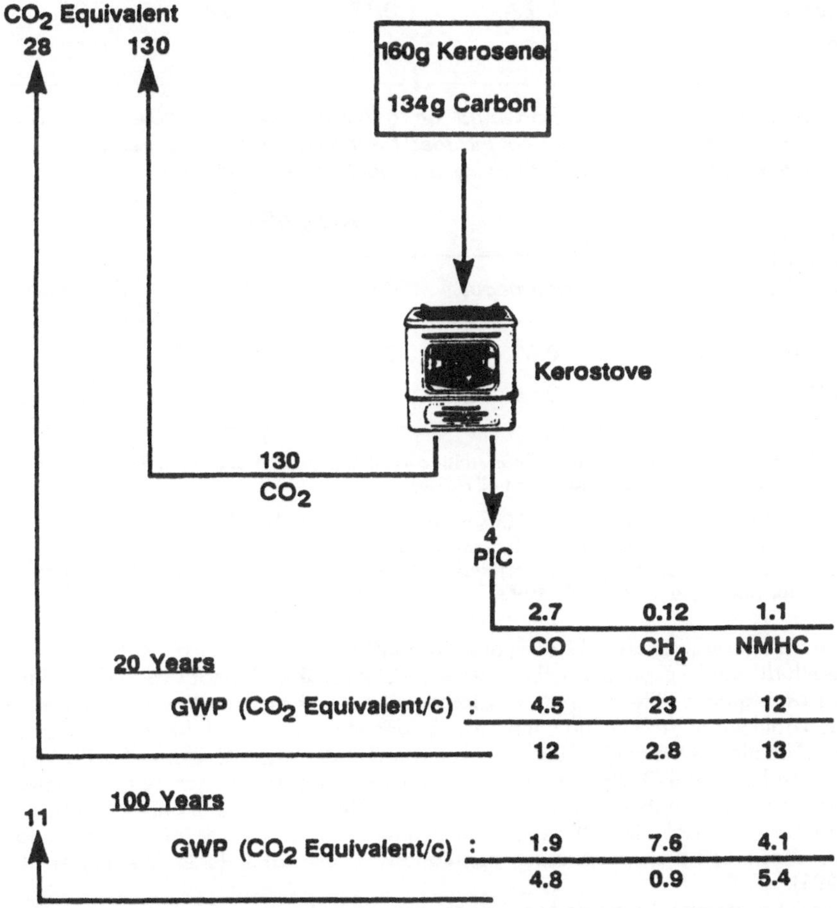

For cooking the same meal as cooked by the woodstove in Figure 4.4, this shows the carbon flows and global warming potentials for a simple wick-style kerosene cookstove. Note that even at a 20-year horizon, the GWP of the PIC is less than one-quarter that of the CO_2. Note also how much less overall carbon is involved because the kerosene stove is more efficient and the fuel has more energy per carbon atom than does wood. Data from Smith et al. (1992).

Figure 4.5 *Flow of carbon through a kerosene-fired stove*

Table 4.3 *Benefits in GWP from fuel switching in cookstoves*

Case A: Where wood is harvested on a completely non-renewable basis, i.e., woodstove CO_2 emissions are included as net atmospheric additions.

Relative GWP in going from wood to:*	*Time horizon*			
	Instantaneous	*20-year*	*100-year*	*500-year*
Kerosene	0.22	0.19	0.23	0.25
LPG	0.14	0.12	0.16	0.17

Case B: Where wood is harvested on a completely renewable basis, i.e., woodstove CO_2 emissions are not included for the woodstove because the CO_2 is completely recycled, but CO_2 from the fossil fuel stoves is included.

Relative GWP in going from wood to:	*Time horizon*			
	Instantaneous	*20-year*	*100-year*	*500-year*
Kerosene	0.69	0.44	0.96	1.5
LPG	0.46	0.27	0.63	1.0

* This does not take into account GGs emitted in other parts of these fuel cycles, e.g., decay of wood residues and emissions at oil refineries.

This GWP includes the production of CO_2 as well as the non-CO_2 GGs: CO, CH_4, and NMHC and is dependent on relative stove fuel efficiencies (wood = 20%; kerosene = 50%; LPG = 70%). Energy per carbon atom is considered to be equal for the fossil fuels and two-thirds as much for wood (see Smith et al., 1993).

Third although charcoal stoves are more efficient and produce less GWP (and health-damaging) pollution than do woodstoves, the charcoal manufacturing process is quite a different matter. Another impact of the generally quite inefficient charcoal kilns in developing countries, besides excessive wood demand, is that a good fraction of the original carbon in the wood ends up as PIC released at the kiln. The resulting total GWP of the charcoal fuel cycle apparently can be many times that of a comparable wood fuel cycle. This effect gives additional impetus to improve charcoal kiln efficiencies and may also argue for more efforts to encourage some charcoal-using populations to use other fuels (Delmas et al. 1991).

More research is being undertaken to verify these findings. The general impact, however, would seem to be that the PIC of biomass stoves is an even greater enemy than has been recognized. We have known that they rob households of some of the energy contained in the fuel and impose health problems on the householders (Smith 1987). Now we have still another incentive to reduce PIC as much as possible (Smith and Thorneloe 1992).

This example illustrates how project evaluations can be affected by consideration of non-CO_2 gases, as well as using GWPs that include estimates of indirect effects. We have not, however, taken such a comprehensive approach in the cost calculations in later chapters of this book. This is partly because many of the projects examined (for example, improved lighting) reduce electricity demand. When generated by fossil fuels, electricity is generally produced in large-scale units that emit small amounts of non-CO_2 greenhouse gases relative to the CO_2 they emit (although if powered by natural gas or coal, there could be significant methane emissions elsewhere in the fuel cycle). This assumption bears examination, however, in future evaluations for large-scale devices and is clearly wrong for some small-scale combustion devices like simple stoves.

References

Delmas, R A, A Marenco, J P Tathy et al., 1991. 'Sources and Sinks of Methane in the African Savanna,' *J of Geophysical Res.* 96(D4): 7287-7299

Ellington, R T, M Meo, and D W Baugh, 1992. 'The Total Greenhouse Warming Potential of Technical Systems: Analysis for Decision Making,' *J Air and Waste Management Association* 42(4): 422-428

Levander, T, 1990. 'The Relative Contributions to the Greenhouse Effect from the Use of Different Fuels,' *Atmospheric Environment* 24A(11): 2707-2714

Pitstick, M E, D J Santini and H Chauhan, 1992. 'Reduction in Global Warming Due to Fuel Economy Improvements and Emissions Controls: New US Light-Duty Vehicles,' *Proceedings of the Intersociety Energy Conversion Engineering Conference*, San Diego, August 3-7

Smith, K R, 1987. *Biofuels, Air Pollution, and Health*, Plenum, NY, USA

Smith, K R, M A K Khalil, R A Rasmussen, S A Thorneloe, F Manegdeg and M Apte, 1993. 'Greenhouse Gases from Biomass and Fossil Fuel Stoves in Developing Countries: A Manila Pilot Study,' *Chemosphere*, 26(1-4): 479-505

Smith, K R, and S A Thorneloe, 1992. 'Household Fuels in Developing Countries: Global Warming, Health, and Energy Implications,' *Proceedings from the 1992 Greenhouse Gas Emissions and Mitigation Research Symposium*, USEPA, RTP, NC, USA

Wilson, D, 1990. 'Quantifying and Comparing Fuel-Cycle Greenhouse Gas Emissions,' *Energy Policy* 18(6): 550-562

The most direct, unambiguous, and accurate reflection of a country's contribution to the greenhouse problem is then given by the total emissions of all the relevant gases (on a CO_2-equivalent basis) over the period of interest. As discussed in Chapter 2, deciding on which gases and time periods to incorporate will be a matter for negotiations, but 100 years of warming into the future seems like a good starting point for discussions.

The responsibility for emissions needs to be tempered by the number of people for which these emissions were released. Thus responsibility is proportional to the natural debt, which we define as the amount of GGs remaining divided by the current population.[m]

m There are several possible formulations of natural debt over a particular historical period (Smith et al. 1993):

ND-1: where total emissions are divided by total person-years;

ND-2: where per capita emissions each year are averaged; and

ND-3: where total emissions are divided by the present population.

For a stable population, all three formulations are identical, but for both increasing and declining populations, the three yield different answers and only the first gives CO_2-equivalent emissions for a person-year of population during the time emissions were taking place. One version of index ND-3 is shown in Table 2.3 of Chapter 2 and is used throughout this chapter.

Before determining responsibility to pay, it is necessary to choose a threshold natural debt.[n] A number of ways to do this exist: for example, based on PQLI, as with ATP; based on the natural debt needed for stable global GG concentration of some specified level; or based on global averages.

The third column of Table 4.2 compares RESP for fossil-fuel CO_2 emissions since 1950 using natural debt at three different thresholds: 5, 10, and 20 tonnes per capita.[o] Note that lower thresholds tend to decrease the relative responsibilities of developed countries, such as the United States, Germany, and the UK, but increase those of developing countries. Only with the lowest threshold does China have any responsibility, and India's responsibility remains at zero for all the thresholds shown.

Obligation to pay

It now remains to combine the practical (resources – ATP) and the ethical (responsibility – RESP) components into an overall obligation-to-pay (OTP) index that is transparent in formulation so that it can be easily manipulated in the course of negotiations. In parallel with Wirth and Lashof (1990), the combination could be a simple sum or the two could be multiplied together.[p]

The second combination (product) implies that if either ATP or RESP is zero, payments are zero. It could be argued, however, that responsibility for a debt does not disappear just because the borrower is unable to pay now. This argument suggests the first formulation where OTP will be greater than zero for a larger group of countries.

Columns 4 and 5 of Table 4.2 compare the additive version of this obligation-to-pay (OTP) index using equal weights for both ATP and RESP and showing also the effect of varying natural debt threshold and income measure (GNP or PPP). The result is more stable than either the ATP or RESP index by itself. The US obligation remains between 35 and 39 per cent, for example. China retains a modest obligation in two scenarios, but India's remains at zero for all combinations.

Smith et al. (1991) compare the multiplicative version of the OTP using equal weights for the two indicators, for both types of income indicators and two different natural debt thresholds. The US contribution is above 69 per

n Parallel to the ATP index, therefore, the final form for determining responsibility (RESP) over some historical period becomes:

$$RESP = \text{Total Emissions} - [POP \times (ND_{threshold})]$$

o Using the same criterion as for ATP (the average natural debt of nations with PQLI of 80–90) reveals an ND threshold of 25 tonnes.

p Sum:

$$OTP = a \times (ATP^{x}) + b \times (RESP)^{y}$$

or products:

$$OTP = (k \times (ATP)^{x}) \times ((RESP)^{y})$$

where a, b, and k are scalers and x and y are exponents that would start out equal to 1.0, but could be changed, if necessary, during negotiations.

cent in all cases, undoubtedly a politically unacceptable result. Most developing countries, on the other hand, have zero OTP.

The present constraint on the maximum UN assessment for any one nation is 25 per cent, substantially less than the 35–39 per cent range determined here for the OTP of the United States. It might be noted, however, that the original limit and US assessment set in 1946 was 40 per cent (UN 1989).

Conclusion

It is widely, although by no means universally, held that, to paraphrase Lord Keynes, economic growth by itself is only a means to certain ends. In other words, after an agreeable quality of life has been thereby achieved, society should consider placing its emphasis elsewhere. More explicitly, after reaching some, admittedly difficult-to-define, level of adequate physical well-being (ethical criterion), individuals should no longer expect special assistance by the broader society to help them develop further economically.[q] This philosophy also is consistent with the physical reality inherent in a finite world; that is, there should be incentives to use finite physical resources in ways that lead to quicker achievement of these minimum levels by humanity at large (efficiency criterion).

One aspect of this approach that has not been well explored is what it implies for the measures of efficiency (indices) that should be used to judge various human activities. Rather than indicators such as income or energy use, which are usually open-ended, it implies the use of thresholds or indicators that actually have fixed ranges, that is, have a maximum corresponding to achievement of the level of adequacy, as already incorporated in the UN scale, for example.

It is thus no accident that our indices both for ability to pay (ATP) and for responsibility contain indicators with thresholds and that ATP is also based on an indicator with finite extent, PQLI. This means that as 100 per cent is neared, the indicator gives little credit for further advancement. Incentive then shifts to promotion of other objectives. If, on the other hand, an open-ended measure such as income is used, an extra 10 per cent looks to be as good for the rich as for the poor, no matter how rich the rich might become.[r]

Implicit in the use of PQLI, therefore, is acceptance that the objective of development assistance and policy should be an improvement in the quality of life. It has long been recognized, however, that there exists a strong

q In this case, by assigning a small or zero obligation to pay for international greenhouse remediation projects.

r In its Human Development Index (HDI), the UNDP (1991) accomplishes this goal by mathematically arranging such that the HDI rises directly with income for those countries below their poverty lines, but slowly above that threshold.

positive relationship between GNP (GDP, PPP) and many measures of quality of life, such as PQLI. As a result, it has been argued that PQLI tells us nothing new and should be rejected as an indicator. There are two counter-arguments: First, although there is a strong overall correlation, the GNP to PQLI ratios are quite different for different countries, an important consideration when assigning international responsibilities and costs. Second, it sends the quite different message that simple increases in per capita income should not be taken as ends in themselves, but as means to improve the quality of life.

Although the index proposed here includes a measure of historical responsibility (based on past greenhouse gas emissions), it counts for only half of the total obligation. The other half is based on current income. Thus, the obligations of countries that have economic problems, such as those in Eastern Europe and the former USSR, will be adjusted accordingly. In addition, only emissions since 1950 are counted, a concession to the political and practical difficulties of determining responsibility previous to the modern era.[s] Combining both indicators also takes into account circumstances in which past emissions may be high, but current income low (for example, Eastern Europe), or vice versa (for example, Norway, which has been blessed with substantial hydropower).

We have now looked separately at indices of both responsibility and resources to determine the relative obligation for the costs of a global programme. Ways to combine the two together would be determined by direct negotiation in international fora, although the simplest combination is presented here as a start. What this chapter does is derive a way of measuring where the world is today, in terms of the present distribution of wealth and greenhouse responsibilities. Before we can judge the distribution of payments for greenhouse remediation projects, however, we need indicators of where the best projects are and where the world ought to be heading (the Who Can? and Who Should? questions of the bottom line on Figure 4.1). This is the task of the next chapters.

References

Agarwal, A, and S Narain, 1991. *Global Warming in an Unequal World: A Case of Environmental Colonialism*, Centre for Science and Environment, New Delhi, India

Arrhenius, S, 1896. 'On the Influence of Carbonic Acid in the Air Upon the Temperature of the Ground,' *Philosophical Magazine* and *J of Science* S15, 41(251): 237–276

s The total fossil-fuel CO_2 emissions for 1950–86 (126 GT [billion tonnes] is more than 70 per cent of the total since 1850 (178 GT in Chapter 3). Given that a larger portion of the earlier emissions have been removed by natural sinks, our cut-off at 1950 still accounts for at least three-quarters of the total.

Feiveson, H A, et al., 1988. 'Princeton Protocol on Factors That Contribute to Global Warming,' Woodrow Wilson School of Public and International Affairs, Princeton University, Princeton, NJ, USA
Grubb, M, 1989. *The Greenhouse Effect: Negotiating Targets*, Royal Institute of International Affairs, London, UK
Gurney, K., 1991. 'National Greenhouse Accounting,' *Nature* 353: 23
Krause, F, W Bach and J Koomey, 1992. *Energy Policy in the Greenhouse*, J Wiley & Sons, New York
Michaelis, P, 1992. 'Global Warming: Efficient Policies in the Case of Multiple Pollutants,' *Environmental and Resource Economics* 2: 61–77
Morris, M, 1979. *Measuring the Condition of the World's Poor*, Overseas Development Council, Elmsford, NY, USA
Mukherjee, N, 1992. 'Greenhouse Gas Emissions and the Allocation of Responsibility,' *Environment and Urbanization* 4(1): 89–98
NZMERT (New Zealand Ministry of External Relations and Trade), 1990. 'United Nations Handbook,' Wellington
Smith, K R, 1977. 'The Interaction of Time and Technology', Ph.D. Dissertation, University of California, Berkeley
Smith, K R, 1989a. 'Developing Countries and Climate Change: Implications for Risk Management,' in D. Street and T. Siddiqi, eds, *Proceedings of the Workshop on Responding to the Threat of Global Warming: Options for the Pacific and Asia*, pp 2–37–2–39, Argonne National Lab/East-West Center, Argonne, IL, USA
Smith, K R, 1989b. 'Have You Paid Your Natural Debt?' Environment and Policy Institute, East-West Center, Honolulu, HI, USA
Smith, K R, 1991. 'Allocating Responsibility for Global Warming: The Natural Debt Index,' *Ambio* 20(2): 95–96
Smith, K R, and D R Ahuja, 1990. 'Toward a Greenhouse Gas Equivalence Index: The Total Exposure Analogy,' *Climatic Change* 17: 1–7
Smith, K R, J Swisher, R Kanter and D R Ahuja, 1993. 'Indices for a Greenhouse Gas Control Regime That Incorporates Both Efficiency and Equity Goals' Chapter 5 of C Suddayas, ed, *Energy Investments and the Environment*, Economic Development Institute, World Bank, Washington DC
Solomon, B D, and D R Ahuja, 1991. 'International Reductions of Greenhouse-Gas Emissions: An Equitable and Efficient Approach,' *Global Environmental Change* 1(4): 343–350
Summers, R, and A Heston, 1988. 'A New Set of International Comparisons of Real Product and Prices: Estimates for 130 Countries, 1950–1985,' *J of the International Association for Research in Income and Wealth* 34(1): 1–26
Summers, R, and A Heston, 1991. 'The Penn World Table (Mark 5): An Expanded Set of International Comparisons, 1950–88,' *Quarterly J. of Economics* (May): 327–368
UN (United Nations), 1989. 'Evolution of the Methodology for the Scale of Assessments and Its Current Application,' Committee on Contributions, A/CN.2/R.532, New York City, USA
UN, 1992. 'Report of the Committee on Contributions,' GAOR 47th Ses., Supp. 11, New York City, USA
UNDP (United Nations Development Programme), 1991. *Human Development Report*, Oxford University Press, NY, USA
UNEP (United Nations Environment Programme), 1992. 'Report of the Seventh

Meeting of the Executive Committee of the Interim Multilateral Fund for the Implementation of the Montreal Protocol,' UNEP/OzL.Pro/Ex.Comm/7/30, NY, USA.
USNAS (US National Academy of Sciences), 1991. *Policy Implications of Greenhouse Warming: Report of the Mitigation Panel*, NAS Press, Washington, DC, USA
Westing, A H, 1989. 'Law of the Air,' *Environment* 31(3): 3–4
Wirth, D A, and D A Lashof, 1990. 'Beyond Vienna and Montreal, Multilateral Agreements on Greenhouse Gases,' *Ambio* 19 (6–7): 305–310
World Bank, 1990. *World Development Report*, Oxford Press, NY, USA
World Bank, 1992, 'Global Environmental Facility: The Pilot Phase and Beyond,' with the UNDP and UNEP, Working Paper #1, Washington, DC, USA

Part II

Resource transfers

5

North–South carbon abatement costs

Peter Hayes[a]

The signing of the Climate Change Convention in Rio de Janeiro in June 1992 was a momentous event. With hindsight, however, it will recede into a waypost on a much longer journey toward international cooperation on climate change. The hard work of achieving abatement still lies ahead. The signatories to the Convention must now negotiate subsidiary agreements or protocols which will determine how much abatement each country must achieve over time; and who should pay for the costs and who should reap the benefits that will flow from this activity. If successful, these protocols will provide substance to the Convention's rhetoric (see the Appendix to this book).

In this chapter, I estimate the cost of abating carbon dioxide from fossil fuel usage over a transition period to a 'sustainable' level of emissions. I do so by aggregating nations into three categories: rich, industrial nations (North), transitional nations (the former Soviet Union and Eastern Europe), and poor and modernizing nations (South)[b]. This approach makes it possible to determine the costs for the South to comply with the Climate Change Convention.

Greatly simplified, this calculation requires at least ten sequential steps plus many intermediate assumptions, explained below. Conservative estimates, combined with reasonable 'medium' assumptions in face of uncer-

---

a Nautilus Pacific Research, Berkeley, California. I thank Nautilus Associates Megan Van Frank, Frank Muller, and Rachel Sommerville for research assistance in completing chapters 5, 6, and 14 of this book; Nautilus economist Lyuba Zarsky and other reviewers; Nautilus staffer Cassandra Morrow for maintaining the greenhouse database; Robert Lingren, Alan Miller, and Irving Mintzer at the Center for Global Change, University of Maryland, for facilitating research for this study; Kirk Smith for sharing his ideas and data; Chieko Umetsu for preparing figures used in this chapter; Robert Hayes for computer support; the Australian Government for permission to draw on a study conducted with its support on these issues; and many colleagues too numerous to list for critical comments and suggestions.

b These blocs of nations are not presented as likely negotiating entities. Indeed, chapter 14 suggests that neither the 'North' nor the 'South' are unified political entities in the context of climate change politics. Rather, the aggregation is necessary to obtain order of magnitude estimates of possible transfers from rich nations to poor ones.

tainty, render this method capable of approximating the rate and magnitude of probable carbon abatement costs.

I begin by providing the reader with an overview of the method. Next, I outline the ten steps that must be taken to estimate abatement and other greenhouse-related costs. Finally, I estimate the overall incremental cost to the North, East, and the South for various abatement costs.

Before discussing abatement costs, however, I must first delineate the critical characteristics of the Climate Change Convention under which protocols must be developed to abate carbon emissions.

Climate Change Convention

The Climate Change Convention resembles an open ended 'General Agreement on Climate Change', rather like the GATT or General Agreement on Tariffs and Trade, while taking the symbolic form of a framework convention (see Appendix). The terms of the agreement have been left largely unspecified in the original text and will be negotiated in protocols over the coming years. Within that general framework, an enormous range of political and economic combinations is possible.[1] This section therefore states a specific scenario that is used in the remainder of this chapter (see box).

In principle, such a scenario is susceptible to varying levels of scientific certainty and political response to climate change.[c] In this chapter, however, I simply assume that states perceive climate change to be sufficiently problematic that they act upon their commitments under the Convention. I assume that the framework Convention adopted in 1992 comes into force in December 1993; that the first of a series of revision conferences to consider specific protocols occurs in early 1994; and that abatement actions commence in earnest in 1995.

The five specific elements of the Climate Change Convention are: (i) an assessment process; (ii) targets and timetables; (iii) national greenhouse strategies; (iv) funding; (v) a revision procedure. Two other elements: (vi) liability and enforcement, and (vii) carbon taxes and tradeable emissions, are shown, but I assume that these items will not be addressed until the first revision conference of parties to the agreement.

---

c These scenarios could be refined along scientific and climatic parameters such as: scenario (1) could be high levels of climate change observed early and conclusively validated by science, associated with a strong international agreement reflected in strong protocols to the Convention. Scenario (2) could be low levels of climate change experienced in the distant future with no scientific consensus, associated with ineffectual protocols. And scenario (3) could be medium levels of climate change in the medium term but still scientifically uncertain that greenhouse gases are responsible, associated with weak protocols.

Climate Change Convention protocols scenario

Assumptions:
Easy start (consensual issues, credit early starters, use soft law, transition periods)
Slow start, CO_2 fossil fuel only (minimum adherence to protocols, ratchet up later)
Modest start (OECD, regional only, regional experiments)
Limited North–South linkage (common understandings)
Limited issue linkage: biomass, sinks, biodiversity, and desertification addressed after carbon emissions from fossil fuel

1 **Administrative and assessment process**
Conference, elected bureau, permanent committees, strong secretariat
A centralized scientific assessment and updating process Protocol on monitoring and data sharing

2 **Targets and timetables**
Into force December 1993, revision December 1994
Overall goal adopted: to stabilize or to not exceed natural carbon dioxide levels by more than 50 per cent in 2025
Sub-targets adopted during protocol negotiations
2.1 OECD stabilize CO_2
2.2 Parties reduce GDP Energy Intensity each year
2.3 Parties stabilize Forest Cover

3 **National strategies**
Each country or regional grouping should prepare a national or regional strategy for addressing greenhouse warming, to be updated periodically.

4 **Mechanism to collect and allocate funds for:**
4.1 Technical assistance
4.2 Training/human resource development
4.3 Information
4.4 GHG abatement project funding and technology transfer issues is addressed after first review of national strategies in 1994.

5 **Revision procedure:** first revised in December 1994 and protocols adopted.

6 **Verification, liability and enforcement** – none at outset; later $100 million per year needed for verification of compliance

7 **Taxes and tradeable emission entitlements**
Test nationally and regionally first.

I assume that the specific elements of a 'winning formula' for effective protocols will combine a set of initial targets for fossil fuel carbon emission reductions with an assessment process, and a requirement that individual countries prepare national or regional strategies to address the problem. Industrial countries are committed to providing national reports and strategies within six months of the Convention coming into force. Developing countries have up to three years to meet this obligation. It is conceivable that the Clinton Administration will commit the United States to reduction targets. This shift would enhance the pace and magnitude of commitments from the OECD bloc as a whole. Meanwhile, the scenario shown here is consistent with positions advanced during the negotiations leading up to Rio, and the terms of initial or subsequent adherence by key developing countries as specified in the Convention itself and the Agenda 21 action plan of the Earth Summit.[2]

If, as James Sebenius argues, the fundamental negotiating task in controlling climate change '*is to craft and sustain a meaningful "winning" coalition of countries backing such a regime*,'[3] then the two conditions for constructing this coalition needed to negotiate effective protocols are: (1) to provide enough gain for each member of the coalition to adhere at all to these protocols, relative to alternatives; and (2), to avoid, accommodate, or neutralize potential blocking coalitions of interest. Blocking coalitions will exploit scientific disagreement, economic interest, and ideology to oppose greenhouse gas reductions. The Convention contains five broad countervailing strategies that may head off potential blocking coalitions in the negotiations over protocols.

1 **An easy start** that prevents blocking coalitions by picking non-controversial subjects for the first negotiations, credits early starters, and uses 'soft law' informal options and transition periods. This logic suggests that protocols should be directed at politically weak or morally suspect greenhouse contributors, especially those located in 'green' countries with strong anti-greenhouse interests.)
2 **A slow start** that creates the institutional basis to ratchet controls upward later, possibly requiring adherents to the initial Convention to agree only to at least one protocol (as did the 1975 Barcelona Convention to protect the Mediterranean). As carbon emissions can be approximated from existing fuel use, these would be difficult to falsify and are easily monitored. Fossil fuel emissions are therefore the most feasible for controls under a protocol, followed by controls related to carbon from burning non-commercial wood fuels.[4] Other uncontrolled greenhouse gases (excepting chlorofluorocarbons which are already controlled in a separate treaty) would be regulated by adding protocols.
3 **A modest start** that includes agreements by regional groupings such as the OECD which complement the universal Convention, thereby demonstrating the economic advantages of controls, testing (on a small or regional scale) schemes such as tradeable emission permits, and setting floors for controls and standards.
4 **A limited North–South and East–West linkage** that defuses ideological blockages which stalled the New International Economic Order. The terms of the protocols that implement the Convention's strong statements as to financing and technology transfer will be crucial. Common understandings of the climate change issue will be extremely important in moderating ideological clashes driven by opposed interests. Dialogue

and information dissemination are crucial in this regard, and include collaborative first–third world research, regional workshops, informal conferences, neutral non-governmental organizations, broad advisory groups, and cross-cutting coalitions of business and green advocates.

5 **A limited issue linkage** that offers limited reciprocity or ecological 'side payments' on climate and development related issues such as desertification and biodiversity but avoids anything that smacks of the rich countries imposing a 'New International Ecological Order' on developing countries.

These are the basic political building blocks of the abatement scenario used in the cost calculations that follow.

Method overview

There are 10 links in the logical chain used to calculate the costs of Southern compliance with the Convention (see box on page 106). Also displayed are the key variables that affect each step. The quantitative goal of the chapter is to estimate a time profile of national greenhouse gas emissions, required reductions, and the costs related to achieving the required reductions. This stream of costs is then present valued, converted to an annuity, and compared with the national or aggregated greenhouse gas reduction obligation-to-pay rankings described in Chapter 6.

Which gases?

For the reasons adduced above and in Chapter 2, only carbon emitted as carbon dioxide from fossil fuel usage is included in this analysis.[d] I have not included costs related to greenhouse gases that are already partly controlled by the Vienna Convention on ozone depletion. Nor do I include costs that will arise from controlling greenhouse gas emissions from agricultural or land use changes. While past estimates and projections have been made for emissions of these gases,[5] such figures are highly speculative and scientifically controversial, making the monitoring and verification of controls on these gases such as methane emissions from paddy fields or cattle virtually impossible.[6] Also, a separate agreement and financing arrangement will

d CO_2ff refers throughout to carbon dioxide emitted from fossil fuel (ff) usage, to remind the reader that other sources of this gas (such as land use changes) are not included in this analysis. Carbon is also denoted as C on occasion in this chapter and refers only to the mass of carbon emitted as carbon dioxide (that is, the mass of the oxygen atoms is not included). Readers can convert the mass of C as CO_2 to the mass of CO_2 by multiplying C by 3.667 – the ratio of the molecular weight of carbon dioxide (44) to the atomic weight of carbon (12).

Method overview

Key variable is shown in [ ]

1 Select cost elements
- Which gases [CO_2 fossil fuel]
- Which losses [coastal protection]

2 Distribute national carbon sink property rights [science, population, territory]

3 Estimate marginal abatement costs
- Nordhaus/US National Academy of Sciences
- Study estimates [RR ratio or ReqRed/CO_2ff Projected, levelized cost abatement $/T C]

4 Project emissions
- Reference scenario [CO_2ff only to IPCC stringent target]
- Efficiency scenario [CO_2ff adjusted for autonomous energy efficiency increase]

5 Estimate required reduction
- Efficiency scenario [adoption rate of efficiency, renewables, energy substitution to force CO_2ff-ReqRed 2025 to equal target emission]

6 Estimate coastal protection costs [time profile costs, discount rate]

7 Present value and annuitize calculated incremental abatement and coastal protection costs
[Discount rate, limited global and zero local benefits counted]

8 Redistribute according to Obligation to Pay Index
- South's OTP = 7% World incremental Cost
- North's transfer OTP to South = South's Incremental Cost - South's OTP
- North's OTP = North's incremental cost + North's transfer [Obligation To Pay index]

9 Evaluate transfer mechanisms
- Carbon tax [projected versus emitted carbon]
- Tradeable permits [savings from trade S/T abated]
- Regional tradeable permits [barriers to trade]

10 Estimate monitoring and verification costs [size, number of sources, gases, science]

likely deal with carbon storage and release in forests, although such was not achieved at Rio in 1992.

A comprehensive agreement that incorporates the relative global warming potential of different greenhouse gases and regulates the net emissions from all sources to achieve the least cost abatement strategy is certainly optimal in the long run.[7] But desirable as it is, achieving such comprehensiveness and flexibility is likely to prove elusive at the outset of the Convention.[8] In addition, the IPCC is revising the radiative forcing equivalence of the different greenhouse gases relative to carbon dioxide. Therefore, I cost only abatement of carbon released during use of fossil fuels which is a substantial portion of the total greenhouse problem. Fortunately, these emissions are not only the most quantifiable; they are also among the least costly. Emissions of fossil carbon are also reasonably reliable (± 5–10 per cent for many big countries although the confidence level is much lower in many small, poor countries) and are estimated regularly through an existing, generally accepted UN office.[9] These emissions are amenable to actions that can be taken and are well understood in the near and medium term, many of which are already economically justified.

Moreover, my modest approach is also conservative in that I estimate only the minimal costs of compliance with the Convention. Later – when the scientific basis improves for the control of greenhouse gases other than carbon dioxide from fossil fuels – so the requisite economic information for computing and allocating related abatement costs will become available. Until then, it is premature to include more than carbon dioxide released from projected fossil fuel usage.

Discounting

Some economists have argued that carbon emissions in the distant future should be discounted as being less significant than those emitted in the near term.[10] There is, however, no ecological argument for discounting future emissions. Indeed, if the absorptive capacity of the atmosphere is finite and if ecological degradation accelerates in non-linear fashion after system thresholds of change are exceeded,[11] then future emissions may be *more* significant than past or near term emissions. (This would imply that negative rather than positive discounting should be used.)

In this study, I did not discount future emissions of CO_2ff. The economic value of the costs of future abatement, however, is discounted by a real discount rate of 5 per cent throughout. Arguably, a higher discount rate (such as the 10 per cent used by the World Bank for project evaluation) is justified in light of the capital scarcity and high social opportunity costs of capital investments in developing countries. Conversely, as economist William Cline has argued, an appropriate annual discount rate for long and very long term cost-benefit analysis might be as low as 1–2 per cent.[12] Moreover, as the US National Academy of Sciences notes, uncertainty as to the impact of greenhouse emissions may increase rather than decrease as more knowledge accumulates, thereby warranting a decrease in standard discount rates to as low as 3 per cent.[13] In short, if it is difficult to estimate the impact and costs of pollution, then it is even more difficult to quantify our ignorance about these impacts over time – a fact that should be reflected in economic analysis.

Of course, investors and consumers may demand much higher returns than the social discount rate adopted in this study, due to market failures, lack of information, and competing investment opportunities. Indeed, these individual preferences and market failures explain why there is a gap between economically justified greenhouse abatement strategies and private investment in energy efficiency and other abatement options. However, social policy should be based on economic, not financial criteria. By the same token, however, it is apparent that many institutional, regulatory, and pricing policy initiatives are necessary to overcome the 'payback' gap that characterizes socially justified strategies.

Calculating cost

With these analytic steps defined, it is possible to specify more precisely the

Incremental cost method

Incremental Cost = Net Present Value P of (A = Incremental Abatement Cost) + (B = Coastal Protection Cost) where:

A = Incremental Abatement Cost = Sum of A given P of:
$IC_{1,2,3}$ = Required Reduction Ratio RR × Marginal Abatement Cost = CO_2ff $Projected_{k,t}$ / $Required\ Reduction_{k,t}$ =

{(Base year $Q_{t,k}$ × $Q_t/cap_{k,t}$ × $Projected\ Population_{k,t}$ × Efficiency Rule 1)}/ (Base year $Q_{t,k}$ × Efficiency Rule 2)} × ($Marginal\ Cost_{k,t,1,2,3}$ at $RR_{k,t}$)

B = Global Coastal Protection Cost = Sum of A given P of:
(Half estimated 100-year coastal protection capital cost)/30y

Key: **A** = Annuity given, P = Present Value of IC = $NPV_{i,ICk,t}$
Base year = 1995, **CO_2ff** = carbon dioxide emissions from fossil fuel use, T C/y
$IC_{1,2,3}$ = Incremental Cost for Marginal Abatement Costs, cases 1, 2, 3
i = discount rate for present valuing, 5% per year
k = country or grouping of countries (North, East, South, World)
$Marginal\ Cost_{k,t,1,2,3}$ = Marginal Cost of CO_2ff Abatement for country k at year t, $/T C abated-$year^{-1}$
Case 1 = Nordhaus schedule, high cost
Case 2 = US National Academy of Sciences schedule, low cost
Case 3 = study estimate, medium cost
n = range of t, 0–30 years from start 1995 to end 2024
P = Net present value of IC for abatement and coastal protection
Q_t = CO_2ff in year t, T C/year
Efficiency Rule 1 for Projected $Q_{k,t}$:Base Year $Q_{k,t}$ plus 40 % of Reference Scenario annual growth in $Q_{k,t}$; equivalent to a global 1%/y autonomous energy efficiency increase
Efficiency Rule 2 for $Required\ Reduction_{k,t}$: Base Year $Q_{k,t}$ – (94%/y of $Q_{k,t-1}$ due to 6%/y adoption of efficiency, renewable or energy substitution options) + (65% of growth in $Q_{k,t-1}$), iterated until $Q_{k,2025}$ converges on target)
Efficiency rule 2 varies for South (98%/y and 45%/y versus North/East of 94%/y and 65%/y to achieve convergence
$RR_{k,t}$ = Required Reduction Ratio for country k in year t, $0<RR<1$
t = year between 1995 and 2025

method for calculating the cost of compliance for a country or grouping of countries with carbon abatement required by my Climate Change Convention scenario. In the algorithm I have used to calculate the costs of the South's compliance (see box), the total incremental cost consists of two elements, calculated separately. These are labelled A, the cost of abating CO_2ff emissions in the South during 1995–2025; and B, the cost of constructing coastal protection in countries vulnerable to sea level rise. Each of the sequential steps presented in this box and the previous one are explained in greater detail below.

Marginal abatement costs

The first step in the calculation is to estimate the marginal cost curve. After reviewing the problems associated with cost data, I construct three cost curves used in the calculations.

Data problems

Complete, disaggregated supply curves for greenhouse gas abatement do not exist for most non-OECD countries in the world. In Figure 5.1, I show earlier (1989) 'indicative' estimates made by the consulting firm McKinsey

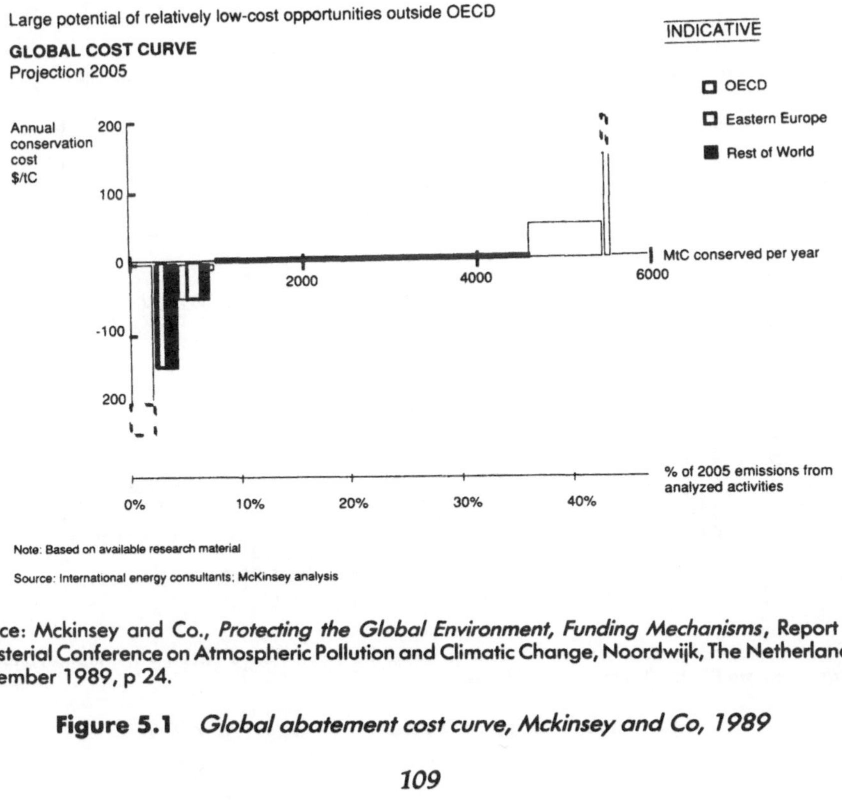

Source: Mckinsey and Co., *Protecting the Global Environment, Funding Mechanisms*, Report to Ministerial Conference on Atmospheric Pollution and Climatic Change, Noordwijk, The Netherlands, November 1989, p 24.

Figure 5.1 *Global abatement cost curve, Mckinsey and Co, 1989*

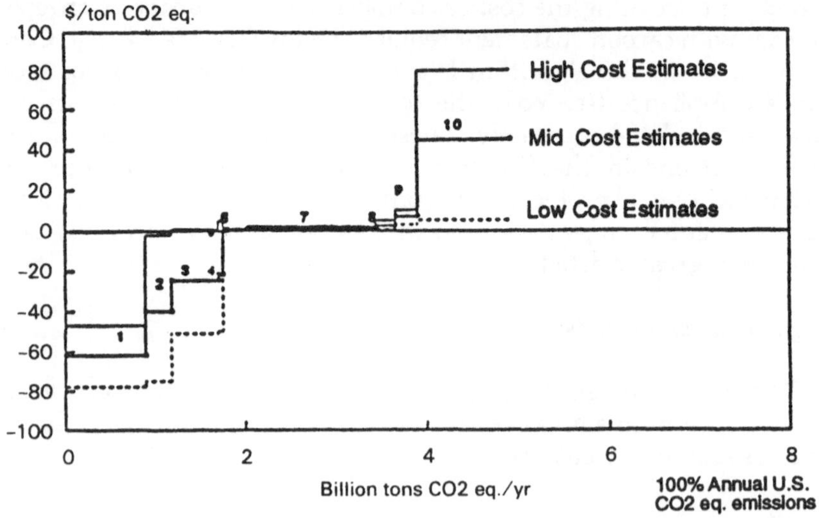

National Academy of Sciences, National Academy of Engineering, Institute of Medicine, *Policy Implications of Greenhouse Warming*, Report of the Mitigation Panel, Committee on Science, Engineering, and Public Policy, National Academy Press, Washington DC, 1991, pp II-13.

Figure 5.2 *US NAS marginal cost estimates*

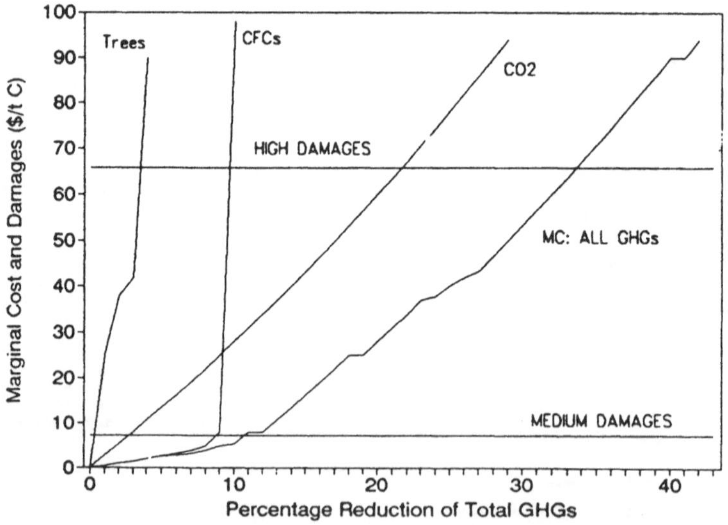

Source: W Nordhaus, 'A Sketch of the Economics of the Greenhouse Effect,' *American Economic Review*, volume 81, no 2, 1991, pp 146–248.

Figure 5.3 *Marginal costs and damages, Nordhaus*

and Co for the Dutch government. They estimated that the global marginal cost curve per unit abatement achieved is negative for the first five per cent abatement; hovers around zero for the next 30 per cent abatement; and then becomes positive (rising to about $50/T-C-y^{-1}$ abated between 35–45 per cent reduction ratio and steeply thereafter). McKinsey's study, however, gives no empirical basis for its estimates which were reportedly based on consultations with practitioners of energy economics.[14]

Figure 5.1 may be compared with Figure 5.2, where I present estimates obtained from the US National Academy of Sciences (NAS) which gives a US-only cost curve for greenhouse gases (in contrast to Nordhaus who gives a global cost curve for greenhouse gas abatement, see below and Figure 5.3).

The NAS estimates are based on detailed end use engineering studies of the potential for existing energy services to be provided from efficiency options, thereby abating greenhouse gas emissions. The resulting supply curve of abatement options and costs is provided in Figure 5.2. (Note that costs in Figure 5.2 are given in $/T CO_2-y^{-1} abated; these are converted to $/T-$y^{-1}$ Carbon as CO_2 in Figure 5.4 which is used in this study.)

Nordhaus's cost estimates related to reduction of CO_2 from use of fossil fuel and other sources and are shown in Figure 5.3.[15] This curve in turn provides the basis for case 1 in Figure 5.4, where the cost curves used in this study are shown. Only the curve related to CO_2ff in Figure 5.3 is incorporated into Figure 5.4.

Cost is always positive in the Nordhaus study – which is contrary to detailed technology-cost driven estimates such as the NAS study for the United States or those by Amulya K.N. Reddy for the Indian state of

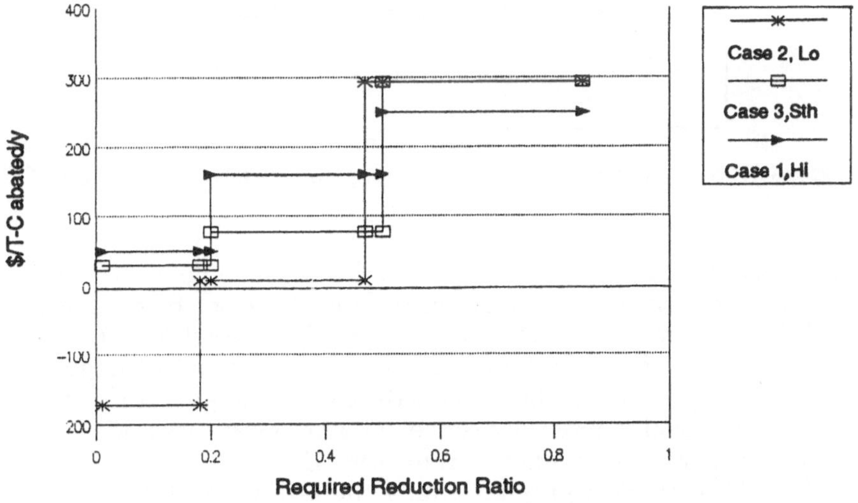

Figure 5.4 *Marginal cost, CO_2 reduction (Dollars per tonne of carbon abated per year)*

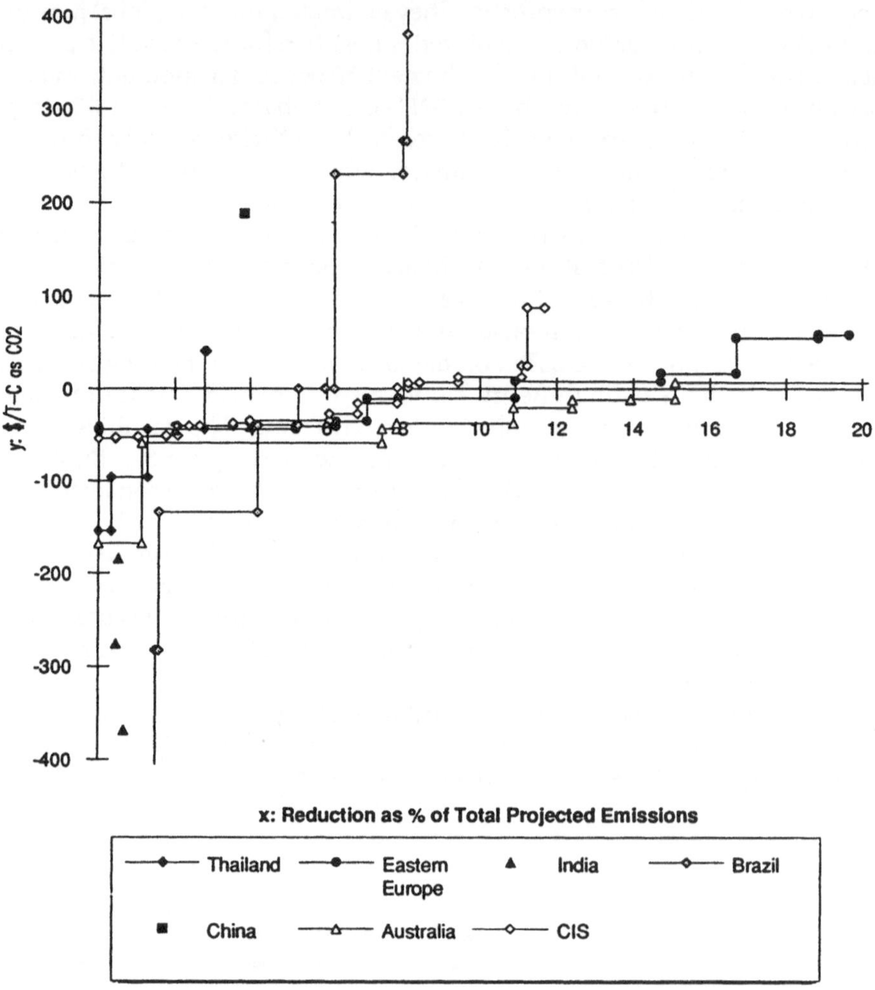

Figure 5.5a *Partial country abatement cost curves (-$400 to +400/T-C as CO_2)*

Karnataka.[16] This assumption derives from the origins of Nordhaus's estimates, in turn, a regression of prior estimates made by macroeconomic modellers as to tax rates required to dampen demand sufficiently to achieve target reduction levels. Such methods assume that current energy markets are reasonably efficient, which makes the cost of energy futures other than 'business-as-usual' inherently positive – in spite of contrary evidence and the practice of many energy utilities in the OECD.[17]

The NAS curve in Figure 5.2 shows the low, mid-point, and high cost estimates at an implementation rate of 50 per cent of the identified technological potential. In Figure 5.4, case 2 is based on the NAS mid-point

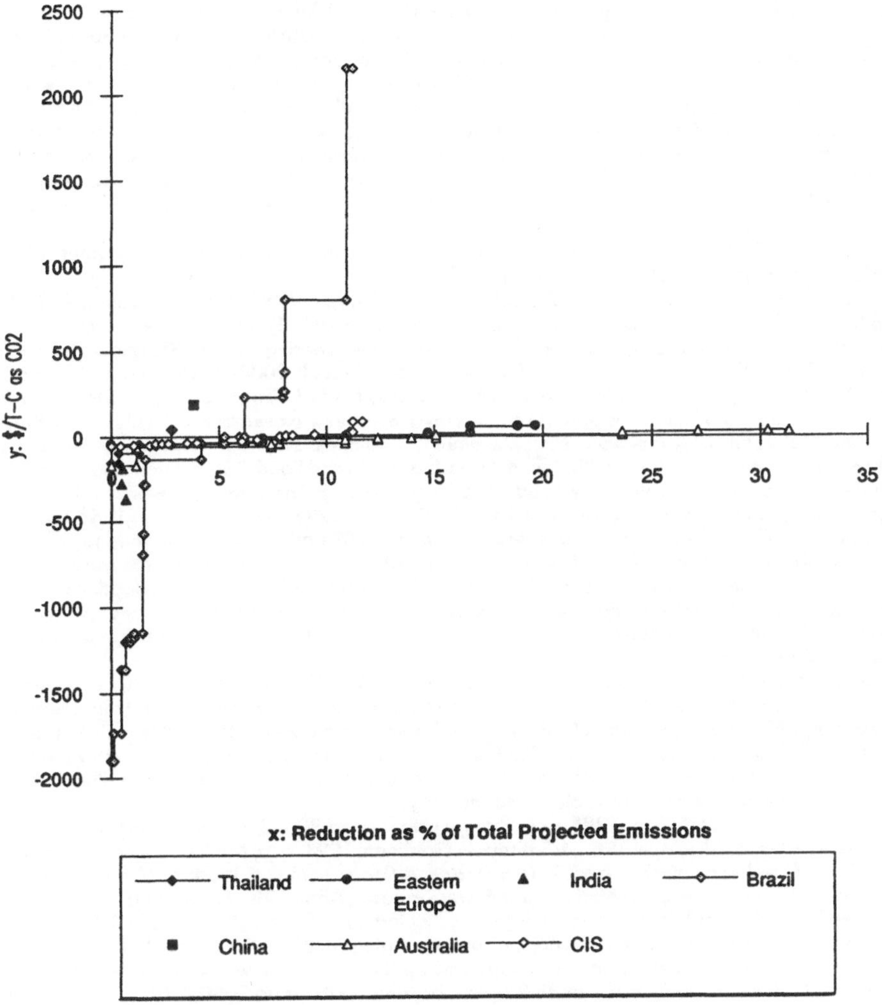

Figure 5.5b *Full country abatement cost curves*

Key: a = The year and amount of total projected emissions; b = The year of abatement cost estimate; c = The year of dollar price; d = Discount rate; e = Time horizon for cost estimate; f = Source; g = Assumptions used in calculations; h = Abatement technology from least cost (left) to highest cost (right).
ST = Short term scenario; MT = Medium term scenario; LT = Long term scenario; n/a not applicable or known; RE = Residential energy IE = Industrial energy; AE = Agricultural energy; CE = Commercial & service sector energy; TE = Transport energy.
EE a: 1985, 356MT-C; b,c: 1988; d:7%; e: savings projected for 2005–2025; g: The share of coal in electricity production decreases from the current 69% to 50% and 30% in 2005 and 2025. f: Chapter 12; h: IE1.Buildings insulation, IE2.Boiler replacement, IE3.Heating improvement, IE4.Cogeneration, IE5.Improved transmission and distribution, IE6.Improved existing industrial equipment, IE7.Ferrous metals, IE8.New electrical motors in industry, IE9.Construction industry improvements.
CIS/USSR a: 2005, 1315MT-C; b: 2005; c: 1988. For cost of carbon saved, rubles figures developed by Makarov, Bashmakov, were used and ruble (where $1 = 6 Rubles) were discounted to correspond

to early 1992's bank exchange rate in Moscow ($1 = 90 Rubles); d: 7%; e: capital cost for 1990–2005, savings projected for 2005; f: Chapter 12; g: CIS/USSR refers to former Soviet Union: h: AE1.Shifting from harvesters to site threshing, IE1.Switching small boilers to high-grade fuels, RE1.Insulation of steam supply network, IE2.Advanced technologies for industrial heating, AE2.Insulation of cattle breeding buildings, IE3.Automation of heating stations, IE4.Efficient centralized boilers, IE5.Change inefficient ovens to large boilers, IE6.Regulated electric drive, IE7.Control and measurement in energy use, IE8.Low capacity multifuel boilers, IE9.Reduction of electric transmission losses, IE10.Replacing wet cement clinker with dry method, IE11.Gas turbine and combined cycle plants, RE2.Efficient lighting, IE12.Improved brick production, IE13.Improved gas compressors in pipelines.

Thailand a: 2001, 34.96MT-C; b: 2001; c: 1991; d: n/a; e: 1991–2001; f: Chapter 11; h: RE1.Refrigerators, RE2.Lighting, RE3.Cooling, RE4.Rice Cookers.

India a: 1987, 144MT-C; b: 1989 (ST), 2000 (MT), 2000 (LT); c: 1989, 15.28Rs/dollar (ST), 1991, 18.81Rs/dollar (MT,LT); d: 5%; e: 1989 (ST), 2000–2025 (MT,LT). For ST, investment in 1989 generates benefit in the same year. For MT and LT, investments during 1989–2000 generate benefit during 2000–2025. Costs for MT and LT were levelized through 2000–2025. f: Prodipto Ghosh, *Analysis of Energy and Technology Options for Reducing Global Climate Change Concerns: India Country Report*. Tata Energy Research Institute, New Delhi, India. November, 1991; Jayant Sathaye & Nina Goldman, *CO_2 Emissions from Developing Countries: Better Understanding the role of Energy in the Long Term*. Volume III: China, India, Indonesia and South Korea. International Energy Studies Group. Lawrence Berkeley Laboratory. July, 1991. g: The total CO_2 emission level was estimated by taking the average of 133 MT-C in 1988 (Gosh, 1991) and 155 MT-C in 1985 (Sayathe & Goldman, 1991). For MT and LT, investment between 1989 and 2000 was assumed to generate benefits of energy reduction from 2000 to 2025. h: IE.MT (Insulation of waste heat recovery systems, replacement of inefficient boilers, better instrumentation and control systems, and adoption of better technologies). IE.LT (Introduction of cogeneration systems, adoption of new energy efficient technologies, and automation of process control). IE.ST (Improved housekeeping, energy audits, and training of personnel).

Brazil a: 2000, 84.6MT-C; b: 2000; c: 1991; d: 12%; e: 1991–2000; f: Chapter 10; h: CE1.Lighting, IE1.Electric ovens & boilers, IE2.Variable speed drivers, IE3.High efficient motors, IE4.House keeping measures, RE2.Air conditioning, IE5.Lighting, IE6.Electrolytic processes, TE1.Improvement in automobiles, TE2.Efficient diesel motors, TE3.Improved urban transportation, TE4.Alternative fuel-natural gas, TE5.Alternative fuel-alcohol, RE3.Efficient refrigeration, CE2.Public illumination, TE6.Highway improvement, RE4.Solar water heating.

China a: 1985, 478MT-C; b: 1985; c: exchange rate for mid 80's (2.9 yuan/dollar) was used for 1985 cost estimate; d: n/a; e: 1985; f: Sathaye & Goldman, 1991 for a; Mark D. Levine et al. *Energy Efficiency, Developing Nations and Eastern Europe, A Report to the US Working Group on Global Energy Efficiency, Developing Nations and Eastern Europe*, April, 1991 for b,c; h: IE.

Australia a: 2005, 52.4MT-C if no change in technology, i.e. 'frozen efficiency'; b: 1991 costs of equipment; c: 1991 A$ N.B. A$1 = US$0.75 approximately; d: 8%; e: 1991–2005 (2030). Costing of measures is based on emission and energy savings throughout the whole life of the equipment concerned, which in some cases is up to 25 years beyond 2005, i.e. to 2030; f: Chapter 13; g: see Chapter 13; h: CE1.Commercial miscellaneous, CE2.Commercial HVAC, IE1.Industrial metal processing, IE2.Industrial electric motors and drives, IE3.Industrial high temperature IE4.Smelting, IE5.Industrial electrolysis, RE1.Residential hot water, CE3.Commercial lighting, RE2.Residential refrigeration, RE3.Residential major appliances.

Figure 5.5b *Full country abatement cost curves (cont)*

cost estimate at a 100 per cent implementation rate (as it was the only one that achieved an 80 per cent reduction level). Also, I excluded the technological options and abatement associated with halocarbons and agriculture as unrelated to CO_2 fossil fuel abatement. As is evident, the NAS identified substantial amounts of abatement that should be available at negative cost (up to about 20 per cent reduction level), offsetting the steep rise in marginal cost at higher levels of abatement.

Some 'point' data are also available to supplement and validate the marginal cost curves presented in Figure 5.4. In Figure 5.5, I summarize the cost curve data provided in the national case studies in Part III of this book. Readers are cautioned that these costs were not calculated over the same time frames, nor using the same discount rates, technological repertoires, etc. (The divergent parameters are stated concisely in the notes to the figure.) Moreover, some of the curves are for one sector only (as in Thailand) whereas others are for all sectors. These curves, therefore, are not strictly commensurable. Nonetheless, they give a summary view of the state-of-the-art of estimating technological costs of carbon emission abatement in developing countries.

As can be seen, the cost ranges from ± $2000/t of carbon abated (in the case of Brazil) and cluster around ± $100/t. The reduction level covered by the curve ranges from a few per cent (in India) to about 32 per cent reduction relative to projected emissions (in Australia). The cost curves in these developing and transitional economies are similar in shape to those of the US National Academy referred to above, but extend to only 10–30 per cent reduction levels.

In general, these curves indicate that carbon savings may be obtained in the South at a significant savings and low net cost, provided that the required reductions do not increase much above the twenty per cent level. Much more research is needed to ascertain the true shape of these curves and the extent of possible reductions in southern economies. Although many studies have begun, usable results will not be available in many southern signatories to the Climate Change Convention until about 1995.[e,18]

Some data are also available for the costs of carbon fixing by reforestation and reversal of deforestation where costs range from a few dollars to upwards of $60 per tonne of carbon fixed (see Figure 5.6); and for country level industrial, building, and transport energy efficiency (with implied costs per abatement of carbon emissions). But virtually without exception, these data are scattered, apply to different years, are incomplete (that is, only some portions of cost, such as capital cost, are given), and are partial (that is, only parts of a sector are covered). In short, these data are almost useless for the purposes of a comprehensive, global analysis.[19] It will take some years before carbon abatement cost data of greater reliability and scope are generated and collated. Until then, the only recourse is to use hypothetical curves supplemented with point data and judgement.

---

e Two sets of studies are in progress as of early 1993 that promise to provide additional data. These are the UN Development Programme studies of least cost greenhouse gas reduction curves in fourteen Asian developing countries; and the UN Environment Programme's greenhouse gas abatement costing studies conducted under the aegis of the UNEP Collaborating Centre on Energy and Environment in Denmark.

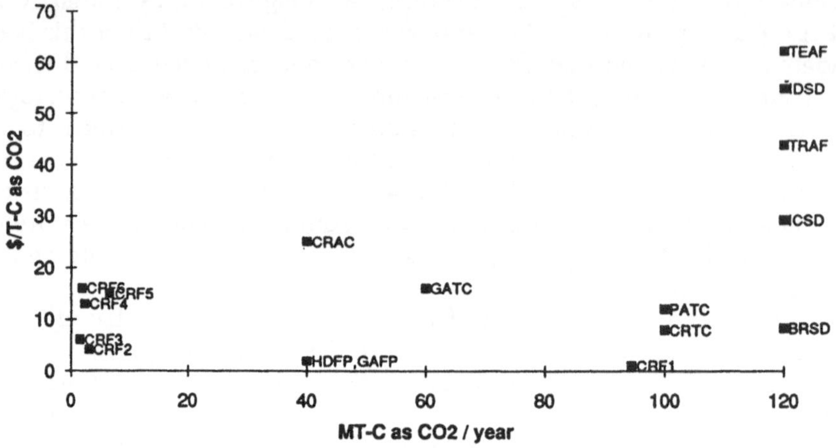

Key: same as Figure 5.5
BRSD = Slowing Deforestation in Brazil; CRF1 = Forest Reserves in Costa Rica; CRF2 = Agroforestry in Costa Rica; CRF3 = Forest Restoration in Costa Rica; CRF4 = Fuelwood Plantations in Costa Rica; CRF5 = Timber Plantations in Costa Rica; CRF6 = Natural Forest Management in Costa Rica; CRTC = Total Carbon Storage Potential in Costa Rica; CRAC = Additional Carbon Storage Potential in Costa Rica; GAFP = Forest Protection in Guatemala; GATC = Total Carbon Storage Potential in Guatemala; HDFP = Forest Protection in Honduras; ICSD = Slowing Deforestation in Ivory Coast; IDSD = Slowing Deforestation in Indonesia; PATC = Total Carbon Storage Potential in Panama; TEAF = Afforestation in Temperate Zone; TRAF = Afforestation in Tropical Zone.
Costa Rica a: n/a; b,c: 1990, 1991 (CRTC, CRAC); d,e: n/a; f: Joel N. Swisher, *Cost and Performance of CO_2 Storage in Forestry Projects*. Civil Engineering Dept. Stanford University, April, 1991; g: b,c are taken as same as year of source publication.
Panama, Guatemala, Honduras a: n/a; b,c: 1988; d,e: n/a; f: Swisher: 1991; g: b,c are taken as year of source publication.
Brazil, Indonesia, Ivory Coast a: n/a; b,c: 1989; d: 5%; e: n/a; f: Joel Darmstadter, *The Economic Cost of CO_2 Mitigation: A Review of Estimates for Selected World Regions*, Energy and Natural Resources Division, Resources for the Future, Washington DC, January 1991; g: b,c are taken as year of source publication. Estimates show cost only, not carbon abatement level.
TEAF, TRAF a: n/a; b,c: 1991; d: 5%; e: n/a; f: J Darmstadter, 1991; g: b,c are taken as year of source publication. Estimates show cost only, not carbon abatement level. The costs of afforestation are estimated for plantation of 30 year-rotation and matured forest.

Figure 5.6 *Carbon abatement by forestry management*

Marginal cost curves

The studies referred to above provide a range of marginal cost estimates drawn from two schools of energy economics that may be termed the top-down, price-driven 'economic pessimists' (high cost) on the one hand, and the bottom-up, end-use oriented, 'technological optimists' (low cost) on the other.[20]

As was noted earlier, I constructed three hypothetical marginal cost curves drawn from this literature to calculate incremental abatement cost (see Figure 5.4).[f] I am not suggesting that any of these curves are 'correct'. Rather, I seek to ascertain the impact of each curve on the estimated transfer justified by the obligation-to-pay index.

I simplified the well-known Nordhaus marginal cost schedule to refer to a required reduction level (the range in the ratio 'RR' defined in the box above summarizing the incremental cost method), using a weighted averaging procedure to simplify the original cost schedule into a simple, three-stage step function. This cost estimate is called Case 1, as are subsequent treatments of incremental abatement costs and distribution of costs based on it. It amounts to a high cost estimate in this chapter (although the low cost case is steeper at higher levels of required reduction).

Similarly, I used the NAS high cost estimate at 100 per cent implementation of technological potential, again employing a weighted averaging procedure to simplify the curve to three abatement levels that are very close to (but not exactly the same as) the ranges used in the Nordhaus marginal cost curve. This curve is called Case 2 and like Case 1, denotes subsequent calculations based on it. It amounts to a low cost estimate in this chapter.

Third, I postulated yet another marginal cost curve that combined elements of Cases 1 and 2. Specifically, the curve for the North and the East were taken from NAS and are identical with those used in Case 2. The curve for the South, however, assumes positive cost for the first two abatement levels, but at a level substantially below those given by Nordhaus in Case 1. The cost for the highest abatement level reverts to the same as Case 2; the first level, \$30/T $C\text{-}y^{-1}$ abated, is based on an estimated reforestation option in the South; the second level, \$77/T $C\text{-}y^{-1}$ abated, is a hypothetical estimated cost of industrial energy efficiency in developing countries. The World Bank's Global Environment Facility, for example, has published

f Estimates of marginal abatement cost at the following required reduction ratios (RR) were developed on a weighted average basis for each tonne-carbon conserved (T-CC) on the high, low, and medium cost curves. In the high cost curve (derived from Nordhaus's global abatement cost curve which excludes de(re)forestation, see Figure 5.3), the following step function was used: $0 \leqslant RR < 0.2$, \$50/T-CC; $0.21 \leqslant RR < 0.5$, \$160/T-CC; $0.51 \leqslant RR < 1.0$, \$250/T-CC. For the low cost curve (from the US National Academy of Science cost curve using their high estimate at 100 per cent implementation rate, see Figure 5.2), the following step function was used: $0 \leqslant RR < 0.18$, \$-173/T-CC; $0.19 \leqslant RR < 0.47$, \$7/T-CC; $0.48 \leqslant RR < 1.0$, \$250/T-CC. In the medium cost curve, a positive cost curve for the South only is substituted for the latter cost curve which is retained for the North and the East. For the medium curve for the South, the following step function was used: $0 \leqslant RR < 0.2$, \$30/T-CC; $0.21 \leqslant RR < 0.50$, \$77/T-CC; $0.51 \leqslant RR < 1.0$, \$294/T-CC. As no national abatement cost curves were available for the South, the last 'medium' curve for the South was based on sectoral point data from energy efficiency projects rather than systematic analysis. In principle, a smooth logarithmic or logistical marginal abatement cost function could be developed to fit empirical and judgemental estimates of the shape of the abatement cost curve. A simple step function is adequate for this study's illustrative purpose.

similar estimates that also hold the abatement costs to be mostly positive.[g,21] The upper level in Case 3 of \$294/T C-y^{-1} abated is the same as in Case 2.

The reason for adopting this composite marginal cost curve, called Case 3 hereafter, will become evident below. But crudely, the curve serves as a medium alternative to the high Case 1 and to the low Case 2.

Three aspects of these marginal cost curves should be kept in mind in subsequent applications. First, it should be noted that the cost in dollars per tonne of carbon abated is an annual, levelized cost of full, life cycle costs including components of capital, operating and maintenance, and salvage cost.[22] Thus, each tonne of carbon abatement paid for in a given year must be paid for again the following year.

Second, the incremental abatement profiles do not account for any benefits that flow from abatement (such as reduced local pollution, preservation of ecosystem values and services, etc.). As noted earlier, it is not possible yet to estimate the benefits of avoided climate change in a quantitatively meaningful fashion. For this reason, it is also highly unlikely that the protocols to the Convention will be based on a global cost-benefit analysis.[23] The analysis herein is therefore a truncated approach that amounts to determining least cost abatement paths for various countries and groupings of countries, and allocating the cost on efficiency and equity criteria.

Third, only technologically defined costs are included here. Many other components of cost are not included. These costs include the impact of carbon emission reductions on trade competitiveness; macroeconomic impacts, especially intersectoral adjustment costs; human resource development costs; and institutional change costs. Little work has been done to measure these categories of cost as of early 1993, especially in developing countries.

A final caveat is in order. No attempt has been made to ensure that the three cost curves used in this study are stated in constant dollars in the same year, using the same discount rate or other underlying assumptions. The curves have been taken 'as given' in the sources and are used in an heuristic exercise to calculate incremental costs. An actual application of this

g The realism of this approach is dubious, however. In China, for example, industrial energy efficiency investments in the second half of the 1980s of 8.2 billion yuan resulted in savings of 125 yuan per tonne of avoided coal production for a total of 28 million tonnes of coal avoided per year by the end of the decade. A rough conversion to levelized cost using the 1989 exchange rate of 4.7 yuan per US\$, assuming a 10 per cent discount rate, and assuming that such investments have a 10 year lifetime suggests a nett savings (negative cost) of about \$20/tonne of carbon abated, relative to providing the same services from expanding coal fuel supplies. Even this figure may be understated as the Chinese estimate of the investment cost attributes it all to energy efficiency, although many such investments are made for non-energy efficiency related reasons. That is, the real savings are probably larger. I am grateful to Mark Levine for providing information on China.

procedure under a protocol of the Convention would have to attend to these methodological details more rigorously.

With the marginal cost schedules defined, it is necessary next to determine the projected emissions that must be abated. I develop two scenarios to this end, a reference scenario, and an efficiency scenario for purposes of abatement cost calculation.

Reference emission projections

Many projections exist for future emissions. These projections are sometimes calculated via large-scale econometric modelling. Simpler methods using projected population, GNP per capita, and emissions per capita are also popular. I used a simple method in this study that applied projected emissions per capita estimated by the IPCC working groups to projected populations in 1995, 2000, and 2025.[24] The intermediate years were then interpolated.

Whichever method is employed contains many variables that are subject to great uncertainty. A GNP-based estimate is subject to enormous variation in projected GNP, even if historical GNP per capita and emissions per GNP rates are employed. Econometrically-driven models simply project the past onto the future (and probably overstate energy demand and thereby emissions).

Simple population-based models are attractive because the emissions per capita can be estimated physically. However, the method suffers from the disadvantage that demographic dynamics are also controversial and projected populations may be wrong; and projected emissions per capita from

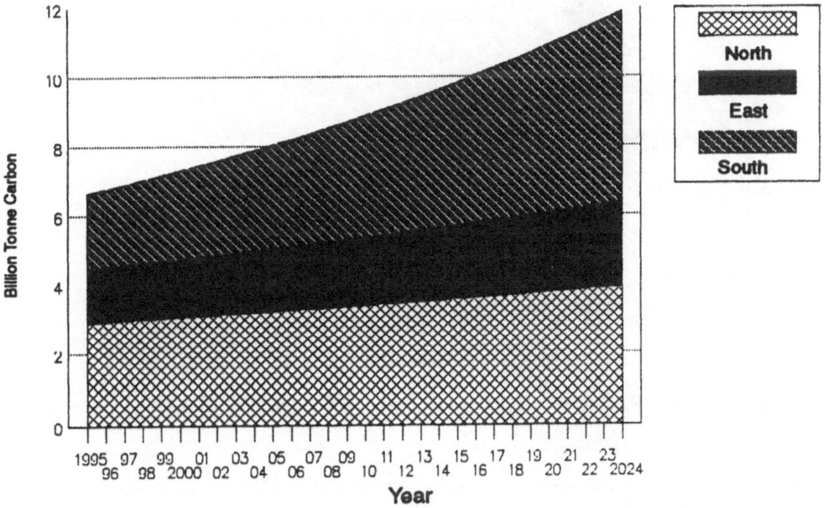

Figure 5.7 *Projected CO_2ff emissions, 1995–2025, reference scenario*

the IPCC are available only at a regional level (and thus are often too high or too low for a given country over the first five years of the projection).

The projected carbon dioxide emissions from fossil fuel usage in the reference scenario are shown in Figure 5.7. Starting at about 6.6 GT/y of C in 1995, the world reaches about 11.8 GT/y of C in 2025 in the unrestrained reference scenario. This projection is consistent with a variety of other IPCC and OECD projections and is adopted in this study as the reference scenario.

Efficiency Scenario

The reference scenario, however, is unsuitable for the purpose of calculating incremental abatement cost. No allowance is made for emissions avoided due to the inexorable increase in energy efficiency due to technical change in product and process design and operation,[h] nor for steps taken each year to increase energy efficiency, use renewable energy, and to substitute lower for higher carbon intensity fuels. This adjustment is now made in the efficiency scenario.

Emission projections

The projected emissions in the efficiency scenario are shown in Figure 5.8.

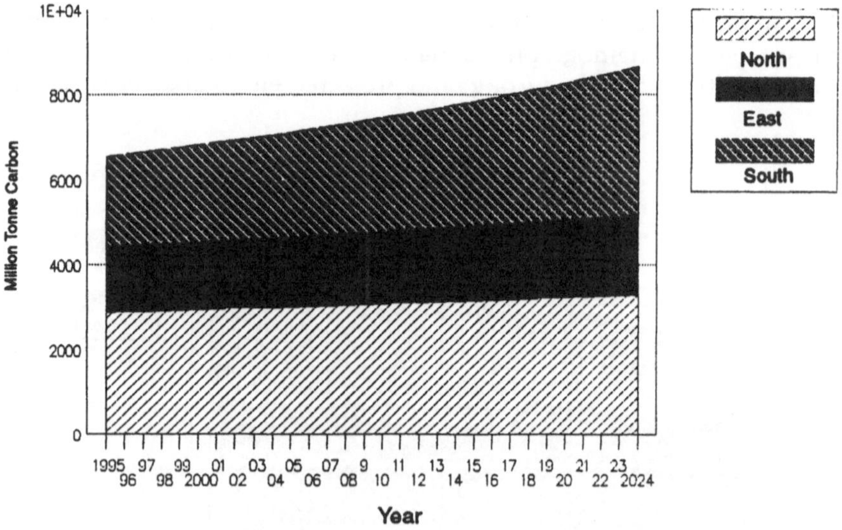

Figure 5.8 *Projected CO_2ff emissions, 1995–2025, efficiency scenario*

h Known as 'autonomous energy efficiency increase' to energy economists. Some hold that this factor is strongly positive (0.5–1 per cent per year reduction); others claim that it is negative due to the impact of energy efficiency investment opportunity costs on general capital productivity.

Starting at about 6.6 GT/y of C in 1995, the world reaches about 8.6 GT/y C rather than 11.8 GT/y of C in 2025 in the unrestrained reference scenario, a fall of nearly 27 per cent. This reduction occurs because of the effect of what is referred to in the box above as *Efficiency Rule 1* whereby the scenario assumes that only 40 per cent of the increment in the annual growth assumed in the reference scenario actually results in emissions. Increased energy efficiency associated with a wide range of managerial and technical measures are assumed to avoid the other 60 per cent of the emissions associated with annual growth in the reference scenario. This step serves as a proxy for the unknown increases in autonomous energy efficiency in each region.[i]

I seek to determine the scale of funding in the South required for carbon abatement because of climate change. I am not suggesting that the North should embark upon an open-ended transfer in the form of compensation for its pre-emption of 'atmospheric space', although some have argued that it should. Only those activities required to abate carbon emissions from fossil fuel usage are included. But not all of the abatement (and therefore cost) in the South that is required to meet emission targets would be incurred as some energy efficiency and abatement would occur regardless of energy efficiency or climate change. In principle, this abatement needs no additional funding from the North and should be removed from the South's marginal abatement cost.

Additionally, this procedure does not cover any costs of 'autonomous', incremental reduction of carbon emissions that arises from economy-wide technical change. In fact, a substantial fraction of this 'autonomous' reduction potential that is reflected in the 'efficiency' projection of emissions will not be realized without additional funding beyond that focused directly on carbon abatement. It is impossible, however, to determine the impact of these two conceivable and opposite adjustments to the incremental cost of abatement in the South. Thus, no deduction is made to the transfer of resources justified by the obligation-to-pay index in order to reflect the net impact of these two flows.[j]

i This absolute decline between reference and efficiency scenario projected emissions amounts to a global 1 per cent/y increase in 'autonomous' energy efficiency (0.6 per cent in the North, 0.8 per cent in the East, and 1.6 per cent in the South). This step reduces the requisite carbon abatement and avoids cost associated with projected emissions that would in fact be avoided by non-energy sector technical change throughout the economy. Otherwise, projected but non-existent carbon emissions would be included and abated, thereby overstating the abatement costs in the scenario.

j The issue also cuts the other way. As will be argued in Chapter 14, it will be impossible in practice to limit the transfer of resources to 'carbon abatement' or 'energy efficiency' for the simple reason that energy efficiency and carbon conservation-related technical change cannot be distinguished from the whole array of technical changes and practices that result in increased factor productivity in an economy.

Required reductions

The target global carbon dioxide emission level in 2024/5 in the efficiency scenario is set at 2.8 GT. This emission is halfway between two IPCC stringent emission reduction scenarios.

The first is IPCC's 'alternative accelerated policies scenario' in which CO_2 atmospheric levels are not allowed to exceed 400 ppmv in 2100, thereby keeping 'equivalent' CO_2 concentration at about 420 ppmv, or a fifty per cent increase of the pre-industrial level of 280 ppmv. (See case e in Figure 5.9). In this case, CO_2 emissions must fall to about 4 GT/y by 2025.[25]

The second is IPCC's tougher projection of the reductions needed to simply stabilize CO_2 concentration at its current level in 2100, that is, 420 ppmv. The projected emission in 2025 that is consistent with this target is 1.5 GT.[26] (See case f, Figure 5.9).

This goal is adopted as a conservative basis for estimating cost. If scientific understanding shows that the reductions are unnecessary, it allows the rate of reduction to slacken early enough so that significant positive cost is not

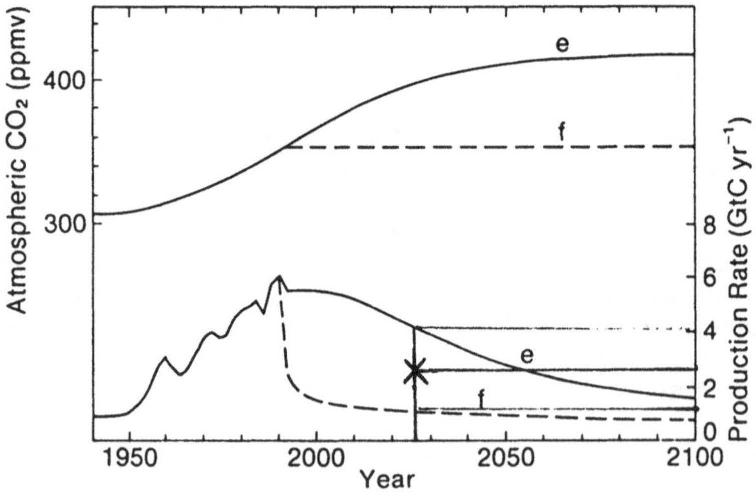

Note: Future CO_2 rates calculated by means of a box diffusion carbon cycle model so as to yield the prescribed atmospheric CO_2 concentrations after 1990.
Case e shows concentration increasing steadily (logistic function of time) to 420 ppmv, a 50% increase over preindustrial level.
Case f shows concentration held constant at 351 ppmv after 1990 (25% increase over preindustrial level).

Source: R Watson *et al*, 'Greenhouse Gases and Aerosols,' in Intergovernmental Panel on Climate Change, *Climate Change, The IPCC Scientific Assessment*, Cambridge University Press, New York, 1990, p 15.

Figure 5.9 *Future CO_2 production rates vs atmospheric concentration*

incurred (the early phase of reductions should be achieved at negative cost). It requires that net terrestrial carbon emissions from biotic sources be reduced to zero by 2025 – something that needs to be done for many reasons other than the increasing greenhouse effect. Finally, it limits the rate of realized global mean temperature increase – assuming that the atmospheric models are correct – to about 1 degree centigrade by 2100 relative to 1980 (or about half a degree centigrade by 2025). Relatedly, the anticipated sea level rise from global warming is also restrained to about 25 cm by 2100 (or about 15 cm by 2030).[27]

By adopting this approach, I am implicitly defining 'sustainability' as accepting ecological damage over the next decade associated with a possible rate of temperature increase of 0.1 degree per decade, and a sea level rise of 0.03 metres per decade. It is sobering to note, as the IPCC Working Group I stated at the Second World Climate Conference in 1990, that: 'If the forcing were then held constant, temperatures would continue to rise slowly, but it is not certain whether it would take decades or centuries for most of the remaining rise to equilibrium to occur.'[28]

Allocation rule

A rule is now needed to allocate the permissible global carbon dioxide emissions from fossil fuel of 2.8 GT among states in 2025. Assuming that the same states exist, the total projected emissions are distributed in proportion to each nation's share of carbon sinks for emissions in 1987. Thus, each nation's target emission rate in 2025 must not exceed its allocated carbon sink property right. This property right was calculated on the basis of current population (for oceanic sinks) and national territory (for land-based sinks).[k,29]

A nation's oceanic sink was obtained by dividing the estimated 1987 oceanic sink by 1987 global population and multiplying by national 1987 population; and a nation's terrestrial sink was obtained by dividing the estimated 1987 terrestrial sink by all nations' land area and multiplying by national land area. A nation's fraction of total sink equals the sum of the two divided by global sink. Converted to a fraction of total sink, the national distribution of emission rights can be used to allocate the permissible emissions in 2025.

---

k There are many different ways to allocate emissions which reflect differing emphases on equity and efficiency goals including: equal emissions; proportional to area; proportional to current emissions; per GDP (corrected for purchasing power); per capita; per adult capita; historical per capita (as in the obligation to pay index in this book); and mixed schemes such as that used in this chapter. Each approach offers different pros and cons in terms of simplicity, equity, efficiency, and pragmatic or subsidiary goals such as accounting for dissimilar geography, resource endowments, and economic structures; population growth; inclusion or exclusion of past emissions; ease of negotiation; draws on precedent, etc. A variety of these allocational bases should be tested in future analyses. The one used in this study is not sacrosanct.

Specifically, the total estimated 1987 carbon sink was about 4.7 GT of carbon, spread equally between land and oceans. Thus, the average terrestrial sink in 1987 was about 0.18 tonnes of carbon per hectare; and the oceanic sink was about 0.48 T C per capita sunk at sea.[30] This approach assumes that all people have a natural right to oceanic sinks as a common heritage; but that nations have a property right to (and responsibility to maintain) national terrestrial carbon reservoirs in forests and soil.

It is noteworthy that estimates of sinks are still highly uncertain. Recent investigations have shown that terrestrial ecosystems of the northern hemisphere may be taking up substantially more carbon than previously thought. Alternatively, the northern oceans may be bigger sinks than the southern oceans.[31] Other scientists have noted that four possible terrestrial biotic carbon exchanges may account for the missing sink.[32] These processes are net deforestation in tropical rainforests, carbon fixing by temperate and high latitude boreal ecosystems, and carbon dioxide fertilization of primary productivity, especially in tropical rainforest. Substantial uncertainty therefore remains as to the size and location of sinks that are assumed here to become national property rights, or the basis of claims to the future permitted level of carbon emissions in 2025.

Each country is then placed on a trajectory that smoothly reduces its annual emissions from 1995 down to its permissible fraction of the 2025 global total. This identity is achieved by imposing two kinds of reductions on emissions, referred to as *Efficiency Rule 2* in the box above. The first applies to the base year emissions; and the second, to the remaining growth increment.

The base year (1995) emissions must be reduced by 6 per cent annually in the North and East, and by 2 per cent per year in the South, whether by offsetting expansion of the carbon sink, or by additional reductions to emissions relative to those projected. (The divergent reduction rates for the North/East versus the South are required to avoid forcing the South to reduce more than their target 2024/5 emission rate.) I assume that existing emissions can be phased out only as fast as existing capital stock is turned over, thereby constraining the rate of abatement of base year 1995 emissions.[33]

Each year, the growth increment must be further reduced by 65 per cent (or 45 per cent in the case of the South) to ensure that the total reductions avoid sufficient emissions by 2024/5 to meet the target levels. In the efficiency scenario's projected emissions, each nation's and national grouping's required annual reduction must increase rapidly each year from an initial global 0.3 GT C abatement in 1995 to 5.7 GT C abatement in 2024/5 (see Figure 5.10). Thus, the absolute required reduction must increase at about 10 per cent per year to achieve this goal.

Concurrently, the two IPCC scenarios that provide the target global emission adopted in this chapter assume that deforestation is halted and that biotic sinks have become major net sinks by 2025. Controls are also required on other greenhouse gases if the 'equivalent' carbon dioxide targets are to be

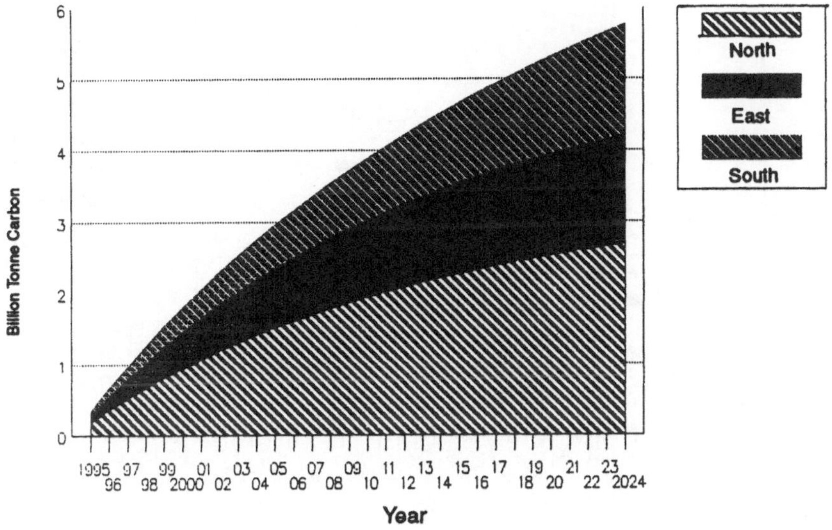

Figure 5.10 *Required reduction, 1995-2025, efficiency scenario*

achieved. Implicitly, therefore, I assume in these scenarios that these measures have been implemented to match the control on carbon dioxide from fossil fuel usage. I do not include the costs of these measures except and insofar as they provide an option to 'sink' fossil fuel carbon emissions. Forestry management, for example, is one of the technological options included in the cost curve for Case 2.

The resulting reductions profile generates CO_2ff emissions in 2005 that are about 30–40 per cent above the Toronto Target emissions that year (that is, 80 per cent of the recorded 1988 emissions). Thus, the required reductions in this chapter are lenient with respect to medium term abatement targets that some industrial countries have already adopted. But the required reductions are stringent by 2025 relative to projections in order to minimize climate change induced damage associated with keeping carbon dioxide concentrations to between today's carbon concentration (already a 25 per cent increase over the pre-industrial level) and a 50 per cent increase in the pre-industrial carbon load of the atmosphere.

Required reduction ratio

With the projected annual emission and calculated required reduction available for each nation or national grouping, it is now possible to compute a ratio of the two, the 'required reduction ratio' or 'RR' factor. This ratio, calculated for each year for each nation and national grouping, is shown in Figure 5.11. As can be seen, the global RR starts quite low (at 0.05 in 1995) and reaches about 0.66 in 2024/5. However, it starts low in the South in 1995

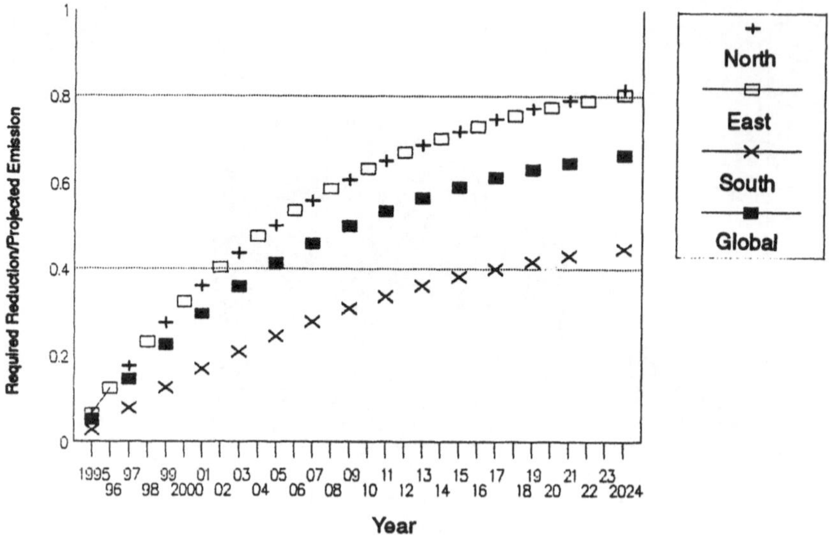

Figure 5.11 *Required reduction ratio 1995–2025*

(at 0.02) and remains relatively low in 2024/5 (at 0.45), whereas it reaches about 0.82 in the North and 0.80 in the East by 2024/5.

Incremental abatement cost

The RR ratio is used to trigger the next abatement level of cost in the marginal cost curves discussed above (see footnote f). The incremental abatement cost for a given year is then calculated for each nation and national grouping by applying the RR ratio for that year by the relevant marginal cost curve.[l] This procedure generates a stream of annual abatement costs for each of the marginal cost curves which are called 1, 2 and 3.

Incremental abatement cost 1

In Figure 5.12, I show the incremental cost calculated for the RR ratio noted above and the Nordhaus marginal cost curve, that is, for Case 1. As can be seen, the global, annual cost (undiscounted) starts at $16.8 billion in 1995

l Thus, countries continue to obtain reduction at the low end of the cost curve over all years in the reduction scenario; the higher costs are only applied for those years when the reduction ratio is high enough to activate higher marginal costs; and these higher costs are only applied to reduction tonnage above the reduction ratio which triggers the higher cost. Thus, average cost of reduction will always be less than the marginal cost in any given year. This approach is akin to a progressive income tax system wherein income tax rates increase at higher income brackets; but the ever increasing tax rates only apply to income within specified ranges.

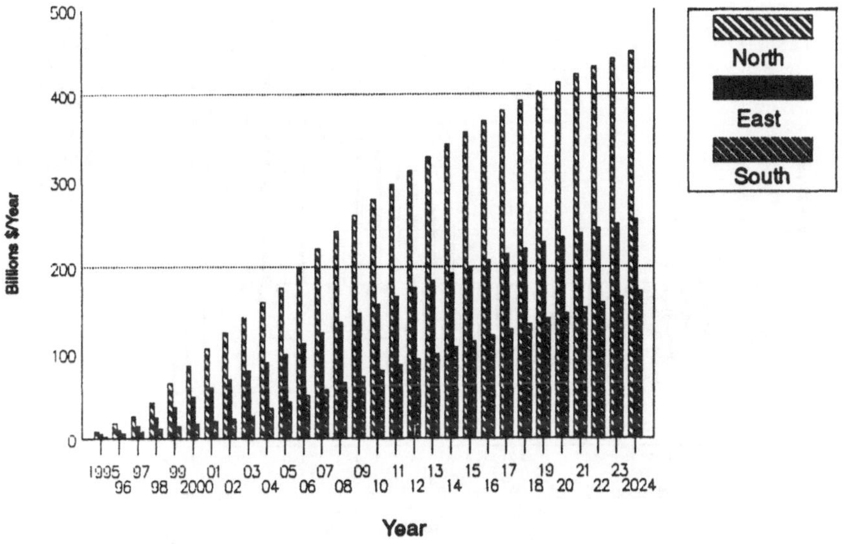

Figure 5.12 *Incremental abatement cost, Case 1 (Nordhaus MC), 1995–2025*

and reaches $876 billion in 2024. The cost is always positive and is substantial for all nations and national groupings.

Incremental abatement cost 2

In Figure 5.13, I show the incremental cost for the North/East and the South respectively. In this case, the incremental abatement cost is calculated for the RR ratio noted above and the US National Academy of Sciences marginal cost curve, that is, for Case 2. As can be seen, the global, annual cost (undiscounted) starts at –$58 billion in 1995 and reaches $276 billion in 2024.

Global incremental abatement cost is much smaller in Case 2 than in Case 1 because the South's cost begins and remains negative until 2024/5. This result in turn is due to the fact that the South's RR of 0.44 in 2020 remains below the 0.47 needed to kick in the positive cost in the marginal cost curve. Indeed, the South's cost remains a negative (undiscounted) $101 billion in 2024/5.

The North and the East, however, pass this point in 2007–8, after which they begin to incur positive cost, reaching an annual, undiscounted cost of $239 and $131 billion respectively in 2024/5.

From a purely economic perspective, therefore, no transfer payment can be justified from the North to the South when a marginal cost curve in the South is assumed to be close to that presented by McKinsey & Co. or the US National Academy of Sciences as portrayed in Case 2. This issue is dealt with in greater detail below and in Chapter 6.

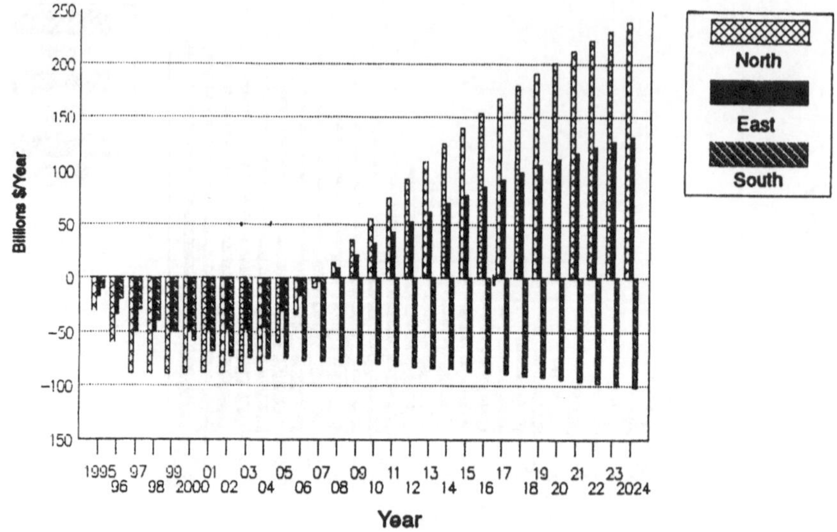

Figure 5.13 *Incremental abatement cost, Case 2 (US NAS), 1995–2025*

Incremental abatement cost 3

To provide an intermediate estimate between the high cost structure assumed by Nordhaus and the low cost structure derived from the US National Academy, a third composite marginal cost schedule was derived that assumed the NAS Case 2 for the North and the East – the same as in the previous run – but substituted my own estimates below Nordhaus but above NAS for the South (see footnote f).

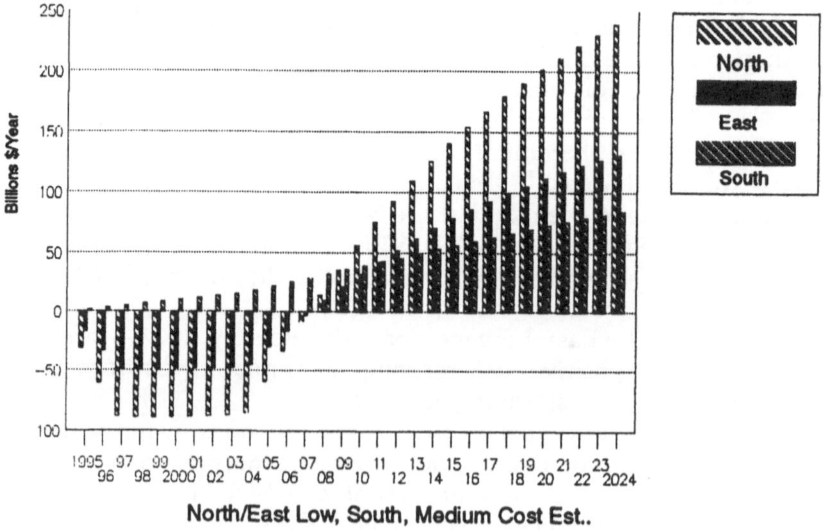

Figure 5.14 *Incremental abatement cost, Case 3, 1995–2025*

Naturally, the North and the East follow the same trajectory as in the previous section (see Figure 5.14). After an initial burst of negative cost, they move into a protracted phase of positive cost that accumulates between 2008 and 2025. This latter phase more than offsets the initial phase of negative cost. In the South, however, cost is positive from the start, building up to an (undiscounted) $85 billion by 2024/5 from $1.7 billion in the base year 1995.

Vulnerable coastal states and sea level rise

The potential impacts of climate change on island and coastal developing states have been widely discussed. Ten per cent of the world's population lives within 20 km of the coast. Ten per cent of the world's coastal zone has a population density greater than 100 people per square km. One study showed that no less than twenty-seven states are highly vulnerable to the impacts of climate change on populations due to inundation of land by sea level rise, given their national abilities to take protective measures against this threat. Among the most potentially affected are Bangladesh, Egypt, The Gambia, Indonesia, Maldives, Mozambique, Pakistan, Senegal, Surinam, and Thailand and a host of other low-lying island microstates such as Kiribati.[34] Areas at high risk in Asia Pacific are shown in Figure 5.15.

Bangladesh is a candidate for the worst case scenario. Nearly 80 per cent of Bangladesh is composed of a complex delta that feeds from three rivers into the Bay of Bengal. Agricultural output from these lands produces about 55 per cent of national GDP, and employs about 85 per cent of the population.[35] Bangladesh is already susceptible to enormous storm surges. A recent cyclone, for example, cost over $3 billion in repairs, ignoring loss of life, production, and environment.[36] One analyst estimates that sea level rise induced by climate change may force the relocation of up 80 million people, reduce rice-producing land by up to 2.6 million hectares and projected rice output in 2010 by 8–15 per cent.[37]

Many problems arise, however, when one attempts to attribute cost to sea level rise. First, it is still scientifically problematic to assign causality for sea level rise to climate change.[38] Indeed, climate scientists are still trying to resolve the many practical problems involved in quantifying the regional and local effects of global climate change on small spatial scales where topography, proximity to the ocean, and local biogeography all affect the local climate. It is still not possible to predict the long run effects of global climate change on the frequency and intensity of storm systems, temperature and rainfall, and monsoons.[39]

Also, relative sea level rise is caused by a number of factors, including some non-climatic ones. Along the coast in Bangladesh, for example, the deltaic plains are subsiding due to tectonic shift or the weight of accumulating sediment, resulting in a tilting from west to east and a shift in the relative sea level with or without climate change.[40]

Finally, in many high risk coastal zones, human habitation and economic

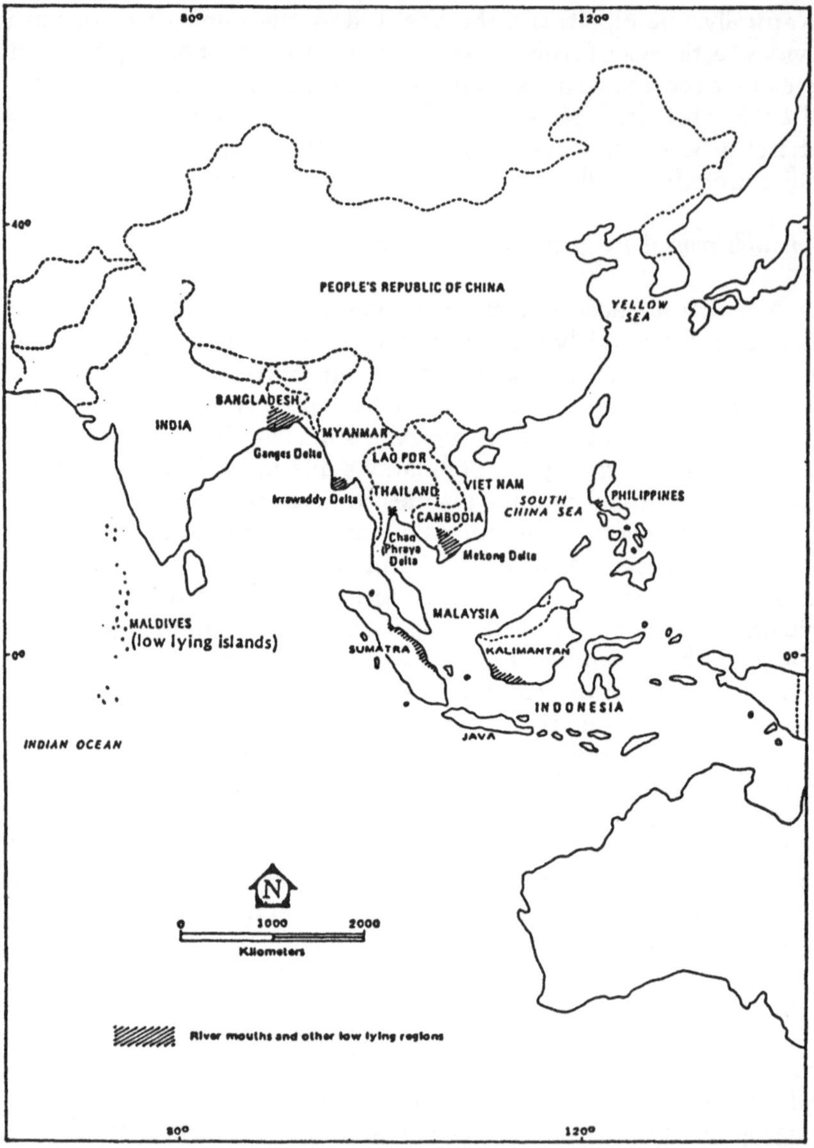

Figure 5.15 *Vulnerable coastal/Island states*

uses are already highly questionable and should change – with or without climate change risks superimposed on existing stresses.[41] Furthermore, if climate change induces people and production to shift out of vulnerable coastal zones, some of the cost will be absorbed by private parties and will not require supplementation by additional public funds.

Finally, data on 'natural disasters' and vulnerability of different social groups are poor and unsuitable for purposes of determining compensation for damages related to climate change.[42] Indeed, a strong argument can be made that reducing stress on the poorest, least adaptable social strata in vulnerable states requires that such compensation be spent on welfare and development, not on climate change mitigation projects.[43]

For all these reasons, it is impossible to quantify the costs of sea level rise due to climate change. Instead, I estimated only the costs associated with a barrier protection against sea level rise for the developing world, although it should be noted that in general, selective retreat is preferable to erecting capital defences against the fury of the oceans.[44]

I adopted a Dutch estimate of $488 billion provided to the IPCC as the global cost of a barrier defence for vulnerable coastal states against sea level rise of one metre in 100 years.[45] This figure does not allow for annual maintenance costs, population relocation costs, losses of production such as fisheries, loss of land values, loss of national existence and associated international environmental refugees, or the increased costs of exploiting continental seabed resources. Given the uncertainty, this figure is used to illustrate how this cost might be allocated according to the obligation-to-pay index. It is not presented as a definitive estimate of the costs of sea level rise induced by climate change. In future calculations, a range of estimates should be used.

I further assumed that half the total 100-year cost is incurred in, and spread equally over, the first 30 years of our scenario. This cost is present valued for each region and is added to the total cost of abatement to produce an overall incremental annual cost consisting of carbon abatement and coastal protection. Of course, abatement will reduce the rate of sea level rise and thus the 100-year cost. However, even if carbon emissions were reduced to levels that restrain atmospheric CO_2 to a 50 per cent or less increase above pre-industrial levels, the IPCC's best guess of the inexorable sea level rise associated with our carbon emission/abatement scenario – the legacy of our ancestors' profligate energy usage – is about 15 cm by 2025. Richard Warrick and Atiq Rahman call this momentum the 'committed' sea level rise.[46] It is therefore prudent to assume that defensive actions will be taken earlier rather than later, especially as many of these works will take up to fifty years to complete.[47]

No account was made in this chapter of other climatic change related costs imposed on the South. In this regard, losses of agricultural, forestry and fishery production are particularly important, especially in the poorest developing countries in semi-arid tropical and sub-tropical regions. However, careful studies summarized by Martin Parry suggest that the net yield and welfare impacts of a range of climate change scenarios on agricultural GDP are likely to be small, and offset by increased output elsewhere in a given nation or national grouping.[48] Moreover, the net impact on the food production system and the geographical distribution of these costs and benefits cannot be determined at this time.[49] For this reason, no compensation for foregone agricultural, fisheries, or forestry production is included. In the future, however, such costs should be estimated using long time horizons and low discount rates.[50]

Discounted incremental cost

In line with the three marginal cost curves introduced earlier, three

Discounted incremental costs, Case 1

A Abatement Costs

Definition: Incremental Cost 1, using Nordhaus marginal cost curve

Net Present Value (not adjusted for Obligation-To-Pay), 1990–2025, $billion/year

North	*East*	*South*	*Global*
2932	1653	867	5452

Annuity (not adjusted for South's Obligation-To-Pay)

191	108	56	355

% of global annuity (not adjusted for South's Obligation-To-Pay)

54	30	16	100

B Global, 100 Year, Instantaneous Coastal Protection Costs
Total: 488 billion

***B1** South's Island/Coastal Protection Costs*
Total cost estimated at 206.6 billion over 100 years
Estimate that 50% of cost incurred between 1995–2025
Annual cost is (0.5 × 206.6 billion)/30 years = 3 billion/year
NPV of South's coastal protection cost = 53 billion

***B2** North's Coastal Protection Costs*
Total cost estimated 227.6 billion over 100 years
Estimate that 50% of cost incurred between 1995–2025
Annual cost is (0.5 × 227.6 billion)/30 years = 4 billion/year
NPV of North's coastal protection cost = 58 billion

***B3** East's Coastal Protection Costs*
Total cost estimated 53.9 billion over 100 years
Estimate that 50% of cost incurred between 1995–2025
Annual cost is (0.5 × 53.9 billion)/30 years = 1 billion/year
NPV of East's coastal protection cost = 14 billion

***B4** Global Coastal Protection Costs*
NPV of global coastal protection cost = 125 billion

***B5** Total Abatement Plus Coastal Protection Costs*

North	*East*	*South*	*Global*	
2990	1667	920	5577	(net present value, billion $)
195	108	60	363	(annuity, billion $/year)
54	30	16	100	(% of total)

distributions of discounted total cost are presented in the following section. In all cases, a real discount rate of five per cent is used to present value streams of cost arising from abatement and coastal protection. No adjustment is made at this stage for obligation-to-pay based on historic contribution to climate change and ability to pay. This redistribution of cost is performed in the next chapter.

Discounted incremental costs, Case 2

A Abatement Costs

Definition: Uses US National Academy of Science marginal cost curves for all areas

Net Present Value (not adjusted for Obligation-To-Pay), 1990–2025, $billion/year

North	*East*	*South*	*Global*
104.0	68.8	-1002.2	-829.5

B Global, 100 Year, Instantaneous Coastal Protection Costs

Total: 488 billion

B1 *South's Island/Coastal Protection Costs*
Total cost estimated at 206.6 billion over 100 years
Estimate that 50% of cost incurred between 1995–2025
Annual cost is (0.5 × 206.6 billion)/30 years = 3 billion/year
NPV of South's coastal protection cost = 53 billion

B2 *North's Coastal Protection Costs*
Total cost estimated 227.6 billion over 100 years
Estimate that 50% of cost incurred between 1995–2025
Annual cost is (0.5 × 227.6 billion)/30 years = 4 billion/year
NPV of North's coastal protection cost = 58 billion

B3 *East's Coastal Protection Costs*
Total cost estimated 53.9 billion over 100 years
Estimate that 50% of cost incurred between 1995–2025
Annual cost is (0.5 × 53.9 billion)/30 years = 1 billion/year
NPV of North's coastal protection cost = 14 billion

B4 *Global Coastal Protection Costs*
NPV of global coastal protection cost = 125 billion

B5 *Global Coastal Protection Costs*

North	*East*	*South*	*Global*	
162.3	82.5	-949.3	-704.4	(net present value, billion $)
10.6	5.4	-61.8	-45.8	(annuity, billion $/year)

C Net present value adjusted by South's Obligation to Pay

No transfer to South is called for on the grounds of Obligation-to-Pay, so the South pays for its own coastal costs in this case.

Case 1: High discounted incremental costs
To calculate the net present value of the stream of incremental cost using the Nordhaus marginal cost curve (see box), table, the flow of discounted abatement cost is added to the discounted value of the stream of costs arising from the construction of coastal protective barriers in the South. The sum of discounted abatement cost ($867 billion) and coastal protection cost ($53 billion) equals total cost in the South with a net present value of $0.92 trillion or 16 per cent of the total global cost of $5.6 trillion (with the North paying $3 trillion or 54 per cent and the East $1.7 trillion or 30 per cent).

Case 2: Low discounted incremental costs
The low ratio of required reduction to projected emissions in the South means that it never reaches a level on the low abatement cost curve that incurs positive cost (see footnote f and Figure 5.11). This outcome contrasts with that in the North which does incur significant positive costs that outweigh the early negative costs, even when discounted (see box). Adding coastal protection costs does not offset the overwhelmingly abatement-related savings in the South although it does increase the North's costs from $104 to $162 billion; and that of the East from $69 to 83 billion.

Consequently, the accumulated net savings in the South ($949 billion of cost) overwhelm the overall positive discounted abatement costs of the North and the East ($162 and $83 billion respectively). This result emphasizes the need for carbon abatement demonstration programmes and research on a scale sufficient to ascertain the shape of cost curves reliably at required reduction ratios of between 0.2 and 0.5 in the South. The empirical cost curves reported in Part III of this book (and summarized in Figure 5.5) mostly address the lower end of the reduction requirement ratio. We remain largely ignorant of the shape of the curve or even its sign above this level in most countries.

There is strong reason to believe that the South's ability to realize its technological potential for low cost abatement – if it exists – will face political and institutional constraints, market imperfections, and high transaction costs. In addition to the impact of outright poverty, these obstacles in the South include competing social priorities, poorly developed capital markets, politicized energy prices, weak state administrative and political structures, powerful private interests, unstable political regimes, high inflation rates, dependent and weak scientific and technological sectors, and short planning horizons.

Case 3: Medium discounted incremental costs
By changing the South's marginal cost to a positive cost, this case adjusts the previous scenario to provide an intermediate cost estimate – above that of the US National Academy of Sciences, but below that of Nordhaus. In this case, the South's total discounted cost for abatement is about $441 billion (see box). The North incurs $104 billion total discounted cost, and the East

Discounted incremental costs, Case 3

A *Abatement Costs*

Definition: Same as Case 2 but South uses a positive cost curve

Net Present Value (not adjusted for Obligation-To-Pay), 1990–2025, $billion/year

North	*East*	*South*	*Global*
104.0	68.8	441.1	613.9

Annuity (not adjusted for South's Obligation-To-Pay)

6.8	4.5	28.7	39.9

% of global annuity (not adjusted for South's Obligation-To-Pay)

16.9	11.2	71.9	100

B *Global, 100 Year, Instantaneous Coastal Protection Costs*
Total: 488 billion

B1 *South's Island/Coastal Protection Costs*
Total cost estimated at 206.6 billion over 100 years
Estimate that 50% of cost incurred between 1995–2025
Annual cost is (0.5 × 206.6 billion)/30 years = 3 billion/year
NPV of South's coastal protection cost = 53 billion

B2 *North's Coastal Protection Costs*
Total cost estimated 227.6 billion over 100 years
Estimate that 50% of cost incurred between 1995–2025
Annual cost is (0.5 × 227.6 billion)/30 years = 4 billion/year
NPV of North's coastal protection cost = 58 billion

B3 *East's Coastal Protection Costs*
Total cost estimated 53.9 billion over 100 years
Estimate that 50% of cost incurred between 1995–2025
Annual cost is (0.5 × 53.9 billion)/30 years = 1 billion/year
NPV of East's coastal protection cost = 14 billion

B4 *Global Coastal Protection Costs*
NPV of global coastal protection cost = 125 billion

B5 *Total Abatement Plus Coastal Protection Costs*

North	*East*	*South*	*Global*	
162.3	82.5	494.1	738.9	(net present value, billion $)
10.6	5.4	32.1	48.1	(annuity, billion $/year)
22	11.2	66.9	100	(% of total)

about $69 billion out of a global cost of $0.6 trillion for carbon abatement only. Thus, the South pays the bulk (72 per cent) of the carbon abatement cost before coastal protection is factored in.

After coastal protection costs are added, the North's total discounted cost increases by $58 billion, the East's by $14 billion, and the South's by $53 billion. In this case, the South shoulders 67 per cent of the total cost, the North 22 per cent, and the East 11 per cent.

Implications for the South

These results are summarized in Table 5.1 and Figure 5.16. If nothing is done to transfer resources and technology, then the high (1) and medium (3) cost cases show that the South would pay between two and ten times more than its 'share' (relative to its obligation-to-pay as determined in Chapters 4 and 6.

Case 1 (high costs) is especially interesting. Here, the northern economies would likely engage in a massive burst of technological innovation in response to rising energy prices and competitive pressures within the North. A dramatic move toward ecologically sustainable growth implies a shift to a 'fourth wave' of industrial capitalism. The changes in store would be on a par with earlier technological revolutions (fossil fuelled factories, mass production lines, information systems).

This transformation would likely reduce the North's marginal production costs in most goods and services relative to the South which would be left far behind in technological and economic terms. Consequently, the North's

Table 5.1 *Summary of incremental abatement and protection costs*

	Region			
	North	*East*	*South*	*Global*
High cost curve: Case 1				
NPV ($billion)	2990	1667	920	5577
Annuity ($billion/year)	195	108	60	363
% of total	54%	30%	16%	100%
Low cost curve: Case 2				
NPV ($billion)	162	83	-949	-704
Annuity ($billion/year)	10.6	5.4	-62	-46
% of total	n/a	n/a	n/a	n/a
Medium cost curve: Case 3				
NPV ($billion)	162.3	82.5	494	738.8
Annuity ($billion/year)	10.6	5.4	32.1	48.1
% of total	22%	11%	67%	100%
Obligation to pay				
% of total	73%	20%	7%	100%

This table shows the incremental cost of carbon abatement and coastal protection costs, before redistribution according to obligation to pay.

Sources: Costs from *Discounted incremental costs* box in this chapter; obligation to pay, see Chapter 6.

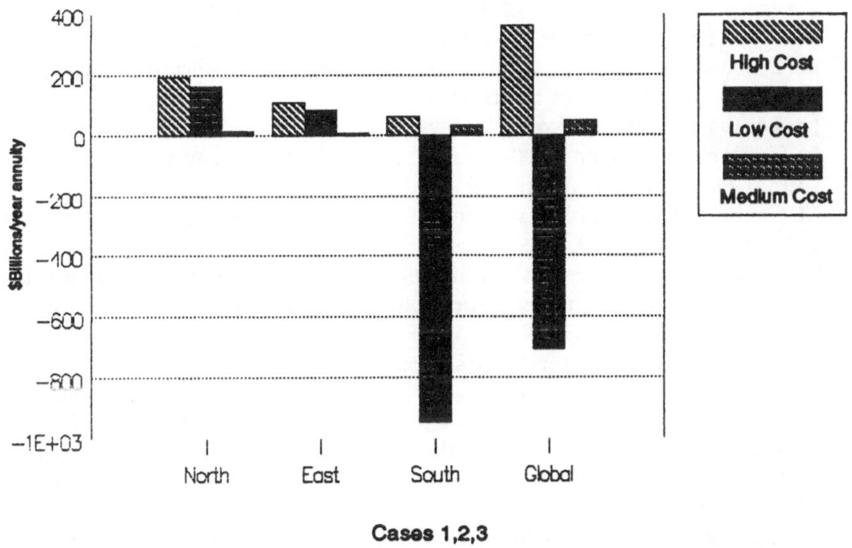

Figure 5.16 *Summary of incremental C-abatement/coastal costs, bill$/y*

carbon abatement costs could be well below those assumed in this chapter, especially in Case 1. Thus, by assuming 'frozen' technological costs over time, I may have overstated greatly the share of global costs that the North would pick up. Only if one assumes that the South achieves low carbon abatement costs that are equal to those of the North (Case 2) can the South come out ahead under its own steam. There is little reason, however, to believe that the South has either the financial resources or the institutional capability to match the North and the East in achieving such a rapid transition.

In the high and medium cost cases, the South is burdened disproportionately (16 and 67 per cent respectively versus its obligation-to-pay of 7 per cent). Dealing with this maldistribution of cost is the subject of the next chapter.

Notes and references

1 For a review of scenario methods in the energy context, see S Encel and N Conner, *Alternatives in Energy Policy and the Use of Scenarios – A Pilot Project*, End of Grant Report No. 813; National Energy Reserch, Development and Demonstration Programme, Canberra, June 1988; and for the same in the context of global climate change, see L Lave and D Epple, 'Scenario Analysis,' in: R Kates, J Ausubel and M Berberian (ed), *Climate Impact Assessment*, John Wiley, New York, 1985, pp 511–528

2 This section is derived from W. Nitze, *The Greenhouse Effect: Formulating a*

Convention, Royal Institute of International Affairs, London, 1990, pp vii–ix. Nitze is a former US State Department official who played a key role in the US participation in the IPCC. In this reference, he predicted accurately the structure and content of the treaty

3 J Sebenius, *Negotiating a Regime to Control Global Warming*, John F Kennedy School of Government paper G-90-10, Harvard University, Cambridge, Massachusetts, 1990. Emphasis in original

4 M Grubb, 'The Greenhouse Effect: Negotiating Targets'; *International Affairs*, volume 66, no 1, 1990, p 85

5 Emission rates, sinks, mean atmospheric residence time of gases, their relative radiative forcing, frequency saturation, chemical synergisms and many other aspects of the gases are still obscure. For estimates and projections of past and projected land use and agricultural emissions, see US Agency for International Development, *Greenhouse Gas Emissions and the Developing Countries: Strategic Options and the USAID Response*; report to Congress, Washington DC, July 1990; and IPCC Agriculture, Forestry and Other Human Activities working group report in World Meteorological Organisation, UN Environment Programme, Climate Change, The IPCC Response Strategies, Intergovernmental Panel on Climate Change, Island Press, Washington DC, 1991, pp 73–128

6 See W Fischer, *A Convention on Greenhouse Gases: Towards the Design of a Verification System*, Forschungszentrum Julich GmbH, Julich, Germany, October 1990

7 For cogent argument for a comprehensive approach, see D Lashof, *Approaching greenhouse gases comprehensively: the value and limits of GWPs*; paper presented to the Workshop on Global Warming Potential Indices, Boulder, Colorado, November 14, 1990; and 'Comprehensive Approach to Environmental Policy,' paper to the Informal Seminar on US Experience, held at US Department of State, Washington DC, 3 February 1990

8 S Barrett, 'Economic Instruments for Global Climate Change Policy', London Business School report to Environment Directorate OECD, Paris, 1990, p 2

9 The reliability of carbon emissions (as against fossil fuel usage) should not be overstated. A number of factors reduce the reliability of fuel usage as the basis for carbon estimates including varying carbon content, combustion efficiency, oxidation time lag, data reporting deficiencies, etc. The ± 5–10% reliability estimate is taken from G Marland and R Rotty, *Carbon Dioxide Emissions from Fossil Fuels: A Procedure for Estimation and Results for 1950–1981*, Energy Office of Energy Research, US Department of Energy, DOE/NBB-0036, Washington DC, June 1983, in CDIAC, *Production of CO_2 from Fossil Fuel Burning by Fuel Type, 1860–1982*, Carbon Dioxide Information Center, Oak Ridge National Laboratory, Tennessee, NDP-006, September 1984

10 W Nordhaus suggests such discounting in his *Contribution of Different Greenhouse Gases to Global Warming: a New Technique for Measuring Impact*, (mimeo), Economics Department, Yale University, February 11, 1990

11 For discussion of possible unexpected, unpleasant effects, see W Broecker, 'Greenhouse Surprises,' in D Abrahamson *et al*, *The Challenge of Global Warming*, Island Press, Washington DC, 1989, pp 196–212; and M Hoffert, 'Climate Sensitivity, Climate Feedbacks, and Policy Implications,' in I Mintzer (ed), *Confronting Climate Change, Risks, Implications, and Responses*, Cambridge University Press, 1992, pp 33–54

12 W Cline, *Estimating the Benefits of Greenhouse Warming Abatement*, Institute for

International Economics report to OECD Environment Directorate, May 1991, pp 22–23

13 As the Academy notes:

> The uncertainties associated with mitigating global climate change and its attendant costs are . . . at least as great as – and probably greater than – the uncertainties associated with other forms of investment that could be undertaken today. Accordingly, the investor averse to risk might conclude that costly mitigation actions should not be undertaken. However, the payoff from mitigation actions now will be greatest if the magnitude of global climate change and the associated costs turn out to be high, even if that is judged to be a contingency of low probability . . . [S]uch investment may still be worthwhile as insurance against an uncertain but possibly costly contingency. How do these considerations influence the discount rate? The precise answer is not at all straightforward, unless the uncertainty itself is related in a particular way to the passage of time. Roughly speaking, however, one can say that where an uncertain outcome (the future payoff from mitigation actions) is negatively correlated with overall economic prospects (as measured by future gross world product per capita), and where the uncertainty grows exponentially with time, some deduction from the discount rate used to evaluate mitigation actions is warranted. How much? That depends in detail on the nature of the uncertainty, an issue that remains to be clarified, and on the degree of our aversion to risk.

US National Academy of Sciences, National Academy of Engineering, Institute of Medicine, *Policy Implications of Greenhouse Warming*, Report of the Mitigation Panel, Committee on Science, Engineering, and Public Policy, National Academy Press, Washington DC, 1991, pp 2–14; see also, F Krause, J Koomey, D Olivier *et al*, *Energy Policy in the Greenhouse, volume 2, Least Cost Insurance Against Climate Risks*, International Project for Soft Energy Paths, El Cerrito, California, forthcoming 1993

14 McKinsey and Co, *Protecting the Global Environment: Funding Mechanisms*, Ministerial Conference on Atmospheric Pollution and Climatic Change, Noordwijk, November 1989, pp 18–24

15 The original regression is found in W Nordhaus, 'To Slow or Not to Slow, The Economics of the Greenhouse Effect,' (mimeo), Economics Department, Yale University, February 5, 1990; The cost schedule used in this study is taken from W Nordhaus, 'A Sketch of the Economics of the Greenhouse Effect,' *American Economic Review*, volume 81, no 2, 1991, pp 146–148

16 Reddy's marginal cost curve for supplying Karnataka's energy services needs is summarised in G Dutt, *End-Use Energy Strategies for Sustainable Development, India as a Case Study*, Office of the Environment, Asian Development Bank, Manila, January 1991, pp 57–59; see also Chapter 8 in this volume

17 See for example R Williams, 'Low-Cost Strategies for Coping with CO_2 Emission Limits (A Critique of 'CO_2 Emission Limits: an Economic Cost Analysis for the USA' by Alan Manne and Richard Richels,' *The Energy Journal*, volume 11, no 3, 1990, pp 35–59

18 See Riso National Laboratory, *Analysis of Abatement Costing Issues and Preparation of a Methodology to Undertake National Greenhouse Gas Abatement Costing Studies*, phase one report, Denmark, August 1992. For a summary of national studies completed or under way as of early 1992, see UN Environment Programme, *Country Study Report*, report to the Intergovernmental Panel on Climate Change, April 14, 1992

19 A selection of the studies reviewed in search of relevant marginal cost data included: P Hoeller, A Dean and J Nicolaisen, 'Macroeconomic implications of reducing greenhouse gas emissions: a survey of empirical studies'; *OECD Economic Studies*, No. 16, Spring 1991, pp 45–78; US Department of Energy, *The Economics of Long-Term Global Climate Change: A preliminary assessment*; report of an Interagency Task Force, DOE/PE-0096P, Washington DC, September 1990; D Dudek and A LeBlanc, 'Offsetting new CO_2 emissions: a rational first greenhouse policy step'; *Contemporary Policy Issues*, volume 8, July 1990, pp 29–42; J Swisher, *A Case Study of Utility Costs for Reducing CO_2 Emissions*, (mimeo), Civil Engineering Department, Stanford University, no date, and his *Prospects for International Trade in Environmental Services: An Analysis of Carbon Emission Offsets*, dissertation thesis, Stanford University Civil Engineering Department, 1991; A Cristofaro, 'The cost of reducing greenhouse gas emissions in the United States', (mimeo), December 4, 1990; S Barrett, 'Economic Instruments for Global Climate Change Policy,' London Business School report for Environment Directorate OECD, Paris, 1990; J Whalley and R Wigle, *The International Incidence of Carbon Taxes*; (mimeo), paper to conference on 'Economic Policy Responses to Global Warming', Rome, October 1990; D Jhirad, 'Power sector innovation in developing countries: implementing multifaceted solutions', *Annual Review of Energy*, volume 15, 1990, pp 365–398; M Levine *et al.*, *Energy Efficiency, Developing National, and Eastern Europe: An Analysis of Key Issues*, report to the US Working Group On Global Energy Efficiency, Lawrence Berkeley Laboratory, March 1991; G Dutt, *End-Use Energy Strategies for Sustainable Development, India as a Case Study*, Office of the Environment, Asian Development Bank, Manila, January 1991; V V Desai, K Nyman, *Industrial Energy Conservation: Notes on Three Country Studies*, Energy Planning Unit, Asian Development Bank, Manila, circa 1986; Energy Research Institute of State Planning Commission, Institute for Techno-Economic Economics and Energy System Analysis of Tsinghua University, *Regional Study of Environmental Considerations in Energy Development Project, People's Republic of China*, Interim Report ADB TA 5357-Regional, Beijing, March 1991; Asian Development Bank, *China, Energy Conservation, Final Report*, Beijing, December 1990; M Philips, *Energy Conservation Activities in Africa and Eastern Europe*; International Institute for Energy Conservation, Washington, September 1990; M Philips, *Energy Conservation Activities in Latin America and the Caribbean*; International Institute for Energy Conservation, Washington, June 1990; M Philips, *Energy Conservation Activities in Asia*; International Institute for Energy Conservation, Washington, September 1990; M Cherniack, *Thailand Electricity Mission*; Mission Report; International Institute for Energy Conservation, Washington, December 1990; E Larson (ed), *Report on the 1989 Thailand Workshop on End-Use-Oriented Energy Analysis*; International Institute for Energy Conservation, Washington, April 1990

20 Good summaries of the varying methods and underlying assumptions of these two schools are found in J Darmstadter, *The Economic Cost of CO_2 Mitigation: A Review of Estimates for Selected World Regions*, Energy and Natural Resources Division, Resources for the Future, Washington DC, January 1991; and M Grubb, *Energy Policies and the Greenhouse Effect*, volume 1, policy appraisal, Royal Institute of International Affairs, London, 1990

21 For the GEF cost data, see Global Environment Facility, *Economic Costs of Carbon Dioxide Reduction Strategies*, Working Paper 3, September 1992, p 53; for the China

estimates, see M Levine, 'China's Energy System: Historical Evolution, Current Issues, and Prospects,' in *Annual Review of Energy and Environment*, volume 17, 1992, p 430

22 For a similar approach to levelized costs, see E Barbier, J Burgess, and D Pearce, 'Technological Substitution Options for Controlling Greenhouse Gas Emissions,' in R Dornbusch and J Poterba, *Global Warming: Economic Policy Responses*, MIT Press, Cambridge, Massachusetts, 1992, p 111. A summary US application of this approach is found in E Rubin *et al*, 'Realistic Mitigation Options for Global Warming,' *Science*, volume 257, July 10, 1992, pp 148–266. For a formal explanation of levelized cost, see J White *et al*, *Principles of Engineering Economic Analysis*, John Wiley and Sons, New York, 1977

23 As argued by N Birdsall and J Dixon, *Some Economics of Global Climate Change: The View from the Developing Countries*, draft paper, World Bank, April 30, 1991, p 2

24 These are given on p xxxiii of the IPCC Policymakers Summary in World Meteorological Organisation, UN Environment Programme, *Climate Change, The IPCC Response Strategies*, Intergovernmental Panel on Climate Change, Island Press, Washington DC, 1991. The IPCC give rates for 1985, 2000, and 2025. These estimates (in T C/capita per year) were interpolated by geometric growth rate between 1985 and 2000 to 1995 as follows: North America, 5.33; Western Europe, 2.22; OECD Pacific, 2.69; Non-OECD Europe, 3.56; Africa, 0.32; Centrally Planned Asia, 0.6; Latin America, 0.61; Middle East, 1.57; South and East Asia, 0.27

25 See R Watson *et al*, 'Greenhouse Gases and Aerosols,' in Intergovernmental Panel on Climate Change, *Climate Change, The IPCC Scientific Assessment*, Cambridge University Press, New York, 1990, p 15. See also Intergovernmental Panel on Climate Change, *Climate Change, The IPCC Response Strategies, op cit* (endnote 24), pp 29–31 for additional information on the assumptions underlying these scenarios, including non-CO_2 greenhouse gas reductions required to achieve these targets

26 *Ibid*

27 For temperature rise, see scenario D in Figure 9 (p xxii) and for sea level rise, scenario D in Figure 14 (p xxxi), in 'Policymakers' Summary,' in Intergovernmental Panel on Climate Change, *Climate Change, The IPCC Scientific Assessment, ibid*

28 J Houghton, 'Scientific Assessment of Climate Change: Summary of the IPCC Working Group I Report,' in J Jager and H Ferguson (eds), *Climate Change: Science, Impacts and Policy*, Cambridge University Press, New York, 1991, p 38

29 See B. Solomon and D. Ahuja, 'International reductions of greenhouse-gas emissions, An equitable and efficient approach,' *Global Environmental Change*, December 1991, p 347

30 1987 population estimates are from Population Reference Bureau, 'Population Data Sheet 1987,' Washington DC, 1987; carbon sink estimates and oceanic/terrestrial spread are from R Watson *et al*, 'Greenhouse Gases and Aerosols,' in Intergovernmental Panel on Climate Change, *Climate Change, The IPCC Scientific Assessment, op cit* (endnote 25), pp 5, 13; land area estimates are from World Resources Institute, *World Resources 1990–1991*, Oxford University Press, New York, 1990, Table 17.1; see also B Bolin, 'How Much CO_2 Will Remain in the Atmosphere?' in B Bolin *et al* (eds), *The Greenhouse Effect, Climatic Change, and Ecosystems*, Wiley and Sons, New York, 1981, pp 94–155; and B Bolin *et al*, 'The

Global Biogeochemical Carbon Cycle,' in B Bolin *et al* (eds), *The Global Carbon Cycle*, Wiley and Sons, New York, 1979 pp 1–56
31 G Pearman, 'Greenhouse Gases: Their Role in Climate Change,' in Association for Science Cooperation in Asia, *Workshop on Greenhouse Gases and Climate Change, An Asian Perspective*, Melbourne, June 17–21, 1991, p 26
32 P Tans, I Fung and T Takahashi, 'Observational Constraints on the Global Atmospheric CO_2 Budget'; *Science*, volume 247, 23 March 1990, pp 1431–1438
33 F Krause, W Bach and J Koomey, *Energy Policy in the Greenhouse, From Warming Fate to Warming Limit: Benchmarks for a Global Climate Convention*, International Project for Sustainable Energy Paths, El Cerrito, California, September 1989, pp I-6-4,5
34 M Ince, *The Rising Seas*, Earthscan Publications, London, 1990, p 59
35 L Edgerton, *The Rising Tide, Global Warming and World Sea Levels*, Island Press, Washington DC, 1991, p 74
36 J Gilbert and M Poletti, 'Policy Related Observations on the Assessment of the Costs of Climate Change,' paper to workshop on a Comprehensive Approach to Climate Change Policy, Centre for International Climate and Energy Research, Oslo, July 1, 1991, p 12
37 M. Asaduzzaman, 'Global Climate Change and Coastal Zone Management in Bangladesh,' in T Siddiqi and D Streets (eds), *Responding to the Threat of Global Warming, Options for the Pacific and Asia*, Argonne National Laboratory and Environment and Policy Institute, East West Center, Workshop Proceedings, ANL/EAIS/TM-17, June 21, 1989, Honolulu, p 5–51
38 See J Godfrey, 'Climate Change and Sea Level Rise,' in Association for Science Cooperation in Asia, *Workshop on Greenhouse Gases, op cit* (endnote 31), pp. 17–21, 1991
39 A Pittock, 'Global Climate Change and the Development of Regional Climate Change Scenarios,' in *ibid*, pp 4.2–3
40 G Quraishee, 'Global Warming and Rise in Sea level in the South Asian Seas Region,' in T. Siddiqi and D Streets (ed), *Responding to the Threat of Global Warming, Options for the Pacific and Asia*, Argonne National Laboratory and Environment and Policy Institute, East West Center, Workshop Proceedings, ANL/EAIS/TM-17, June 21, 1989, Honolulu, p 5–27
41 See A Wijkman and L Timberlake, *Natural Disasters, Acts of God or Acts of Man?*, Earthscan, London, 1988
42 GEMS Monitoring and Assessment Research Centre, 'Natural Disasters,' *Environmental Data Report*, Blackwell, Oxford, 1991, pp 361–382
43 See K Smith, 'The Risk Transition'; *International Environmental Affairs*, volume 2, no 3, 1990, pp 227–251; and K Smith, 'Risk Transition and Global Warming'; *Journal of Energy Engineering*, volume 116, no 3, December 1990, pp 178–188
44 Ministry of Transport and Public Works, Tidal Waters Division, *Rising Waters: Impacts of the Greenhouse Effect for the Netherlands*; Ministry of Transport and Public Works, The Hague, January 1991, p 31
45 This estimate is found in Appendix D, 'A World Wide Estimate of Basic Coastal Protection Costs,' and are summarized on p 15 of Report of the Coastal Zone Management Group, *Strategies for Adaptation to Sea Level Rise*, Intrgovernmental Panel on Climate Change Response Strategies Working Group, November 1990; See also Table 5.5 in Coastal Zone Management Working Group report in World Meteorological Organisation, UN Environment Program, *Climate Change, The IPCC Response Strategies, op cit* (endnote 24), p 153

46 R Warrick and A Rahman, 'Future Sea-Level Rise: Environmental and Socio-Political Considerations,' in I Mintzer (ed), *Confronting Climate Change, op cit* (endnote 11), pp 106–107

47 As argued by the Villach Experts Conference, 'Developing Policies for Responding to Climatic Changes,' Policy Issues Workshop, November 9, 1987, Bellagio, Italy, p 25

48 M Marry, *Climate Change and World Agriculture,* Earthscan Publications, London, 1990, pp 108–119

49 M Parry and M Swaminathan, 'Effects of Climate Change on Food Production,' in I Mintzer (ed), *Confronting Climate Change, op cit* (endnote 11), p 124

50 See W Cline, *The Greenhouse Effect,* Institute of International Economics, Washington DC, 1992

6

North–South transfer

Peter Hayes

In the previous chapter, I calculated the incremental costs to different nations and groupings of nations arising from the carbon abatement scenario and protection against sea level rise. In this chapter, I redistribute these incremental costs based on historic contribution to climate change and ability to pay. As might be expected, the North is obliged to pay substantially more than it would if it ignores its historic contribution and its greater ability to pay than the South. And unless the South's cost is reduced to its obligation-to-pay, it could pay much more than it should – from 58 to 90 per cent more according to the following analysis.

Redistribution in accordance with this indice is then compared with the pragmatic distribution rule known as the UN scale of payments, and some benchmarks. As the obligation-to-pay index does not diverge much from the UN scale of payments, the latter could be very useful in judging national claims for exemption from the former in the climate change context.

Next, I examine how the substantial funds involved might be collected, generated, and transferred by a carbon tax, tradeable permits, or sale of abatement services. This analysis shows that the carbon tax and tradeable permits are feasible instruments to achieve the requisite financing in the South, but that the sale of abatement services would likely only supplement the former two mechanisms. Finally, I direct the reader to a summary of the major uncertainties that affect each of these links in the logical chain presented in this analysis.

Obligation to pay indices

'Obligation to pay' (OTP) is a composite indice based on two constituent measures: 'ability to pay' (ATP) and 'historic contribution to climate change' (HCCC). The indice enables the issue of who should pay (who created the problem and who can afford to pay for the cleanup) to be separated from where abatement should be conducted (the cheapest sites). Here, this concept is used to determine the distribution of the costs of abatement and coastal protection that were calculated in the previous chapter, at high, medium and low abatement cost curves. First, I will briefly review the conceptual components of 'obligation-to-pay' as advanced in Chapter 4.

Ability to pay (ATP) is best formulated as follows:

$$\text{ATP} = \text{GNP} - [\text{Population} \times (\text{GNP/capita})_{\text{T'HOLD PQOL}}]$$

where: ATP = ability to pay;

GNP = Gross National Product;

$[\text{Population} \times (\text{GNP/capita})_{\text{T'HOLD PQOL}}]$ = a basic need adjustment using a threshold physical quality of life or PQOL estimate set at GNP = 1986US$1,800/capita.[a,1]

This criterion is useful because it clearly shows who has gained economically from past pollution of the atmosphere. It also identifies who can afford to shoulder the burden of greenhouse gas abatement in the future. This ATP index does not include any special weighting for populations who are especially at risk from climate change, such as those of developing coastal and island states.

The second index that is used to determine overall obligation to pay is historic climate change contribution (HCCC) defined for a given nation as:

$$\text{HCCC} = \text{national cumulative GHG} - [\text{Population} \times \text{ND}_{\text{T'HOLD}}]$$

where: GHG = greenhouse gas emissions integrated contribution to atmospheric warming using a 100-year time horizon from year of emission.[b,2] Here, carbon dioxide from fossil fuel usage is taken as a proxy for greenhouse gas emissions;

ND = natural debt, the national borrowing of the atmosphere's absorptive capacity in cumulative 1950–86 GHG emissions/1986 population;[c]

$\text{ND}_{\text{T'HOLD}}$ = a threshold 'natural debt' defined as the cumulative per capita emission that is attributable to the universal human right to an equal portion of atmospheric absorptive services such as the carbon sink. Any emission above this per capita level is treated as an excess borrowing that counts in the HCCC.

The HCCC of a given nation is weighted relative to its 1986 adult population on the grounds that current populations should carry the burden of

---

a Empirically, PQOL plateaus at this point with further increase in GNP per capita, suggesting that it is a good basic needs indicator (see Chapter 4). This factor is based on 1978 data and should be validated against the Human Development Index produced by the UN Development Programme in 1990 and updated each year.

b Due to data inadequacies, Smith *et al* compute this index for CO_2 only, and only from 1956 to 1986.

c Smith *et al* compute this measure only for fossil fuel carbon emitted between 1956 and 1986 due to inadequate information on earlier historical emissions for fossil fuel and other biotic sources of carbon.

ancestral cumulative damage and that national responsibility for climate change is proportional to national natural debt.

As was argued in Chapter 4, the HCCC is an important ethical and political concern that should be reflected in determining obligation to pay. Some critics contend that this criterion places too much emphasis on moral responsibility and not enough on current emissions.[3] However, allowing past decision-makers to avoid liability for their historical contributions to cumulative and irreversible environmental degradation such as climate change fails to provide current decision-makers with an incentive to protect the rights of future generations. The current generation of leaders cannot disavow its obligation to pay off its natural debt from the immediate past at the same time as it claims to be adopting the principle of intergenerational equity. Moreover, the political leaders of the South are not about to let the North occupy all the global atmospheric commons without first obtaining significant compensation – a point stressed repeatedly by the Group of 77 in the negotiations over the Convention.[4] A transparent, quantitative index of historical responsibility will facilitate greatly the ongoing negotiations over protocols on this issue. It is therefore the necessary starting point of meaningful bargaining over this issue, even if political-economic power ultimately elbows aside much of the moral imperative represented by the HCCC index.

These indexes can be combined into a total obligation-to-pay indice by either multiplication or addition. Each index could be weighted, for example, from concern with equity. Here, however, the indexes are simply added without weighting.[d] Of course, politics would determine how the indexes would be combined in an actual negotiation. As both indexes give similar distributions of responsibility, the selection of one rather than both, or the best combination of the two, would be a pragmatic political question.

In Table 6.1 and Figure 6.1, I show the per centage distribution of national and aggregate obligation to pay. The South's obligation to pay is about 7 per cent, the North's about 73 per cent, and the East's about 20 per cent. With this indice, the incremental costs calculated in Chapter 5 can be redistributed between nations in accordance with their global obligation to pay, a procedure followed in the next sections.

If the obligation to pay indice were adopted in a protocol to the Climate Change Convention, then it should be recalculated periodically (say every five years) to reward those who have decreased their historic greenhouse gas contribution relative to projections; and to penalize those who have increased their historic greenhouse gas contribution relative to projections. A similar adjustment to obligation-to-pay should be made for each nation's achieved GNP per capita growth rates relative to those projected in the initial calculation of a nation's obligation to pay.

---

d When multiplied, the two indices give some countries (especially the United States) such a high OTP that the index would not facilitate negotiations.

Table 6.1 *Obligation to pay, combined index*

Country		*% of total*	*% of group subtotal*
North			
United States	US	37	50
Germany (United)	UG	8	11
Canada	CA	3	4
Belgium	BE	1	1
United Kingdom	UK	6	8
France	FR	4	6
Japan	JAP	8	10
Italy	IT	3	4
Australia	AU	1	2
Greece	GR	0	0
Netherlands	NE	1	2
Spain	SP	1	2
Portugal	PO	0	0
Turkey	TK	0	0
Subtotal, North	NO	73	100
'East'			
Soviet Union	SU	16	77
Rest of East Europe	EE	5	23
Subtotal, East	EA	20	100
'South'			
China	CH	1	22
India	IND	0	0
Indonesia	INDO	0	0
Rest of South	RoS	5	78
Subtotal, South	SO	7	100
TOTAL	WORLD	100	-

National obligation to pay = (% world ability to pay, at threshold income of (1986) US$1,800/capita) + (historic contribution of cumulative emissions, 1950–86) – (1986 population (national) × natural debt threshold of 10 tonnes of carbon per capita (1986))

At first sight, these adjustments appear to increase the relative obligation-to-pay of the developing countries (which will rapidly increase their share of historic contribution to atmospheric carbon loading, and their ability to pay) and reduce the obligation of the developed countries (which will have an ever smaller share of the accumulated carbon contribution to the atmosphere and likely a lesser GNP growth rate than in the South). In reality, both adjustments would be sensitive to the effect of population growth on the 'basic needs' allowance in the obligation-to-pay indice. Therefore, the 'redistributive' impact of regular recalculation of the obligation-to-pay indice

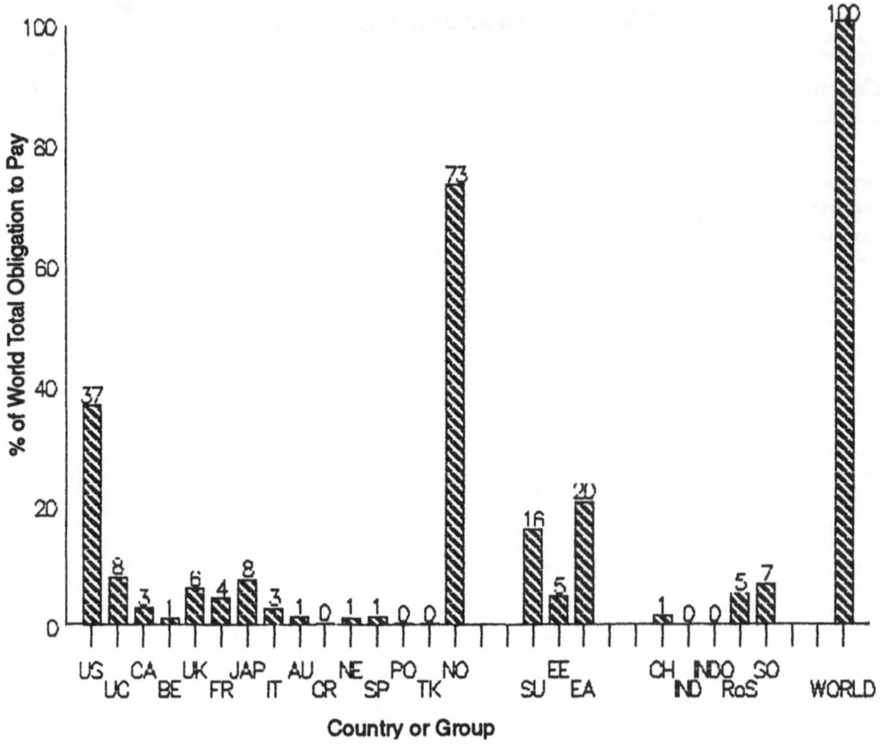

Figure 6.1 *Obligation to pay: ability to pay + historic resp – natural debt threshold*

is indeterminate. High population growth in the South, for example, would increase its permitted 'basic needs' emissions that would not count toward its obligation to pay. It could (depending on the impact of population growth on GNP growth) also keep per capita income in the South below the basic needs threshold per capita income (US$1,800/capita) that determines ability to pay in this study.

If the emissions index in the obligation-to-pay indice is reanalysed using this study's projected population in and cumulative emissions up to 2025 for the North, East and South, then the North's total obligation to pay (updated for cumulative contributions from 1950 all the way to 2025) would increase by about 7 per cent, the East's by about 8 per cent, and the South's by about -14 per cent relative to the distribution of the obligation-to-pay indice in 1986/7 (58, 23, and 18 per cent respectively).

Conversely, if the basic needs allowance in the emissions component of the obligation-to-pay indice is pegged to its 1986 level (that is, a 1986 population level is used rather than the 2025 figure), then the shift in relative responsibility due to the increases in cumulative contribution from 1987 to

2025 is from the North (–5 per cent) to the East (+1 per cent) and the South (+4 per cent).

How population growth is treated in the two indexes that constitute the obligation-to-pay indice is therefore critical to the impact of its periodic adjustment. The choice of which population to use is political, and there being no 'right' answer. In any case, whatever its direction, the overall redistribution of obligation-to-pay arising from periodic adjustment is relatively small, even after thirty years of additional emissions. I therefore proceed using the current estimate of the obligation-to-pay index stated above in redistributing incremental costs.

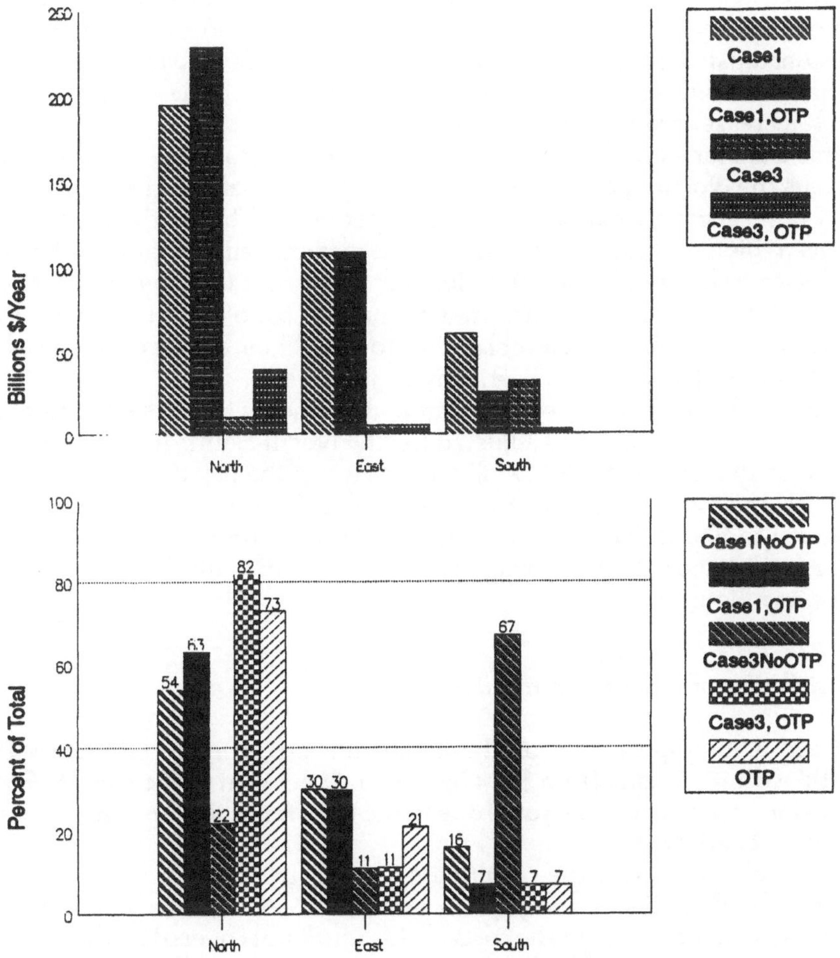

Figure 6.2 *Cost with and without OTP redistribution, Cases 1 and 3*

Redistribution of incremental cost

In this section, I take the estimates of incremental cost developed in chapter 5 and redistribute between regions those costs that are 'excess' to their obligation-to-pay. As there are three incremental cost cases, so there are three calculations of redistribution of cost that follow.

Redistribution of high incremental cost

At the high abatement cost curve, the present value of the total cost is estimated at $5.6 trillion. The South's incremental cost, therefore, should be reduced to only 7 per cent of this global cost or $390 billion, or by 58 per cent of its unadjusted incremental cost. The difference - $529 billion - is treated as the responsibility of the North and is transferred to the latter's account which increases from 2.9 to 3.5 trillion dollars as a result.

This transfer - an annuity of $34 billion over the 30 year scenario period - boosts the North's payment from 58 to 63 per cent of the global total (still short of its strict obligation-to-pay of 73 per cent). The residual difference between this distribution of global cost and that implied by the obligation-to-pay indice now lies between the North and the East (and is not addressed further here). The latter incurs an incremental cost of $1.7 trillion, that is, about 9 per cent more than its obligation-to-pay. Thus, another transfer, this time from the North to the East, may be justified.

Expressed as an annuity of thirty annual payments that are equivalent to the total present values and adjusted for the North-South transfer, the bill at high abatement cost is $229 billion per year in the North, $108 billion in the East, and $25 billion in the South, for a global annual cost of $362 billion (rounded to $363 billion in box on page 151). The annual cost in billions of dollars redistributed by the obligation-to-pay indice is displayed in the box above and Figure 6.2.

Redistribution of medium incremental cost

In Case 3, the application of the obligation-to-pay indice results in the North's cost increasing from $104 billion (or 16 per cent of the total) to $605 billion (or 82 per cent of the total) due to the transfer of $442 billion from the South to its account.

In Figure 6.2, I show the resulting overall distribution of annual cost in billions of dollars. The North pays an annual bill of $39 billion (of which $29 billion is a transfer to cover the costs of the South not covered by the South's own obligation to pay). The East is responsible for about $5 billion per year while the South is responsible for about $3 billion per year of the global annual total of $48 billion.

Distributed cost, Cases 1 and 3

I. Redistribution of cost, high cost curve, case 1

Definition: Incremental Cost 1, using Nordhaus marginal cost curve and net present value of incremental cost to each area adjusted to account for South's excess of cost above its obligation-to-pay between 1995–2025, in $billion/year

Global Cost, total nett present value (NPV)[a] = 5577 billion

South's obligation-to-pay = 7% of global cost = 390 billion (NPV)
South's obligation-to-pay as % of South's total abatement and coastal protection cost = 42%
North-to-South transfer to cover the difference between South's obligation-to-pay and its total cost = 529 billion (NPV)
North-to-South transfer expressed as an annuity = 34 billion/year

North	*East*	*South*	*Global*	
NPV of cost, adjusted for transfer according to obligation-to-pay				
3520	1667	390	5577	billion $ (NPV)
Annuity (adjusted for South's obligation-to-pay)				
229	108	25	363	billion $/year
% of global annuity (adjusted for South's obligation-to-pay)				
63	30	7	100	

II. Redistribution of cost, medium cost curve in south, case 3

Definition: Incremental Cost 3, using this study's estimated abatement cost curve in South, and US National Academy of Science cost curve for North and East, in net present value of incremental cost to each area adjusted to account for South's excess of cost above its obligation-to-pay between 1995–2025, in $billion/year

Global Cost, total nett present value (NPV)[a] = 739 billion

South's obligation-to-pay = 7% of global cost = 52 billion (NPV)
South's obligation-to-pay as % of South's total abatement and coastal protection cost = 10%
North-to-South Transfer to cover the difference between South's obligation-to-pay and its total cost = 442 billion (NPV)
North-to-South Transfer expressed as an annuity = 29 billion/year

North	*East*	*South*	*Global*	
NPV of cost, adjusted for transfer according to obligation-to-pay				
605	83	52	739	
Annuity (adjusted for South's obligation-to-pay)				
39	5	3	48	billion $/year
% of global annuity (adjusted for South's obligation-to-pay)				
82	11	7	100	

III. % obligation-to-pay based on historic emissions and ability to pay

North	East	South	Global
73	21	7	100 (% of total)

a From box on page 132
Source: Box on page 132 and text

Redistribution of low incremental cost

As I stated in Chapter 5, Case 2 provides no economic grounds for a transfer to the South. Expressed as an annuity, therefore, the North pays $10.6 billion per year in this scenario; the East $5.4 billion per year; and the South gains from a negative annual cost of $62 billion per year (see box, page 133 and Figure 5.13). On this basis, the North would not transfer additional resources to the South for the simple reason that the South should abate to the extent projected out of economic self-interest, without regard for the climate *per se*.

This result implies that developing countries can reap the advantage of being latecomers to industrialization. Rather than waiting for long-lived capital stocks to turn over, they may be able to install modern, resource saving technologies as they industrialize.

Conversely, the notion of enormous savings being reaped in many developing countries is incredible when juxtaposed against their evident inability to reduce the enormous waste in their existing economies, let alone massively abate their future emissions. Even if these low and negative cost technological opportunities exist, most developing countries are unable to finance the front end investments needed to tap the potential. Admittedly, the obligation-to-pay index does not provide an economic rationale for a transfer from the rich to the poor under these assumptions. Equally, only the rich states can provide the substantial concessional financing needed in the developing countries.

Developing countries face not only an absolute scarcity of investment resources with which to respond to climate change. They will also incur costs from intersectoral adjustment, trade impacts, human resource development, institutional change, and highly priced information, all of which will inhibit their ability to increase their energy efficiency. Overcoming these barriers to abatement will impose costs that are not reflected in the low abatement cost curve used in this case.[5]

How much financing might be needed to meet these challenges so that the developing countries reduce their emissions as described in Chapter 5? The simple answer is that no one knows. One study states that developing countries need to redirect at least 2 per cent of their gross domestic products – or about $58 billion in 1989 – toward human development priorities such as education, health care, and social services that are integral to sustainable development.[6] At the June 1992 Earth Summit, UN officials estimated the cost of implementing a global sustainable development strategy as about $600 billion per year, of which about $125 billion must be financed externally in the form of technical and economic assistance.[7] Only a portion of these costs would be attributable to carbon abatement activities; but not much abatement may be achieved unless such broad based development occurs.

The reader can imagine the size of this daunting task by supposing that within about three decades, all developing countries must have the same proportion of scientists and technicians in their population as do the rich countries today – arguably a prerequisite of realizing substantial carbon abatement. To do so, the developing countries would have to increase their number of scientists and technicians at 10 per cent per year, from today's 36.7 million to about 579 million in 2025.[e,8] This growth in human resources would demand tens, perhaps hundreds of billions of dollars of investment in education and training each year.

In Chapter 14, I return to these issues that take us well beyond a technological approach to determining requisite transfers from the rich to the poor states. These difficulties also reveal the limits of a method that fuses a moral argument with a technological and economic approach to determining who should pay for a greenhouse gas regime. The obligation-to-pay method supplies some minimum estimates of the possible transfers but it may not be applicable in all cases and almost certainly must be supplemented with other approaches on pragmatic and political grounds.

Benchmarks

According to this approach, substantial transfers to the South may be required to fund the technological and economic costs not covered by its own obligation to pay. These fall within the range $29–34 billion per year for thirty years, depending on the underlying marginal cost assumptions. Thirty billion dollars per year is a reasonable mid-point estimate of the justified, minimal and additional financing needed by the South to achieve its required reduction targets. To this amount should be added a substantial sum to

e Currently, the industrial countries have 81 scientists and technicians per 1,000 people while the developing countries have only 9. Current developing country population is a little greater than 4 billion. By 2025, it could reach 7 billion or more. Between 1995 and 2025, therefore, an average of 18 million new scientists and technicians must be trained *each year* to reach the target of 542 million by 2025. That is [{(7.15 billion developing country population in 1995) × (target fraction of 81 scientists and technicians/1000 persons)} – {(4.1 billion developing country population in 2025) × (current fraction of 9 scientists and technicians/1000 persons)}]/30 years. This calculation does not allow for brain drain and retirement.

'kickstart' the sustainable development process by training the scientists and technicians who will be needed to implement an abatement strategy in the South.

Thirty billion dollars or more per year is a lot of money. For example, current official development assistance (ODA) for all energy investment in the South currently amounts to about $10 billion per year. Total ODA ran at about $30 billion per year during the 1980s (reaching $46.9 billion in 1989 for the OECD).[9] Total foreign direct investment to all developing countries was about $13 billion per year in the same period.[10] Enabling the South to participate in a global climate change agreement would result in transfers on a scale that would create a new foundation for the political-economic interdependence of the North and the South, on a scale with current aid and foreign investment.

Conversely, world and national GDP growing at 3 per cent per year will increase by 240 per cent over the same period, rendering the annual transfer cost a declining portion of donor country GDP. The transfer to the South of about $30 billion per year pales into insignificance compared with agricultural production subsidies ($50 billion per year in the EC); military spending in the North or the South; or Third World debt (which resulted in a South-to-North net financial flow of $42.9 billion in 1989). There is little doubt that the North can afford to pay $30 billion per year even if it would be difficult to muster the political will needed to do so.

UN scale of payments

Negotiations on the allocation of payment are likely to begin with precedent. A commonly used focal point for bargaining is the UN scale of payments. Slightly modified, this scale formed the basis of contributions to the costs of the ozone agreement.[f] A number of other UN trust funds for environmental purposes have also used weighted contributions based on the global assessment scale of the UN General Assembly. In this system, countries are rated according to a number of economic, geographic and demographic factors. The only limit is a 25 per cent ceiling for the US contribution since 1972. The use of differential scales that modify even the UN sliding scale allows countries to participate in an environmental agreement without incurring an insupportable cost. Under the Montreal Protocol, for example, Singapore contributes $1,500 annually but has the same membership rights as the United States which pays $300,000 per year.

Under the UN scale, the North pays about 77 per cent, the East about 14

f Admittedly, the United States signed the ozone agreement explicitly stating that the use of the UN scale did not set a precedent to be employed in other agreements such as the Climate Change Convention. Nonetheless, it remains a likely starting point for negotiations.

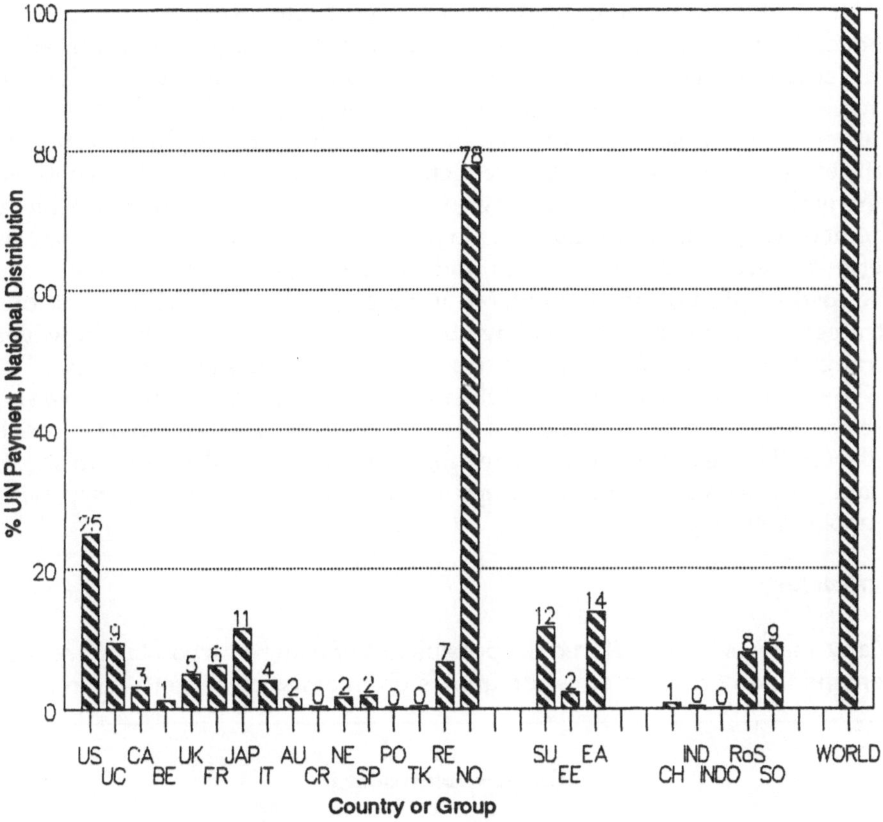

Figure 6.3 *UN scale of payments, %*

per cent, and the South about 9 per cent of total UN cost (see Table 6.2 and Figure 6.3). It is interesting to compare this with the obligation-to-pay index referred to earlier. The North's obligation-to-pay is about 73 per cent of global cost; the East's about 20 per cent; and the South's about 7 per cent. Negotiations on protocols to the Climate Change Convention are likely to proceed pragmatically by countries making bids to vary their contribution relative to the UN scale until consensus is reached. The obligation-to-pay index therefore provides a convenient method to analyse the validity of claims for special treatment with respect to the UN scale.

Financing mechanisms

Three instruments to finance the South's incremental costs beyond its obligation-to-pay have been canvassed widely. These are carbon taxes, tradeable permits, and trade in abatement services. Below, I examine the role that each might play in transferring financial resources on a large scale.

The reader should note that I treat all three mechanisms as a means to an end: each of them is a way to ensure that reduction targets are achieved at least cost, from a global perspective. Minimizing the global cost of reduction (that is, efficiency) *per se* is not the overriding priority in this study, however. Rather, I aim to see how carbon emission reductions might be achieved efficiently in an equitable carbon reduction strategy. Equity has been defined here with respect to an overarching goal, the creation of a strategy to achieve the necessary emission reductions in accordance with obligation-to-pay as defined above. Taxes, permits and trade, therefore, are enlisted to minimize the costs of promoting an equitable and ecologically sustainable emissions strategy. As is proper, means have been subordinated to ends – however strange this relationship appears to some in the economic profession.

The reader should be alert that the numbers that follow are indicative and are used to explore the contours of the terrain rather than to provide precise answers. The results should not be interpreted as the 'real' figures. In short, I aim only to sketch the landscape rather than to produce a map with pinpoint accuracy.

Carbon tax

Many analysts have explored the possibilities of using a carbon tax to reduce demand to accord with emission quotas or an emission target.[11] Here, I do

Carbon tax transfer

A. Assuming high abatement cost curve and redistributed cost, case 1 (billions $)

Annuity of North-South transfer according to obligation-to-pay index = 34 billion/year

Net present value of carbon tax on projected northern emissions to finance the transfer annuity, after north has achieved own required reductions = 518 billion

Annuity of carbon tax revenue to fund transfer = 34 billion

Carbon tax level to achieve transfer annuity = 13$/T-C emitted

B. Assuming medium abatement cost curve and redistributed cost, case 3 (billions $)

Annuity of North-South transfer according to obligation-to-pay index = 29 billion/year

Net present value of carbon tax on projected northern emissions to finance the transfer annuity, after north has achieved own required reductions = 445 billion

Annuity of carbon tax revenue to fund transfer = 29 billion

Carbon tax level to achieve transfer annuity = 11$/T-C emitted

Table 6.2 *UN scale of payments*

Group/Country		*% of total UN payments*	*% of total of group/country*
North			
United States	US	25.00	32.16
Germany United	UG	9.36	12.04
Canada	CA	3.09	3.98
Belgium	BE	1.17	1.51
United Kingdom	UK	4.86	6.25
France	FR	6.25	8.04
Japan	JAP	11.38	14.64
Italy	IT	3.99	5.13
Australia	AU	1.57	2.02
Greece	GR	0.40	0.51
Netherlands	NE	1.65	2.12
Spain	SP	1.95	2.51
Portugal	PO	0.18	0.23
Turkey	TK	0.32	0.41
Rest of OECD/North	RE	6.56	8.44
Subtotal, North	NO	77.73	100.0
'East'			
Soviet Union	SU	11.57	83.78
Rest of East Europe	EE	2.24	16.22
Subtotal, East	EA	13.81	100.00
'South'			
China	CH	0.79	8.60
India	IND	0.37	4.03
Indonesia	INDO	0.15	1.63
Rest of South	RoS	7.88	85.75
Subtotal, South	SO	9.19	100.00
TOTAL	WORLD	100.73	-

Column one does not sum to 100 per cent due to rounding error and errors in source data

Source: United Nations, Administrative and Budgetary Coordination of International Atomic Energy Agency, A/45/798, November 27, 1990, pp 23–27

not examine the pros and cons of different levels and types of carbon taxes. Rather, I determine the level of carbon tax in the North that would raise the necessary revenue needed for the transfer to cover the costs of the South that are the North's obligation-to-pay.

In the box opposite, I show the carbon tax needed in our scenario to raise the revenue in the North to finance the transfer to the South. To calculate the tax, the North's total 'post-reduction' emissions (projected emissions minus required reductions, in the efficiency scenario in Chapter 5) are multiplied by the tax rate per tonne of emitted carbon in each year. The carbon tax that amasses an annuity based on the present value of this stream of future tax revenues, that equals the transfer implied by the obligation-to-pay indice, is the tax adopted.

Two tax levels are derived, one for the high marginal abatement cost, Case 1, and one for the medium abatement cost, Case 3. (Each case generates a different estimate of the target transfer annuity, namely, $34 and $29 billion respectively). Carbon taxes of between $13 and $11 per tonne of carbon emitted achieve these target annuities. These tax levels are quite small compared with the carbon taxes that are computed by economists as necessary to dampen carbon emissions to a significant degree (according to which taxes of $20–120/T-C abated are required for reduction ratios of from 20 to 80 per cent).[12]

Tradeable permits

Many analysts have argued for tradeable permits instead of or to supplement carbon taxes.[13] Based on limited US experience, tradeable permits are argued to create flexible market incentives to reduce emissions at low administrative cost and lesser social cost.[14] The system requires that total emissions be limited or that a cap be placed on the total number of permits issued. A mix of buyers and sellers must also exist with access to markets for permits. Trade must not be allowed above the total limit or without permits if permits are to maintain their value.

In principle, tradeable permits favour technological innovation due to the continuing incentive for users to avoid having to pay for additional permits. Tradeable permits would permit states to enter voluntarily into an arrangement whereby one reduces its emissions in return for value provided by another which continues to emit, provided the sum of the two national limits combined is not exceeded. Whereas the major deficiency of a carbon tax is that policymakers will not know the tax level needed to achieve a given reduction level, a major inadequacy of a tradeable permit system is that the price of the permits – and thus the cost of achieving limits on emissions – cannot be determined in advance. The initial allocation of tradeable permits as well as the administrative and enforcement mechanisms needed to implement a global regime based on permits (in the absence of world government) are two other major difficulties with the scheme.

I do not address the political feasibility of allocating permits in this study, although I did suggest a combined set of criteria in Chapter 5 with which to set a limit on national emissions. Thus, the cap on emissions was determined by setting emissions at a target level that was defined as 'sustainable' in terms of atmospheric carbon concentration and rates of temperature and sea level rise. I assume here, therefore, that the South is simply issued with an additional amount of saleable emission permits over and above those needed to meet its target required reduction trajectory (that is, it received an 'excess entitlement' that it can sell without being required to offset each sale with equal and additional abatement in the South); and that the same volume of emissions are deducted from the North's permits, forcing the North either to abate emission further, to meet its reduction targets, or to buy permits from the South.

I suppose further that the North would only buy permits from the South when a major price differential makes it attractive relative to its own marginal abatement cost in a given year of the abatement scenario. The upper limit on the permit price is set therefore by the North's marginal abatement cost. In principle, the South could sell excess permits at any price below that upper limit as it obtains marginal rent for any price above zero. (Of course, the lower the price, the more sales are needed to generate the requisite financing of the South's 'excess' incremental costs relative to its obligation to pay.) However, I assume that the South sells its permits only when the price offered is above its own marginal abatement cost, thereby setting a lower limit for the permit price.[g]

This assumption reflects my belief that the supply of low-cost abatement in the South is not unlimited, and that development pressures in the South will push it to exceed its own emission permits in our abatement scenario. Thus, even if it is allocated 'excess permits' at the outset, the South may have difficulty in financing its requisite emission reductions via tradeable permits because each 'excess' permit that it sells incurs an opportunity cost worth its own marginal abatement cost. For illustrative purposes, the permit price is assumed to settle at the South's marginal abatement cost plus half the difference between the North's and the Souths' marginal abatement cost in any year in the scenario.

Tradeable permits, Case 1

In Figure 6.4, I show the high marginal cost curve from Case 1 in Chapter 5 in which the North and the South face the same shaped curve, but travel over it at different rates due to divergent required reduction ratios in each year.

In Case 1, both the North and the South pay the same $50/T-C abated until 1997, so I assume no trade in emission permits before this year. (It is as cheap for the North to reduce another tonne of carbon emissions as it is to buy a permit to emit that tonne from the South). But then the North moves up to $160/T-C abatement cost while the South remains at only $50/T-C abated until it too attains $160/T-C abated in 2002 when it reaches the 20 per cent required reduction ratio. From 1998 to 2002, therefore, I posit that the price of a permit to emit carbon will fall half way between the South's

g In the real world, transfers might be funded by taxes, tradeable permits, and trade in abatement services at the same time. Assuming markets for permits and abatement services were competitive, then the price of abatement services would set the clearing price of excess permits. If abatement services and excess permits were in scarce supply, however, then sellers might obtain an economic rent by selling services and excess permits at a clearing price tending toward the buyer's marginal cost of abatement.

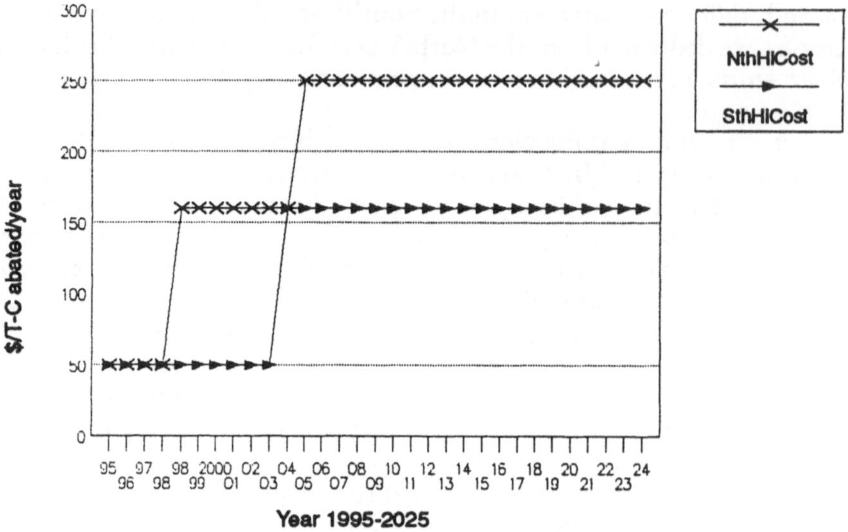

Figure 6.4 *Tradeable permits transfer, Case 1*

(\$50) and the North's (\$160) marginal abatement cost – that is, at \$105/T-C emitted.[h]

Sales fall to zero over the next two years as the cost curves have converged at \$160. Then in 2004, the North moves rapidly up to a \$250/T-C-y-1 abatement cost passing a required reduction ratio of 50 per cent. Consequently, the permit price increases to halfway between the North's own marginal abatement cost and that of the South (\$160/T-C-y-1 abated) or \$205/T-C emitted. As the South's cost never exceeds \$160/T-C-y-1 abated in the rest of the scenario, the permit price does not change thereafter.

Having determined the permit price path, it is possible to derive the volume of permits that must be sold to finance the transfer annuity to the South. The volume of permits that generates a present valued stream of tradeable permits sold at the posited, upwardly ratcheting permit prices that equals the transfer annuity of \$34 billion is 280 million tonnes per year in each 'trading year' (see box, page 161). Relative to the North's average annual required reduction of 1.7 billion tonnes per year in our abatement

h In a freely traded permit market, the clearing price (ignoring transaction costs and market imperfections) would tend toward the South's marginal abatement cost. However, in that world the South would also not be availed of a major 'excess entitlement' from which it would receive the full rent from each sold permit with which to fund the transfer justified by its relative obligation to pay. Rather, it would receive the permit price minus its own abatement cost to offset the sold emission permit. In a freely traded permit market, therefore, a much larger volume of permits would have to be sold to effect the same transfer to fund the South's abatement costs in accordance with the obligation-to-pay index than is the case with our 'excess entitlement' allocation of permits to the South.

scenario, Case 1 offers a plausible way to transfer resources to the South required to effect its abatement commitments.[i]

Tradeable permits, Case 3

Case 3 is a little more complicated as the North and the South are on different cost curves, but the principle is the same. The North's abatement cost is far below that of the South until 2004 when the North's required reduction ratio reaches 47 per cent and its abatement cost jumps to $294/T-C-y^{-1}$ abated – well above the South's $77/T-C-y^{-1}$ abated (which it never exceeds in the rest of the scenario). At the 'split the difference' permit price of $147/T-C-y^{-1}$ emitted, about 200 million tonnes of permitted emissions sold each year by the South to the North would fund the obligated annual transfer of $29 billion (see Figure 6.5). However, unlike Case 1, the underlying cost curves in Case 3 ensure that no trade occurs for the first

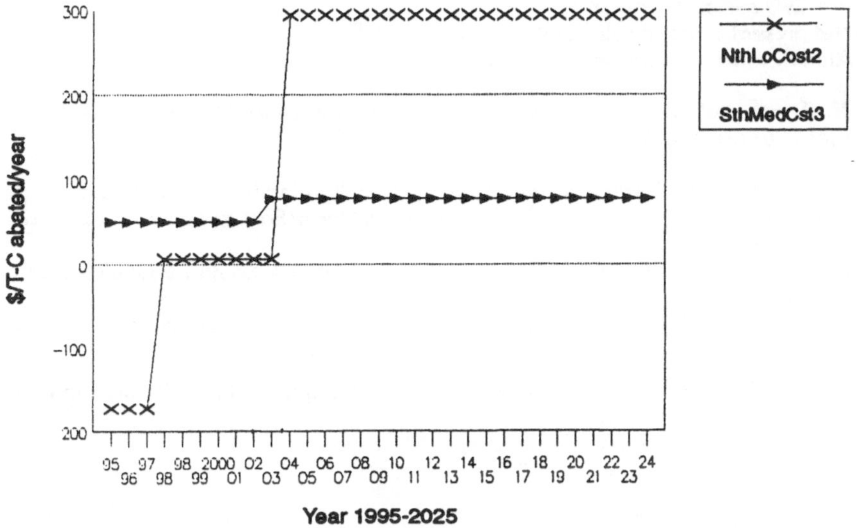

Figure 6.5 *Tradeable permits transfer, Case 3*

---

i The South might not be issued with an 'excess entitlement' but only with its own permitted emissions consistent with its abatement profile in this study's scenario. It could trade these permits in any volume demanded by the North. The permit price would fall between the North and the South's marginal abatement costs. However, the South would be obliged to offset its sales with additional abatement in the South in order to meet its abatement commitments. This deduction from revenue from permit sales would reduce the transfer achieved per permit sold by the South to the North. To achieve the resource transfer in accordance with the obligation-to-pay index, it would have to sell similar volumes of permits to those estimated in the abatement services trade analysed in the next section. As noted there, the South is not likely to have such large volumes of saleable emissions. It follows that an 'excess entitlements' scheme without offsetting abatements in the South would be needed if only a tradeable permits scheme is used to transfer resources in accordance with obligation-to-pay.

Tradeable permit transfer, Cases 1 and 3

I. Tradeable permits, case 1, high cost curve

Annuity of North-South transfer according to obligation-to-pay index, assuming Nordhaus cost curve in all areas = 34 billion/year

phase 1:[a] North and South have same marginal abatement cost, no trade between 1995–1997

phase 2:[a] North buys South's abatement between 1998–2002 at $80/T-C abated

phase 3:[a] North and South have same marginal abatement cost, no trade between 2003–2004

phase 4:[a] North buys South's abatement between 1998–2002 at $205/T-C abated

Volume of sales to reach annuity of transfer according to South's obligation-to-pay = 290 million T-C/year

Net present value of sales at this level = 516 billion

Annuity of sales at this level = 34 billion/year

II. Tradeable permits, case 3, medium cost curve in South, low cost curve in North/East

Annuity of North-South transfer according to obligation-to-pay index, assuming medium cost curve in South, low cost curve in North/East cost curve in all areas = 29 billion/year

phase 1:[a] marginal abatement cost in North is less than in South, no sales by South to North

phase 2:[a] North buys South's abatement between 1998–2025 at $147/T-C abated

Volume of sales to reach annuity of transfer according to South's obligation-to-pay = 250 million T-C/year

Net present value of sales at this level = 443 billion

Annuity of sales at this level = 29 billion/year

a See Figures 6.4 and 6.5

decade, indicating that tradeable permits may not set in motion a scheme that achieves transfers justified by the obligation-to-pay indice – an essential characteristic of a workable scheme.[15]

These two cases demonstrate how the relative cost curves drive the hypothetical North-South trade in permits. It is likely that a big relative cost difference would impel massive innovation in carbon abatement technologies in the North if the South's abatement costs set the price of traded permits well above the North's marginal abatement cost. If the North's and South's marginal abatement costs are approximately equal, trade will be very slow. Either way, however, the fact that the North's marginal abatement cost eventually greatly surpasses that of the South suggests that buying permits from the South would become increasingly attractive to the North.

How exactly the market would operate and where the clearing price would settle would depend on the rate of innovation driving down the permit price, and the oligopolistic behaviour of the rent-seeking southern owners of the permits and monopsonistic behaviour of the rent-avoiding northern buyers.[j]

Emission abatement services

Another possible financing technique involves the South selling low cost abatement services to the North. This scheme assumes that the South has a large stock of inefficient energy-using equipment in its buildings, industries, and transport systems plus forestry related carbon fixing potential that exceeds its own responsibility to reduce emissions. This 'surplus' abatement potential endows the South with cheap abatement options relative to those in the North. Ironically, the highly wasteful energy economy of many southern states endows them with competitive abatement opportunities if an international market can be created for such activities.

Such services can be provided by private entities and do not require interstate agreement as is usually assumed to be required for tradeable permit schemes. (Indeed, the first such project by a US utility which invested in carbon fixing reforestation in Guatemala, was led by private organizations.) Rather, it requires: (1) that abatement targets be adopted by states and that these states devolve responsibility to meet the abatement commitments onto public and private entities within these countries; and (2), that parties to the Convention adopt rules recognizing that abatement paid for by such an emission-reducing entity, but achieved at a saving relative to its own abatement costs in another country, can be debited from the emissions of the investor's country.[16]

Here, I determine the volume of abatement services that would have to be sold by the South to the North to finance the South's incremental costs. Table 6.3 shows the calculation for the high and medium marginal cost curves used to calculate the transfers.

The abatement services market would be analogous to that in tradeable emission permits. Polluters seeking a cheap abatement option relative to either the emission permit price or their own abatement cost would bargain with sellers of abatement services over the cost savings represented by their marginal abatement cost differential. However, abatement prices would be pegged to only this cost savings rather than the North's total marginal abatement cost. Thus, a smaller value is obtained by the South from the North for each sale of a tonne of carbon abatement versus a tonne of carbon

---

j I have not analysed the possibility that the South could buy tradeable permits from the North early in Case 3 when its abatement cost is way above the North's. Doing so might be sensible for the South even if it were allocated additional emission rights. The market should clear in either direction.

permitted emission. Here, I posit that the abatement services price would settle at half the difference between the North's and the South's marginal abatement cost in our scenario.

In the high cost, Case 1 – when the North and the South start out on the same cost curve – there is initially no cost difference, followed by a four year stretch of relative cost advantage to the South and a subsequent twenty year stretch of even greater advantage after 2005 (see Table 6.3).

Table 6.3 *Abatement services transfer, Cases 1 and 3*

A.		CASE 1, HIGH MARGINAL COST TRANSFER OF 34 BILL $/Y Sale of Abatement Services to fund North's Transfer to South, and N-S transfer = ½ price difference at time of sale	
	A1.	Assume sale occurs when MCsth<MCnth, and transfer = ½ of versus North, with saving split 50:50 between North/South	
	A2.	Annuity-OTP N-S Transfer =	34.0 Bill$/y
	A3.	Phase 1, no cost difference, no sale	0.0 $/T C, 1995–1997
		Phase 2, transfer is ½ price diff.	30.0 $/T C, 1998–2002
		Phase 3, no cost difference, no sale	0.0 $/T C, 2003–2004
		Phase 4, transfer is ½ price diff.	45.0 $/T C, 2005–2025
	A4.	Required Sales to generate Annuity OTP?	
		NPV trade 523.2	
		Annuity Trade 34.0	
		Sales volume = 1.2 BillT/y	
	A5.	Total South Req Reduction is only	24.7 Bill T C 1995–2025
	A6.	Trade to fund total transfer 1995–2025 is 30 y x sales/y =	35.9 Bill T C 1995–2025
		or 1.5 times South's required reduction over same period	
B.		CASE 3, MEDIUM MARGINAL COST TRANSFER OF 29 BILL $/Y Sale of Abatement Services to fund North's Transfer to South, and N-S transfer = ½ price difference at time of sale	
	B1.	Assume sale occurs when MCsth<MCnth, and transfer = ½ of versus North, with saving split 50:50 between North/South	
	B2.	Annuity-OTP N-S Transfer =	29.0 Bill$/y
	B3.	Phase 1, no cost difference, no sale	0.0 $/T C, 1995–2004
		Phase 2, transfer is ½ price diff.	70.0 $/T C, 2005–2025
	B4.	Required Sales to generate Annuity OTP?	
		NPV trade 446.1	
		Annuity Trade 29.0	
		Sales volume = 1.0 BillT/y	
	B5.	Total South Req Reduction is only	24.7 Bill T C 1995–2025
	B6.	Trade to fund total transfer 1995–2025 is 30 y × sales/y =	30.0 Bill TC 1995–2025
		or 1.2 times South's required reduction over same period	

With the price trajectory specified, the volume of abatement services needed to fund the South's annual 'excess' incremental cost of $34 billion can be calculated. After discounting and annuitizing the volume of trade at the posited prices, the required sale of abatement services is no less than 1.2 billion tonnes per year. Thus, the South would have to find more than 150 per cent more abatement potential than its own required reduction at these relatively low costs over the thirty year scenario. Applied to the medium cost, Case 3, the same method demands only a slightly smaller volume of trade to achieve the smaller obligated funding in the South of $29 billion.

It is apparent that unlike tradeable permits - and granted all the assumptions stated earlier as to the South's net profit earned on sale of abatement services - the sale of abatement services cannot be relied on to finance the whole of the South's 'excess' incremental costs. It is highly unlikely that such large volumes of abatement opportunities exist in the South, abatement that must be achieved to the credit of other nations and over and above the reductions required on the South's own behalf.[k,17] However, the forestry related carbon fixation opportunities in the South (either forestry maintenance or reforestation) may be much greater than those obtained by abating carbon dioxide from fossil fuel, and entrepreneurs may identify many new ways to provide competitive abatement services in the South.

I conclude that trade in abatement services – especially if supplemented by a scheme related to the South's forestry carbon fixation – would be useful within and between states in responding to emission limits or targets.[18] Northern marginal abatement costs will also increase faster in the North than in the South as Northern states move up the required reduction curve calculated in Chapter 5 (see Figure 5.11). With each year that passes, therefore, it is likely that the price of tradeable permits and abatement services will increase substantially, offering increasing opportunities for the South to reap benefits from their initial endowment of tradeable permits that could thereby finance their own required reductions.

Using these instruments in combination to implement a global carbon reduction strategy may also be productive, as would introducing them at different phases of a strategy.[19] Carbon taxes might be spent best at a national level, thereby stimulating a new wave of technological innovation and driving down the marginal cost of abatement for everyone. Trade in abatement services might be used effectively early on in a reduction strategy to enable the North to take advantage of lower marginal abatement costs in the South, even if the South has few or no excess permits. And tradeable permits might be used later rather than earlier, thereby avoiding the possible perverse outcome that the South would sell 'excess' emission rights at the

k One reviewer questioned this conclusion, stating: 'Is it really a problem if the sales indeed were greater than the South's required reduction, since the required reduction would be rather small anyway? There may be a lot of opportunities to sell [abatement services], depending on how carbon storage in forests is treated.'

outset of a reduction strategy to finance development that expands greenhouse emissions – which is what the world is trying to avoid.[1]

Conclusion

I have shown that the incremental cost of abatement and coastal protection in the South that is justifiably the responsibility of the North is of the order of $30 billion per year. To these costs should be added a substantial sum for human resource development needed to realize the South's carbon abatement potential, perhaps as much again; plus approximately $100 million per year for technical assistance, training, and information provision (at least for the first five years of the Convention), and another $100 million for the costs of monitoring and verifying compliance with emission commitments by parties to the treaty (see Chapter 14).

These cost estimates are indicative and should be treated with caution. Because these figures are based on 'soft' estimates, these conclusions are not robust. They are based on many intervening assumptions and variables, changes in which would require radical revision in the numbers cited (see box). Many of the potential uncertainties may operate in opposite directions, thereby cancelling out each other. Some (such as marginal cost estimates and

Summary of uncertainties

- Present and future per capita carbon intensity (+20%)
- CO_2 fossil fuel usage estimate error (±5–10%)
- Aggregation errors
- Population projections (±10%)
- Sink estimates (±50%)
- Allocation rule for national distribution of sinks (±25–50%)
- 2025 permissible emissions target uncertainty (±10–20%)
- Take-up rate of efficiency, substitution, renewables (±10–20%)
- Cost of autonomous energy efficiency increase (±20–50%)
- Levelized marginal cost schedules (differences of ±100%) in current estimates; impact of 100% from innovation and diminishing returns on future cost ignored)
- Benefits (avoided coastal and agricultural costs) (±25–50%)
- Discount rate (1–2% vs 5% vs 10%)
- Obligation to pay index used to determine the South's obligation (±10% due to periodic readjustment)
- Treatment of non-CO_2 fossil fuel greenhouse gases
- Trade impacts of abatement costs
- Intersectoral adjustment, human resource development, and institutional change costs

Note: author's estimates of the range of positive and negative uncertainty for variables is shown in (±)

1 I am indebted to Joel Swisher for this insight.

addition of non-CO_2 greenhouse gases) may combine to enhance each other's effect on the incremental cost, increasing it dramatically.

It is hard, however, to see that revisions will decrease the cost estimates provided here by a substantial amount. These estimates can be used reasonably as baseline figures of the likely magnitude of the financial flows involved in implementing the emission reduction commitments should the Climate Change Convention be successful.

The marginal abatement cost estimates, trade impacts, and human resource development costs are three particularly potent variables that need much more research before these estimates of incremental cost can be treated with confidence. It is also important that the emissions data base used in this chapter be updated and that the capital requirements be linked to a macroeconomic model that will enable the direct costs to be matched with indirect trade effects on the economic welfare of different nations and national groupings. Developing such a modelling capability to support policy making and the negotiations on the protocols of the Convention is of great importance.

Enough is known now, however, for action to start. Bilateral and regional activities can commence immediately (see Chapter 14). Indeed, such initiatives are a pressing priority as they would demonstrate the feasibility and logic of carbon taxes, tradeable permits and abatement services.

Notes and references

1 See UN Development Programme, *Human Development Report 1990*, Oxford University Press, New York, 1990

2 See K Smith *et al.*, *Indices for a Greenhouse Gas Control Regime that Incorporate Both Efficiency and Equity Goals;* report to Environmental Policy and Research Division, World Bank, Environment and Policy Institute, Honolulu, Hawaii, May 21, 1991; and Chapter 4 of this book

3 M Grubb, J Sebenius, A Magalhaes and S Subak, 'Sharing the Burden,' in I Mintzer (ed), *Confronting Climate Change, Risks, Implications, and Responses*, Cambridge University Press, New York, 1992, p 321

4 See T Hyder, 'Climate Negotiations: The North/South Perspective,' in I Mintzer (ed), *Confronting Climate Change, ibid*, p 328

5 W Nitze, 'Criteria for Negotiating a Greenhouse Convention that Leads to Actual Emissions Reductions,' *International Challenges*, volume 11, no 1, 1991, p 15; see also A K N Reddy, 'Barriers to improvements in energy efficiency,' *Energy Policy*, December 1991, pp 953–961

6 World Resources Institute, *World Resources, 1992–93*, Oxford University Press, New York, 1992, pp 8; 236–7

7 G Piel, 'Agenda 21: A New Magna Carta,' *Earth Summit Times*, September 14, 1992, p 11

8 Scientists and technicians from UN Development Programme, *Human Development Report, 1992*, Oxford University Press, New York, 1992, p 190; population data from World Resources, *World Resources, 1992–93, op cit* (endnote 6), p 76

9 L Lunde, *The North/South Dimension in Global Greenhouse Politics, Conflicts, Dilemmas, Solutions*, Report 9, Fridtjof Nansen Institute, Lysaker, Norway, 1990, p 17

10 Table 1475, US Department of Commerce, *Statistical Abstract of the United States*, recent annual editions, Washington DC

11 See the survey in S Barrett, 'Economic Instruments for Global Climate Change Policy', London Business School for Environment Directorate, OECD, Paris, 1990; a more accessible version is found in Barrett's 'Global Warming, The Economics of a Carbon Tax,' in D Pearce (ed), *Blueprint 2, Greening the World Economy*, Earthscan Books, London, 1991, pp 38–39. See also Price Waterhouse Government Liaison Services, *Consultancy Report to Department of Arts, Sport, The Environment, Tourism and Territories on Carbon Tax*, Canberra, ACT, June 1991

12 For a summary of eight carbon tax studies, see S Barrett, 'Global Warming, The Economics of a Carbon Tax,' in D Pearce (ed), *Blueprint 2, ibid*, pp 38–39

13 See J Epstein and R Gupta, *Controlling the Greenhouse Effect: Five Global Regimes Compared*; Brookings Occasional Papers; The Brookings Institution, Washington, DC, 1990; D Victor, 'Limits of market-based strategies for slowing global warming: The case of tradeable permits'; *Policy Sciences*, volume 24, 1991, pp 199–222; A Markandya, 'Global Warming, The Economics of Tradeable Permits,' in D Pearce (ed), *Blueprint 2, op cit* (endnote 11), pp 53–62

14 For reviews of this experience, see E Meidinger, 'On Explaining the Development of 'Emissions Trading' in US Air Pollution Regulation'; *Law and Policy*, volume 7, no 4, October 1985, pp 447–477; and T Tietenberg, 'Transferable Discharge Permits and the Control of Stationary Source Air Pollution: A Survey and Synthesis'; *Land Economics*, volume 56, no 4, November 1980, pp 391–416

15 M Grubb and J Sebenius, *Participation, Allocation and Adaptability in International Tradeable Emission Permit Systems for Greenhouse Gas Control*, forthcoming in OECD workshop proceedings on greenhouse gas tradeable permits, Paris, June 1991, p 4.

16 J Swisher and G Masters, *International Carbon Emission Offsets: A Tradeable Currency for Climate Protection Services*, Technical Report #309, Department of Civil Engineering, Stanford University, Stanford, California, 28 February 1989

17 Joel Swisher, personal communication, June 22, 1992

18 W Makundi *et al* provide detailed estimates of current, committed and delayed carbon uptake and emissions in southern forests in *Carbon Emissions and Sequestration in Forests: Case Studies From Seven Developing Countries*, Lawrence Berkeley Laboratory, LBL-32119, UC-402 (draft), August 1992

19 See J Swisher and G Masters, 'A Mechanism to Reconcile Equity and Efficiency in Global Climate Protection: International Carbon Emission Offsets,' *Ambio*, volume 21, no 3, April 1992, pp 154–159

7

Insuring against sea level rise

Michael Wilford

The Reports of the Intergovernmental Panel on Climate Change (IPCC) recognize that the human settlements most vulnerable to climate change are those which are especially exposed to natural hazards, that is, to coastal or river flooding, severe drought, landslides, severe wind storms and tropical cyclones. The IPCC also recognized that the most vulnerable populations are in developing countries; and that in small island countries as well as in coastal lowlands, inundation due to sea level rise and storm surges are a particular hazard (see box on page 170).[1]

The IPCC Working Group II (Impacts) used various scenarios based on a number of different scientific studies.[2] The main scenario upon which Working Group II based its assessments was:

1 an effective doubling of CO_2 in the atmosphere between now and 2025–2050;
2 a consequent increase of global mean temperature in the range of 1.5°C to 4–5°C;
3 a sea level rise of about 30–50 cm by 2050 and of about 1 metre by 2100, with a rise in the temperature of the ocean surface layer of between 0.2°C and 2.5°C.[a,3]

This scenario pre-dated, but is in line with, the assessments of IPCC Working Group I.

In assessing the effects of climate change on the oceans and coastal zones the IPCC reports point out[4] that global warming would not only accelerate sea level rise, but would also modify ocean circulation and change marine ecosystems, with considerable socioeconomic consequences. These effects will add to the present trends of rising sea level and damage to coastal resources from pollution and over-harvesting. A 30 cm sea level rise will

a Readers might note that 1 metre is the high estimate of sea level rise from the 1990 and 1992 assessments by Working Group I, Intergovernmental Panel on Climate Change. Their best estimate was 30 cm by 2050 and 70 cm by 2100. Their low estimate was 15 cm by 2050.

Vulnerability of the Maldive Islands to sea level rise

The current environmental problems of the Maldives are in large part the result of the high density population (650/km^2) which is aggregated onto relatively few islands within each atoll. The problems of the capital island, Malé, have reached a critical level in several areas increasing the island's susceptibility to episodic events such as storm generated high waves. In other areas and atolls the problems are also locally severe increasing the susceptibility of such areas to future climatic change and sea level rise. This increased susceptibility is due to:

- coral mining for construction and road surfacing;
- land reclamation, particularly on the seaward edges of islands;
- construction of coastal infrastructure including sea walls, breakwaters, jetties, piers, groynes and harbours;
- aquifer depletion and saline intrusion.

All of the current environmental problems are exacerbated by:

- high population growth;
- a lack of mechanisms within government for taking environmental problems into consideration in the planning process;
- a lack of guidelines and procedures for the evaluation of environmental issues;
- a lack of an adequate in-country data base covering many physical and biological parameters;
- a shortage of trained manpower at all levels.

The current, environmentally unsound, practices will increase the susceptibility of the Maldives to changes predicted to occur as a consequence of global warming and the greenhouse effect. Assuming a sea level rise of 12–18 cm by the year 2030 one might anticipate profound effects on those islands of the Maldives which have been structurally modified, since the normal processes of sand genesis, deposition, removal and flux between sinks have been altered by changes to the micro-climate and current regimes.

The impact of 'high waves' will be greater with greater mean sea level and such increases must be taken into consideration in planning future coastal infrastructure. Changes to aquifer volumes may be expected under higher sea levels; however, such changes will be less important on islands where the aquifer is not currently over-exploited. Saline intrusion will be exacerbated in those aquifers which are heavily used for human consumption.

Increased temperatures (of perhaps 1.5°C by 2030) will affect the human environment, agricultural production, and marine ecosystems. Given the country's proximity to the equator, the Maldives can expect a lower than average temperature rise which may have little impact on the human environment but may be expected to result in some increased demand for air conditioning. Agricultural production and terrestrial ecosystems are likely to be less affected than marine organisms such as corals, many of which are currently growing at temperatures close to their upper thermal tolerance limits.

Perhaps the area of greatest current concern in the Maldives is the possibility of an increased frequency of storm-generated swells and high waves, particularly given the experiences of the country during 1987. Analysis of the meteorological patterns in the Indian Ocean is urgently required to predict the possibility of an increased frequency of such events.

Situations of high sea level at the coast of the atolls are caused by storm surges and waves setup. A degree of coastal flooding due to high tides has been experienced in the past at various places. Recent flooding has been made more noticeable by its impact on construction such as sea walls and houses near the shore and on low-lying reclaimed land. The July 1988 high water situation at Thulhadhoo (Malosmadulu Atoll) was caused by high southwesterly waves (5 m high, periods 12–15 seconds) in association with high spring tide and a southwesterly wind. The damage caused was enhanced by the absence of beaches and the presence of vertical low seawalls which magnified overtopping and flooding. These events are a reminder that occasional natural events of long distance swells and high water levels due to wave surge and/or high tides, would in themselves cause little damage to the Maldives atolls, were it not for the mismanagement that has taken place in recent years. This mismanagement includes stripping the islands of the natural defences afforded by the outer reefs, the reef flats and the beaches.

Social impacts arising from changes to island stability and/or habitability are likely to be extensive given the nature of Maldivian society which is characterized by generally low mobility and strong attachment to individual atolls and islands. Economic impacts will be most intensely felt if the tourist industry is adversely affected. The present structure of the tourist industry is based on 'resort islands' which are essentially self-contained and as a consequence pack considerable infrastructure on the land and coastline of very small islands. The present tourist industry is concentrated in the Central Maldives, hence increasing the risk to this sector of the economy.

Excerpt from J Pernetta and G Sestini, *The Maldives and the Impact of Expected Climatic Changes*, UNEP Regional Seas Report 104, UN Environment Programme, Nairobi, 1989, pp 27–8; 38.

threaten low islands and coastal zones. A one metre rise would render some island countries uninhabitable.

The Coastal Zone Management Subgroup also emphasized that:

> Sea level rise could increase the severity of storm-related flooding. The higher base for storm surges would be an important additional threat in areas where hurricanes, tropical cyclones and typhoons are frequent, particularly for islands in the Caribbean Sea, the south eastern United States, the tropical Pacific and the Indian sub-continent . . . Many small island States are . . . particularly vulnerable. This is reflected in their very high ratios of coastline length to land area. The most seriously threatened island States would be those consisting solely, or mostly, of atolls with little or no land more than a few metres above sea level. Tropical storms further increase their vulnerability and, while less in magnitude than those experienced by some of the world's densely populated deltas, on a proportional basis such storms can have a much more devastating impact on island nations.[5]

These predictions of the increasing impact of severe storms have not been lost upon the insurance industry.

In a review of the effect of climate change on insurance one of the world's largest reinsurers has recently written:[6]

> Detailed measurements in the Pacific show that the areas with water temperature at the surface above 27°C have expanded by about one-sixth in the last two decades. While substantial fluctuations from one year to the next

> and additional factors such as El Niño make it impossible so far to prove the effect of such higher temperatures on the frequency of tropical cyclones, super-hurricanes 'Gilbert' and 'Hugo' may certainly be regarded as a clear sign of an increase in hurricane intensity. According to estimates, hurricane activities in the Caribbean will increase considerably in the next two to three decades, the loss potential going up by more than 50%. Together with the rising level of the sea, this also means a much greater risk of storm surges in densely populated coastal regions in the tropics and sub-tropics.

These insurers predict that there will be a global increase in catastrophic losses from a current annual average of about US$20 billion to about US$100 billion per year in overall economic terms.[7] And that, it may be assumed, does not take into account the potential longer term consequences of climate change.

Insurability of losses

In theory, any loss which is not inevitable is insurable under a commercial contract of insurance. A classic definition of the necessary legal elements of such a contract includes the requirement that the event insured against 'should be one which involves some amount of uncertainty'. The definition continues: 'There must be either uncertainty whether the event will ever happen or not, or if the event is one which must happen at some time there must be uncertainty as to the time at which it will happen.'[8] An element of uncertainty is therefore a necessary legal requirement for a loss to be insurable.

In commercial terms, more will be required if a risk is actually to be insurable on the world insurance markets. The requirements for the insurability of a flood or inundation risk commercially have been stated[9] to be:

1 a sufficient demand for insurance of the risk and sufficient variation in the insurer's portfolio of such risks to ensure an adequate spread of risk;
2 the random occurrence of losses, but neither too often nor too seldom;
3 the ability to restrict accumulation of losses;
4 the traceability of the causes of a loss;
5 the availability of statistical records of losses from the risk, covering sufficiently long periods of time.

The randomness or uncertainty of the event to be insured against is here again an essential element. But for commercial insurers, it is also essential to be able to select the risks that they insure and to be able to calculate realistically what premium they should charge, in order to accumulate over the long term, sufficient funds to meet losses and to make a profit.

The IPCC Working Group I concluded that even if action is taken to limit emissions, there is enough momentum in the global climate system for a rate of accelerated sea-level rise to be inevitable.[10] Gradual flooding and inundation which may be attributable to an *inevitable* rise in sea level will certainly not be capable of being insured on the world insurance markets.

Even in the sphere of catastrophe insurance (which in this context refers to insurance against damage caused by storm, windstorm, hurricane, tropical cyclone, typhoon, as well as flood and inundation) events that up to the present have been regarded as fortuitous, and therefore insurable, may cease to be so regarded in the future. One insurer, in considering the impact of global warming and climate change upon catastrophe insurance, has remarked: 'There should be further examination of what constitutes an "abnormal" loss, and we will have to alter our view of what is perceived as normal.'[11]

In considering the potential for heavy increases in catastrophe losses as a result of climate change, it is important to bear in mind that the current levels of catastrophe loss represent in turn a dramatic increase over the losses of two decades ago. Considering the major natural disasters over the last three decades and extrapolating figures to 1990 prices, it can be shown in the 1980s compared with the 1960s, that: overall economic loss from major natural disasters increased by a factor of 3.1; insured losses increased by a factor of 4.8; and major catastrophes increased by a factor of 5.0.[12]

The world's insurance and reinsurance markets will find it difficult to cope with such rapid change in the incidence of natural hazards. For example, the reinsurers of first layer catastrophe insurance covers granted to United Kingdom companies for worldwide exposures paid out between 1979 and 1988 in claims resulting from 'abnormal' weather experienced in the UK alone, more than twice the total amount received by these reinsurers in premium.[13]

In stable climatological conditions, an insurer or reinsurer can assess the probable 'return period' of a known catastrophe peril by reference to historical events and statistical records. The premium for the risk can then be calculated with some reasonable measure of accuracy. Rapid changes in climate make the calculation of premium for natural catastrophe risks a very much more difficult and uncertain exercise. In a 1963 study, for example, it was calculated on the basis of existing records that the return period for a flood which inundated over 800 square kilometres of the east coast of England in 1953 was 1 in 200 years. Yet between 1963 and 1989 (26 years), the 1 in 200 year storm surge level occurred eleven times.[14]

Given the difficulties currently facing insurers and reinsurers in basing premium calculations on past climatological records and those posed by predictions of global warming and sea level rise, insurers and reinsurers will likely become more selective and restrictive in their offered cover.

Options for insurers

The options available to insurers and reinsurers in such circumstances include:[15]

1 impose upon the insured, in the case of original insurances, substantial deductibles per policy or risk;
2 impose a limit figure on the amount of their liability;
3 charge sufficient premium to enable adequate reserves to be accumulated;
4 exclude particular risks (such as, for example, wind storm, flood, inundation) either absolutely or in particular localities;
5 avoid insuring any catastrophic risks in particular localities or limit overall catastrophe exposure in particular localities;
6 withdraw from the insurance or reinsurance of catastrophe risks altogether;
7 redefine categories of loss which in insurance terms are at present regarded as 'abnormal' or fortuitous, and therefore ordinarily insurable, and those which are regarded as 'normal' or non-fortuitous, and ordinarily not insurable.

Additionally, insurers and reinsurers can assist and encourage the adoption of loss-prevention measures either by advice or by the introduction of conditions into policies of insurance or reinsurance. But faced with the probability of heavy increases in catastrophic losses from so-called natural causes, insurers and reinsurers will be forced to take some, if not all, of the measures mentioned if they are to continue to provide an effective and responsible service to their insurees and if they are to remain in business as insurers.[16]

This logic suggests strongly that insurers will not cover a substantial part – possibly the most substantial part – of the potential loss and damage resulting from climate change and sea-level rise. In short, such risks will be uninsurable on the world's insurance and reinsurance markets. This outcome has indeed been recognized by the insurance industry itself. One company recently stated that:

> On the level of direct insurance and reinsurance, the insurance industry fortunately has a wide range of 'tools' and instruments for supervising and controlling its exposure to catastrophe risks.
>
> These underwriting facilities . . . must however be used in full and without delay if insurers are to bring the current adverse development to a halt. At the same time these activities must be accompanied by structural loss prevention measures and suitable precautions taken by the authorities.
>
> Wherever the national insurance industry supported by worldwide reinsurance capacity (which, ultimately, is only available if premiums, terms and conditions are adequate) is not able to cover an extreme catastrophe risk, *it may become necessary for the government to assume a share in catastrophe covers, at least for losses exceeding a certain level, or to guarantee indemnification in all events, as some governments already do in the case of the risk of flood and inundation.*[17]

It is evident that even in the case of fortuitous catastrophic losses, the world insurance industry will not be able to bear the burden of providing a global spread of compensation without state intervention. And in the case of those losses which the insurance industry will have to redefine as 'normal' – for example, flooding or inundation resulting from inevitable sea level rise – State funding will be the only source of compensation of those who are affected by it.[18] Since the least developed countries are those most vulnerable to future catastrophic loss and what may come to be regarded as the inevitable consequences of climate change, any state compensation regime will have to be internationally funded.

Two internationally funded compensation schemes operate at present. These deal with oil pollution and nuclear damage. Neither case, however, is analogous exactly with loss consequent upon climate change and sea level rise. On the contrary, the incidence of oil pollution or nuclear damage is more likely to be traceable to an identifiable source and referable to a discoverable cause.

Loss or damage of the nature associated with climate change, whether of a catastrophic or 'abnormal' nature or of a 'normal' or inevitable nature, is highly unlikely ever to be conclusively attributable to any legally definable source or cause. Thus, if a compensation scheme is to form any part of a Climate Change Convention, it could probably not be based upon ordinary legal criteria of liability or responsibility. Rather, it would have to be based on a criterion of responsibility in its broadest sense.

Oil pollution[19]

Compensation for loss or damage resulting from the escape or discharge of oil from ships is governed by two conventions. The first lays down a regime of civil liability for oil pollution. The second sets up an international compensation fund to cater for those instances where the civil liability regime does not afford compensation to the injured party.

The first Convention, the *International Convention on Civil Liability for Oil Pollution Damage 1969*, contains uniform international rules and procedures under which shipowners are made strictly liable for certain types of oil pollution damage. It imposes upon the bulk carriers of persistent oils the obligation to effect insurance against the liabilities imposed by the Convention. The Civil Liability Convention came into force in 1975 and there were amendments to its financial provisions by Protocol in 1976.

The *1971 International Convention on the Establishment of an International Fund for Compensation for Oil Pollution Damage* ('the IOPC Fund') provides for the payment of compensation to any person suffering oil pollution damage, as defined in the Civil Liability Convention, if that person is unable to obtain full and adequate compensation for one of the following reasons:

1 that no liability for the damage arises under the Civil Liability Convention, because the shipowner can invoke one of the exemptions from liability set out in that Convention;
2 the shipowner is financially incapable of meeting his obligations under the Civil Liability Convention and his insurance is insufficient to satisfy the compensation claims;
3 the damage exceeds the shipowner's limit of liability under the Civil Liability Convention.

The IOPC Fund also provides an indemnity to shipowners against a proportion of their liabilities under the Civil Liability Convention. It is financed by levies on oil importers. By Article 10 of the Fund Convention, contributions to the fund are levied on any person who has received 'contributing oil' (crude oil and heavy fuel oil) in a quantity exceeding 150,000 tonnes in one calendar year in a Contracting State.

Under Article 11 initial contributions are payable by such importers when a State becomes a member of the IOPC Fund. Annual contributions to the fund are levied to meet the anticipated payments of compensation and indemnity by the IOPC Fund in the ensuing year and to meet also the administrative expenses (Article 12).

The functioning of the IOPC Fund is controlled by an Assembly, an Executive Committee and a secretariat headed by a Director. The Assembly comprises representatives of the governments of all Member States. The Executive Committee is elected by the Assembly and must approve settlements of claims against the Fund. The amount of annual contributions to the Fund is decided annually by the Assembly.

Contributions are levied directly from and paid by the individual importers and not by the Contracting States, unless a Contracting State declares that it assumes that obligation (Article 14). The Contracting States are required to communicate annually to the Director of the IOPC Fund the name and address of any person in that State who is liable to contribute to the Fund as well as details of the quantities of 'contributing oil' received.

Contributing oil is counted for contribution purposes each time it is received at ports or terminal installations in a Contracting State after carriage by sea. The place of loading is irrelevant: the oil may be imported from abroad, carried from another port in the same State or transported by ship from an off-shore production rig. Oil received for transhipment to another port or received for further transport by pipeline is also considered as 'received' for contribution purposes. Fears expressed at the time the Fund Convention was adopted that the Secretariat would have difficulties in collecting contributions have proved to be unfounded.[20]

In 1984 two Protocols were adopted amending the Civil Liability Convention and the Fund Convention. They increased the limits of the shipowners' liability and also increased the limit of compensation payable by the IOPC Fund in respect of any one incident from an aggregate amount of

60 million SDRs (Special Drawing Rights) to 135 million SDRs. This latter figure will be increased automatically to 200,000,000 SDR when there are three Member States of the 1984 Fund whose combined quantity of contributing oil received during a given year in their respective territories exceeds 600 million tonnes.

The Protocols also extend the scope of the Conventions to oil spills from unladen as well as laden tankers and redefined the term 'pollution damage'. The Protocols introduce the concept of compensation for impairment to the environment, although they provide that such compensation shall be limited to loss of profit and to costs to reasonable measures of reinstatement actually undertaken or to be undertaken.[21]

The 1984 Protocols are not yet in force. The 1984 Protocol to the Fund Convention enters into force when ratified by at least eight States and when the total quantity of contributing oil received during a given calendar year in all the ratifying States is at least 600 million tonnes. As of 1 May 1991, forty-five States had ratified the Fund Convention.

Nuclear damage[22]

A different approach from that adopted by the Oil Pollution Fund Convention has been adopted in the case of nuclear damage. The 1960 Paris Convention on Third Party Liability in the field of Nuclear Energy,[23] concluded under the auspices of the OEEC (now OECD) was the first international convention to regulate the liability for risks arising out of the peaceful use of nuclear energy.

The purpose of the Paris Convention is to harmonize national legislation with regard to third party liability and insurance against atomic risks and to establish a regime for liability and compensation in the event of a nuclear incident. The Convention generally applies only to nuclear incidents occurring, and damage suffered, in the territory of Contracting States. In 1963, a supplementary convention to the Paris Convention was adopted ('the Brussels Supplementary Convention'[24]).

Article 3(b) of the Brussels Supplementary Convention (as amended by the subsequent 1964 Protocol) provides that the contracting parties undertake that compensation in respect of damage caused by a nuclear incident (other than one occurring entirely in a territory of a non-contracting State) shall be provided up to an amount of 120 million units of account per incident as follows:

1 up to an amount of at least 5 million units of account, out of funds provided by insurance or other financial security, such amount to be established by the legislation of the contracting party in whose territory the nuclear installation of the operator liable is situated;
2 between this amount and 70 million units of account, out of public funds

to be made available by the Contracting Party in whose territory the nuclear installation is situated;

3 between 70 million and 120 million units of account, out of public funds to be made available by the Contracting Parties according to the formula for contributions specified in Article 12 of the Brussels Supplementary Convention.

Article 12(a) of the Brussels Supplementary Convention provides that:

> The formula for contributions according to which the contracting party shall make available the public funds referred to in Article 3(b)(iii) shall be determined as follows:
>
> (i) as to 50%, on the basis of the ratio between the gross national product at current prices of each contracting party and the total of the gross national product's current prices of all Contracting Parties as shown by the official statistics published by the Organisation for Economic Co-operation and Development for the year preceding the year in which the nuclear incident occurs;
>
> (ii) as to 50%, on the basis of the ratio between the thermal power of the reactors situated in the territory of each Contracting Party and the total thermal power of the reactors situated in the territories of all the Contracting Parties . . .

The Brussels Supplementary Convention therefore creates an international pool out of which the highest layer of compensation is payable in the event of damage from a nuclear incident, and levies contributions toward that pool from Contracting States according to a formula based upon GNP and total capacity of their nuclear reactors.

Unlike the Paris Convention, the *1963 Vienna Convention on Civil Liability for Nuclear Damage*[25] is potentially of worldwide geographical application. The Vienna Convention also establishes a regime for liability and compensation for nuclear incidents, but it permits the Contracting State within whose territory the installation is situated to limit the liability of the operator.

By Article VII, the installation State undertakes to require the operator to maintain insurance or other financial security covering his liability for nuclear damage in such amount as the State shall specify. But the installation State is bound to meet compensation claims up to the operator's limit of liability, if the operator's insurance or other financial security is inadequate to meet such claims. There is, however, no provision similar to the Brussels Supplementary Convention either in the Vienna Convention or elsewhere for any international pooling arrangement to pay compensation above the operator's limit of liability.

Implications

I conclude that the preceding analysis poses at least seven implications for the ongoing negotiations under the rubric of the Climate Change Convention.

First, it is clear that the consequences of inevitable gradual sea level rise due to climate change induced by global warming will not be insurable on the world's insurance markets. Second, even in the case of the 'abnormal' catastrophic consequences of climate change and sea level rise - floods, inundation, storms, windstorms, hurricanes, tropical cyclones, typhoons - insurers already acknowledge that there will be a substantial shortfall in their coverage of losses in the future. Insurers are already forced to decline to insure certain risks in particular areas and increasingly they will be forced to refuse to insure risks in the geographical regions most prone to catastrophic loss. As a result they will have to impose deductibles, limit the total amounts of their liability on individual risks and limit their liabilities in the aggregate.

If as predicted the incidence of catastrophes resulting from the combination of sea level rise, severe storms and storm surges increases substantially, a large proportion of catastrophic losses will be uninsured and indeed uninsurable on the world's insurance markets. The insurance industry recognises that even in the case of catastrophe insurance, government intervention will be necessary.

Third, losses of such magnitude cannot be carried by the governments of those countries most vulnerable to these hazards - the small island and low-lying coastal developing countries. There is a need for an internationally funded insurance pool.

Fourth, the oil pollution and nuclear damage Conventions referred to in this paper provide examples of international insurance pools that have been set up in other contexts. One is funded by the oil industry, the other by States on a basis which takes account of gross national product and nuclear energy capacity.

Fifth, the basis for contribution to an internationally funded insurance pool in the present context, and the criteria to be satisfied in claiming from such a fund, should form part of the negotiations leading to the Climate Change Convention.

Sixth, such an international insurance pool might be funded by the developed countries on a contribution basis related to gross national product and/or greenhouse gas emission levels and/or by the industries of such countries responsible for such emissions.

Finally, the Ministerial Declaration at the Second World Climate Conference recommended, inter alia, that stepped-up financial contributions be provided by developed countries to address the particular problems and needs, including funding, of low-lying coastal and small vulnerable island countries.[26] Such countries must be the principal beneficiaries of an internationally funded insurance pool set up to cover losses resulting from sea level rise and related catastrophic events stemming from climate change.

The insurance scheme proposed by AOSIS

It was in the light of these conclusions that a proposal for an insurance scheme was put forward by the members of the Alliance of Small Island States (AOSIS) in the negotiations for a Climate Change Convention. The proposal was outlined in an Insurance Annex which was introduced into the draft text in the course of the negotiations and remained in the draft until the last meeting of the International Negotiating Committee for a Framework Convention on Climate Change in New York in May 1992. It was not directly included in the Climate Change Convention, but was reflected in Articles 3 and 4 (see Chapter 1). The text of the proposed insurance scheme is set out in the appendix to this chapter.

The AOSIS Insurance Annex explained that under the insurance scheme, no contributions would become payable by the industrialized countries for at least ten years – the period within which many scientists believe that it may be possible to establish with more certainty the extent to which global warming and sea level rise will increase if greenhouse gas emissions remain unchanged. Further, no claims on the Insurance Pool would arise unless the rate and absolute increase of global mean sea level rise reach certain levels. These figures might be set to reflect a rate of increase beyond which the vulnerable ecosystems of the insured countries could not easily adapt and an absolute rise beyond which significant damage to small islands and low-lying coastal areas would occur.

The Insurance Annex adopted a funding method for the International Insurance Pool that is similar to that of the Nuclear Damage Convention, but recast in the context of global warming and sea level rise by reference to GNP and total CO_2 emission levels of the industrialized developed countries. The scheme would therefore offer incentives to the industrialized developed countries to limit their CO_2 emissions so as to mitigate the rate and extent of global warming and consequent sea level rise.

The scheme would also offer the possibility of building up a long term fund. After a minimum of ten years, a single contribution of a per centage of GNP (adjusted to reflect respective CO_2 emissions) would be made by the industrialized countries. Depending upon the rate and magnitude of global mean sea level rise, no claim on the Pool might arise for several decades, even if 'business as usual' continues. A fund of US$1 billion established in 2003 and invested at a real interest rate of 5 per cent would produce a fund of $4.3 billion in thirty years, and $7 billion in forty years.

What losses would the Insurance Pool cover? The Working Group set up by the Intergovernmental Panel on Climate Change to consider response strategies (Working Group III), and in particular its Coastal Zone Management (CZM) Subgroup, categorized the responses to sea level rise in three groups: retreat; accommodation and protection. To quote the CZM report:

> *Retreat* involves no effort to protect the land from the sea. The coastal zone is

abandoned and ecosystems shift landward. This choice can be motivated by excessive economic or environmental impacts of protection. In the extreme case, the entire area may be abandoned.
Accommodation implies that people continue to use the land at risk but do not attempt to prevent the land from being flooded. This option includes erecting emergency flood shelters, elevating buildings on piles, converting agriculture to fish farming, or growing flood- or salt-tolerant crops.
Protection involves hard structures such as sea walls and dikes, as well as soft solutions such as dunes and vegetation, to protect the land from the sea so that existing land uses can continue.

In general, losses resulting from the retreat option would seem to fall naturally within the scope of the Insurance Pool, whilst expenditure incurred in implementing protection responses would fall under other mechanisms to provide resources to developing countries such as the proposed International Climate Fund. More difficult is the question of whether, and to what extent, losses or expenditure resulting from the adoption of implementation responses should be covered by the Insurance Pool, or whether these should be dealt with in another context. I suggest that much of this latter category of expenditure should be covered by the Pool. This approach accords with ordinary commercial insurance principles that expenditure incurred or loss suffered in responding to an immediate threat of the insured peril should be treated as a loss by that peril. Measures taken in advance of any immediate threat, even if taken on reasonable grounds, would not be covered by the Pool, but would be dealt with under other funding mechanisms.

In principle, the main criterion for entitlement to claim should be proved loss attributable to sea level rise. Apart from inundation and flooding, the most widely predicted consequences of climate change so far as small islands and low-lying coastal countries are concerned are increased incidence and intensity of hurricanes, typhoons, severe storm surges and coral bleaching. It may not be possible yet to prove conclusively that such events are attributable to climate change due to global warming. I suggest therefore that such claims against the Pool should include, at minimum, loss or damage resulting from sea level rise, together with expenditure incurred in connection with immediate accommodation responses, and that the criteria for entitlement to claim against the Pool should be set by reference to the rate of global mean sea level rise over a given period as well as to the total rise from a given date.

For many developing countries, traditional insurance definitions of physical and economic loss would not reflect the impact of sea level rise. The total inundation of a small island with a minimal economic life of its own would result in the inhabitants' loss of their homeland (quite apart from the costs of resettlement). It might also involve loss of development potential; and even loss of the potential economic benefit of its surrounding 200 mile Exclusive Economic Zone (which is defined with reference to terrestrial landmarks). Losses suffered by small island and low-lying coastal countries should be assessed therefore by reference to 'total economic value' – that is, values arrived at not only by reference to actual use value, but also to option and existence values.

The design of the Pool took into consideration the necessity for setting an overall limit to its liability and to structure an equitable system of distribution such that its resources would not be completely absorbed by a single massive loss suffered by one large country. Moreover, the question was also considered as to whether losses suffered which might otherwise be claimable against the Pool could have been mitigated by measures which reasonably could have been taken by the claimant at an earlier stage.

Loss or damage to commercially insured property would not be recoverable under the scheme. I do not expect that this exclusion would discourage insured countries from arranging commercial insurance where available at reasonable rates for two reasons:

1 claims would not be recoverable in full where the totality of claims against the Pool in any one insurance period exceeded the funds in the Pool;
2 in assessing claims, the authority administering the Pool would take into account whether steps could have been taken earlier to avoid or mitigate the loss, including taking out commercial insurance if it could have been obtained at reasonable premiums.

I consider that these factors would encourage also the insured countries to take preventative or mitigating measures where possible, even where no commercial insurance was available.

The Climate Change Convention

As I noted above, the Insurance Annex proposed by AOSIS remained in the draft Convention text until the last negotiating meeting before it was signed in Rio. At that time, all annexes dealing with matters of substance were removed on the grounds that insufficient time remained to reach agreement on matters of detail before Rio; and the view of some countries (although not the AOSIS countries) that the Convention should not in any event be concerned with detailed mechanisms such as the Insurance Pool.

Nonetheless, the Climate Change Convention signed at Rio did refer to insurance. When it comes into force, the parties are bound by Article 4.8 to give 'full consideration to what actions are necessary' in regard to insurance to meet the specific needs and concerns of the most vulnerable developing countries. Small island and low-lying coastal states head the list of vulnerable countries. Also, under Article 4.4, the developed country parties are obliged to assist the vulnerable developing country parties to meet the costs of implementing adaptation to those adverse effects. It is precisely to these adaptation costs that the AOSIS insurance proposal is directed.

The AOSIS countries therefore intend to develop the outline insurance proposal into a fully researched, viable insurance scheme and to put it before the parties to the Convention when protocols related to financial issues are negotiated.

Notes and references

1 The IPCC First Assessment Report, August 1990; see also R Warrick and A Rahman, 'Future Sea Level Rise, Environmental and Socio-Political Considerations,' in I Mintzer (ed), *Confronting Climate Change, Risks, Implications, and Responses*, Cambridge University Press, New York, 1992, pp 97–112

2 Report of IPCC Working Group II, Policymakers' Summary, June 1990

3 IPCC Working Group I, *Scientific Assessment, Climate Change 1992, The Supplementary Report to the IPCC Scientific Assessment*, Cambridge University Press, 1992, pp 5–19

4 IPCC First Assessment Report, Overview, August 1990, p 8

5 'Strategies for Adaptation to Sea Level Rise', Report of Coastal Zone Management Subgroup of IPCC Working Group II, June 1990, p 5

6 *Munchener Ruckversicherungs-Gesellschaft*, 'Windstorm', Munich, 1990, p 111

7 *Ibid.* p 4

8 *Prudential Insurance* v. *IRC* [1904] 2 KB 658 at p 613; see also *Soya GmbH Kommanditgesellschaft* v. *White* [1980] 1 Lloyd's Rep. 491, [1982] 1 Lloyd's Rep. 136, [1983] 1 Lloyd's Rep. 122

9 *Munchener Ruckversicherungs-Gesellschaft*, 'Flood Inundation', Munich, 1973, p 7

10 Report of IPCC Working Group I, Policymakers' Summary, p 23

11 J Hindle 'Warming Signals', *The Review*, March 1989, p 35

12 *Munchener Ruckversicherungs-Gesellschaft*, 'Windstorm', Munich, 1990 p 4

13 J Hindle, *op. cit.* (note 11)

14 G Dimmock, 'The Reinsurers' View', in 'The Greenhouse Effect – Implications for Insurers', Insurance and Reinsurance Research Group Limited, February 1989, p 119 and J Hindle, *op. cit.* citing C T Suthons, 'Frequency of Occurrence of Abnormally High Sea Levels on the East and South Coasts of England, 1963'

15 *Munchener Ruckversicherungs-Gesellschaft*, 'Windstorm' and 'Flood Inundation', *op. cit.*; J Hindle and G Dimmock, *op. cit.*; F Nierhaus and G Berz 'Natural Catastrophes – Elements of Risk', *The Review*, December 1989, p 17; A F Dlugolecki, 'Natural Catastrophes arising from the Greenhouse Effect', *Journal of the Insurance Institute of London*, 1990, vol. 78, p 49; S Hitchcock, 'Changing Weather Patterns and their Effect on Reinsurers', Paper to Insurance Institute of London, March 1989

16 *Ibid.*

17 *Munchener Ruckversicherungs-Gesellschaft*, 'Windstorm', *op. cit.* (note 6) p 11

18 Following disastrous floods in Queensland, New South Wales and Victoria in 1990, the President of the Insurance Council of Australia was reported as stating that few householders in flood-prone areas could obtain insurance cover for flood damage and that the Australian Government should 're-think the

question of catastrophe reinsurance'. The National Insurance Brokers' Association of Australia was also reported as having called upon the Federal Government to provide catastrophe reinsurance above whatever level is freely available in the private insurance market: *The Re Report*, London Issue 90–19, 17 September 1990

19 Mans Jacobsson, Director, International Oil Pollution Compensation Fund, 'The International Conventions on Liability and Compensation for Oil Pollution Damage' in *Oil Pollution Claims and Liability*, Legal Studies and Services Limited, London, November 1989; for a summary of the IOPC, see M Jacobson, 'The International Convention on Compensation for Oil Pollution Damage,' *Siren*, (UN Environment Programme, Nairobi), no 42, September 1989, pp 20–24

20 *Ibid.* p 8

21 Article 1.6 of the 1984 Protocol to the Civil Liability Convention

22 P Sands, *Chernobyl: Law and Communication*, Grotius Publications, Cambridge 1988, p 51

23 *Ibid.* where the text of the Convention is set out at p 53

24 *Ibid.* where the text of the Convention is set out at p 68

25 *Ibid.* p 96; the text of the Convention is set out at p 97

26 Ministerial Declaration at the Second World Climate Conference, Geneva, November 1990, paragraphs 26 and 27

Appendix: Scheme proposed by AOSIS for inclusion in the Climate Change Convention

I The Parties recognise that:

1 There should be established, as an integral part of the Framework Convention on Climate Change, an International Climate Fund to finance measures to counter the adverse consequences of climate change, and a *separate* International Insurance Pool to provide financial insurance against the consequences of sea level rise.

2 Revenue for the Insurance Pool should be drawn from mandatory sources, in particular, developed country assessments.

3 The financial resources of the Insurance Pool should be new, additional and adequate.

4 The Insurance Pool should be under the control and direction of the Conference of Parties.

5 The resources of the Insurance Pool should be used to compensate the most vulnerable small island and low-lying coastal developing countries for loss and damage arising from sea level rise.

II. The Parties further recognise that the formulation of an Insurance Pool scheme involves consideration of the following main questions:

- methods of funding an international insurance pool;
- classification of the types of loss to be covered by the Insurance Pool;
- criteria for establishing entitlement to claim against the Pool;

- methods of evaluating loss resulting from sea level rise;
- limitations on the amount of compensation payable by the Pool.

III The parties accordingly agree as follows:

1 The financial burden of loss and damage suffered by the most vulnerable small island and low-lying developing countries (Group 1 countries) as a result of sea level rise shall be distributed in an equitable manner amongst the industrialised developed countries (Group 2 countries) by means of an Insurance Pool.

2 The Insurance Pool shall be funded by contributions levied on Group 2 countries.

3 The administrating authority ('the Authority') for the scheme shall be a body controlled on an equitable basis by the Group 1 and Group 2 countries within the framework of the Conference of Parties.

4 The contributions referred to in paragraph 2 shall be calculated according to a formula modelled on the 1963 Brussels Supplementary Convention on Third Party Liability in the field of Nuclear Energy, as follows:

(a) As to 50%, on the basis of the ratio between the gross national product at current prices of each Group 2 country and the total of the gross national product of all Group 2 countries in the year prior to the year in which the contribution was levied ('the contribution year');

(b) As to 50%, on the basis of the ratio between the total emissions of CO_2 of each Group 2 country and the total CO_2 emissions of all Group 2 countries in the year prior to the contribution year.

5 Ten years from the date on which the Convention enters into force the Group 2 countries shall contribute to the Insurance Pool an agreed per centage of the total gross national product of all Group 2 countries in the year prior to the contribution year, apportioned as in paragraph 4, provided that over the 10-year period the rate of global mean sea level rise will have reached an agreed figure. If the rate of global mean sea level rise has not reached the agreed figure by the end of the 10-year period, a review shall thereafter be carried out at five-yearly intervals and the obligation of the Group 2 countries to contribute to the Insurance Pool will not arise until the year following the review in which it is established to the satisfaction of the Authority that the rate of global mean sea level rise has reached the agreed figure or that absolute global mean sea level rise has reached an agreed figure.

6 The insurance fund so constituted shall be invested by the Authority in interest-bearing securities as determined by the Conference of Parties.

7 No right to claim against the Pool in respect of loss or damage in any area of a Group 1 country shall arise until:

(a) It shall have been established to the satisfaction of the Authority that the rate of global mean sea level rise and the absolute level of the global mean sea level rise has reached agreed figures;

(b) It shall have been established to the satisfaction of the Authority that the relative mean sea level rise for any insured area in a Group 1 country has reached an agreed level above base levels determined for each area insured (such relative mean level figures having been determined within 10 years of the Convention coming into force);

(c) One year shall have elapsed from the date upon which the figures referred to in sub-paragraph (a) shall have been established as having been reached (that date plus one year being 'the inception date').

8 In the first instance those areas of Group 1 countries which would be directly affected by sea level rise to a level of an agreed number of centimetres above the base levels referred to in paragraph 7(b) shall be valued for insurance. Marketed assets shall be valued on the basis of gross domestic product for the insured area in question. Non-marketed interests shall be valued on the basis of formulae to be agreed.

9 The insured values covered shall be negotiated between the Authority and the Government of each Group 1 country in accordance with valuation principles to be agreed. The same 'policy conditions' shall be applicable to all Group 1 countries.

10 All assets and interests intended to be insured under the scheme shall be listed by Group 1 countries for registration with the Authority. Records of assets and interests registered for insurance shall be kept up to date. Valuations of assets and interests registered for insurance shall be carried out in accordance with the agreed formulae and shall be assessed as soon as possible after the setting-up of the Authority and in any event within 10 years of the Convention coming into force. Revaluations shall be carried out periodically as appropriate.

11 The first period of insurance shall commence on the inception date as defined in paragraph 7(c) and shall cover an agreed period following the inception date. Loss or damage occurring within the first and each following period of insurance, if accepted as a valid claim by the Authority, shall be paid out of the Insurance Pool as accumulated at the closing date of the period of insurance.

12 If the funds in the Pool are insufficient to meet all valid claims, the claims shall be paid out on an equitable basis. If, after payment out of all valid claims in full, any surplus shall remain in the Pool, the surplus shall be carried over to the credit of the following insurance period.

13 Prior to the closing date of the first period of insurance and of each subsequent period, the Conference of the Parties shall, after consultation with the Authority:

(a) fix the length of the next period of insurance;

(b) estimate the probable extent of claims on the Pool during the next insurance period;

(c) determine the level of contributions to be levied on Group 2 countries sufficient to meet the estimated claims after taking account of any surplus carried forward from the preceding period.

14 Claims against the Insurance Pool in respect of insured assets and interests shall be dealt with by the Authority. The Authority shall investigate the cause of any claimed loss, prepare estimates, determine whether the claim comes within the terms of the insurance, evaluate the extent of loss and assess the amount of the claim recoverable by reference to the insured value of the asset or interest and any applicable limits.

15 All assets in insured areas of Group 1 countries, whether commercially insured or not, shall in the first instance be valued for insurance, but no claims shall be accepted by the Pool in respect of property which at the time of loss or damage occurs is insured commercially, whether by a private insurance company or otherwise.

16 In assessing claims against the Pool, the Authority shall determine whether the

loss or damage claimed could have been avoided or mitigated by measures which might reasonably have been taken at an earlier stage. In determining whether measures could or could not reasonably have been taken at an earlier stage, account shall be taken, amongst other things, of the availability of funds, both domestic and international, which would have enabled mitigating or preventative measures to have been taken, and the availability of commercial insurance on reasonable terms.

17 If differences of opinion arise between the Authority and the participating countries, every effort shall be made to negotiate a resolution, but if this is not achievable disputes shall be submitted to an [the] arbitration tribunal under a special arbitration scheme [the Convention].

Part III

National greenhouse gas reduction cost curves

8

Integrating ecology and economy in India

Jayant Sathaye[a] and Amulya Reddy

Introduction

Recent years have witnessed a growing concern regarding the accumulation of greenhouse gases (GHGs) in the earth's atmosphere. This concern has led to many in-depth studies of the phenomenon. Individual country studies have ranged from simple (albeit data-intensive), inventories of GHGs to evaluation of policy options to stabilize or reduce emissions in some future year. Impact studies have focused on better understanding the effects of gases on atmospheric temperature, monsoon patterns and sea level rise.

The Intergovernmental Panel on Climate Change estimates that the global temperature would increase up to 3.5°C by 2100 under the most likely scenarios.[1] The results of these models and other studies prompted calls for an international treaty which nations could adopt to restrict the growth of emissions. Such a treaty was put forward at the 1992 UNCED meeting in Rio de Janeiro. The text of the treaty was debated for many months prior to the convention. One of the critical divisive issues was the sharing of burden among the various parties. This issue lies at the heart of the debate among nations on climate change. Burden sharing is difficult to resolve since the emissions burden that each nation shoulders is different for each gas. And, it depends on the historical cumulative emissions that a country may have emitted through its use of various fuels. The burden varies by the type and extent of impact that a nation may have to bear as well.

The cumulative share of carbon dioxide emissions from the developing countries between 1870 and 1986 is estimated to be only 15 per cent but, with 76 per cent of the world's population, their share of energy-related carbon dioxide emissions in 1986 was about 27 per cent. This share is

a Jayant Sathaye's background work for this paper was supported by the US EPA, Office of Policy, Planning and Evaluation, Climate Change Division through the US Department of Energy under Contract No. DE-ACO3-76SF00098.

increasing since their modern[b] energy growth is faster than for other countries. IPCC scenarios of future emissions indicate that worldwide emissions of carbon and other gases would continue to increase even if the developed countries were to reduce emissions from their current levels.[2] These shares suggest that the developed countries should shoulder a greater responsibility for historical emissions but the opportunities for reducing incremental emissions may be more abundant in the developing countries.

Unlike other airborne pollutants, such as SO_x and NO_x, which are emitted in trace amounts, are very reactive and can be scavenged at the source, GHGs are emitted in far more diluted and much larger amounts, which poses a problem in attempting to physically remove the gases. The main alternatives consist of eliminating or reducing the emissions of these gases, or growing biomass to sequester carbon dioxide.[3] Much of the current deforestation occurs in developing countries, and reversing or slowing this process would aid in reducing future emissions.

Given the above background, developed countries have insisted that reducing GHG emissions from the developing countries is vital. If necessary, such emissions reduction could be accomplished with assistance from the Global Environment Facility (GEF).[c] On the other hand, the developing countries have argued that the growth of emissions is an unavoidable consequence of economic growth, as was the case with the industrialized countries during their past. Hence, global environmental protection should not be allowed to penalize development. Is this difference reconcilable?

For a signatory to the Climate Convention, the adoption of policies and strategies to restrain emissions growth is an important goal to pursue. A nation would find it easier to follow such options to the extent that these are adoptable without hindering its current or anticipated development trends. Many studies have argued that alteration of growth patterns will lower social welfare and add to the cost of future socio-economic development. We will cite several studies for India to illustrate the opposite view that the adoption of such policies need not reduce social welfare. Indeed, accelerated adoption of certain energy and forestry policies, some of which are already being promoted and implemented, will lead to reduced carbon emissions and/or increased carbon sequestration at no additional cost to the nation. The pursuit of such policies will shift the business-as-usual growth to basic-needs-oriented development.

Adoption of such policies may be slowed or thwarted by many barriers in developing economies. In the case of India, scarcity of capital and of hard

---

b Modern energy denotes all energy derived from non-biomass sources. Carbon refers to carbon dioxide throughout this chapter.

c GEF is a pilot programme providing grants and low-interest loans to developing countries to help them carry out programmes to relieve pressure on global ecosystems. The Facility is a cooperative venture among national governments, the World Bank, the United Nations Development Programme and the United Nations Environment Programme.

currency are twin dilemmas which often limit adoption of the most efficient policies. Lack of institutions to facilitate the adoption of high energy-efficiency technologies is another barrier. Reddy (1991) lists the many barriers to improving energy efficiency.[4]

If adoption of such policies were to increase social welfare and achieve reduction of GHG emissions at no extra cost, then is a nation justified in seeking international support for implementation of these options? As we illustrate for India, even if the life cycle cost of abatement projects is less, the up-front costs of these projects may make them prohibitively expensive to pursue. As provided for in the Convention, India could justifiably seek support for projects meriting such assistance through the Global Environment Facility.

This chapter addresses three main issues dealing with these topics: emissions inventory and the uncertainty of estimates; energy efficiency, fuel substitution and the economics of GHG abatement; and emissions and sequestration from biomass growth.

Emissions inventory

The Climate Change Convention calls for each Party to prepare an emissions inventory. GHGs included in the inventory are carbon dioxide, methane, carbon monoxide, nitrous oxide and CFCs. Carbon dioxide and methane are emitted in larger amounts than the other GHGs, and CFCs are regulated under the Montreal Protocol. Several other different inventories have been prepared for India by experts both national and foreign. Emission estimates of CO_2 from energy sources are in relatively close agreement. Estimates of CO_2 from non-energy sources, and of methane from all sources, vary widely and have been debated in many fora.

In Table 8.1, we show the estimates of annual greenhouse gas emissions from anthropogenic activities in India for 1986. Carbon, emitted as carbon dioxide, emissions total 164 teragrams.[d] Of these, emissions from forestry and land use changes amount to 20 teragrams. A more recent estimate places carbon emissions (as CO_2) from fossil fuel use at 133 teragrams in 1988 compared to 139 in 1986, as shown in Table 8.1.[5]

Methane emissions are estimated at 55 teragrams in Table 8.1. Boden et al.[6] estimate emissions of this gas from livestock and rice paddies to be 10 teragrams, compared to 48.2 in Table 8.1. The lower figure is the result of new emissions measurements which reflect smaller rice biomass from Indian paddy fields and the fact that areas emitting high methane flux are a fraction of total paddy area; and above-surface biomass weight is reportedly smaller than elsewhere.

d Teragram = 10e12 grams or one million tonnes.

Table 8.1 *Estimates of annual GHG emissions from anthropogenic activities in India, 1986 (teragrams of gas)*

	CO_2	CO	CH_4	N_2O	CFCs
Energy					
1 Coal production	-	-	1.7	-	-
2 Coal combustion	378 (103)	6.6	0.04	0.03	-
3 Oil combustion	114 (31)	3.9	0.01	0.01	-
4 Gas combustion, flaring	18 (4.9)	0.01	0.002	0.002	-
5 Gas venting, leakages	-	-	-	-	-
Industry					
1 Cement manufacture	18 (4.9)	-	-	-	-
2 CFCs (CFC-11 Equiv.)	-	-	-	-	0.01
3 Landfills	-	-	1.7	-	-
Agriculture and forestry					
1 Animal husbandry	-	-	10.4	-	-
2 Rice cultivation	-	-	37.8	-	-
3 Fertilizer use	-	-	-	0.04	-
4 Biomass combustion	-	55.6	3.5	0.09	-
5 Deforestation, land use changes	73 (20)	-	-	0.03	-
Total	601 (164)	66	55	0.2	0.01

Amount of carbon in the CO_2 shown in brackets

Source: Ahuja, D (1990). *Climate Change Technical Series: Estimating Regional Anthropogenic Emissions of Greenhouse Gases*, US EPA Report No. 20P-2006.

The emissions inventory is for a single year and does not provide guidance on trends or future growth of these emissions. Oak Ridge National Laboratory has tracked historical emissions for several countries, including for India (Figure 8.1). Carbon dioxide emissions from India have increased at 5.7 per cent annually since 1950 as India climbed from thirteenth to fifth place in the world as a national contributor. With increased shares of oil and gas, the share of CO_2 emissions from coal has declined from 87 per cent in 1950 to 71 per cent in 1989.

We have selected future scenarios from two authoritative reports for carbon dioxide emissions from energy and forestry sources. In Table 8.2, we show emissions estimates for commercial energy sources for 1985 and 2025, and for biomass sources for 1986 and 2011. Emissions from modern energy sources increase at a rapid pace in these scenarios but those from biomass increase much more slowly in either scenario.

Energy efficiency and fuel substitution

The production and use of modern energy in India generated 115 million

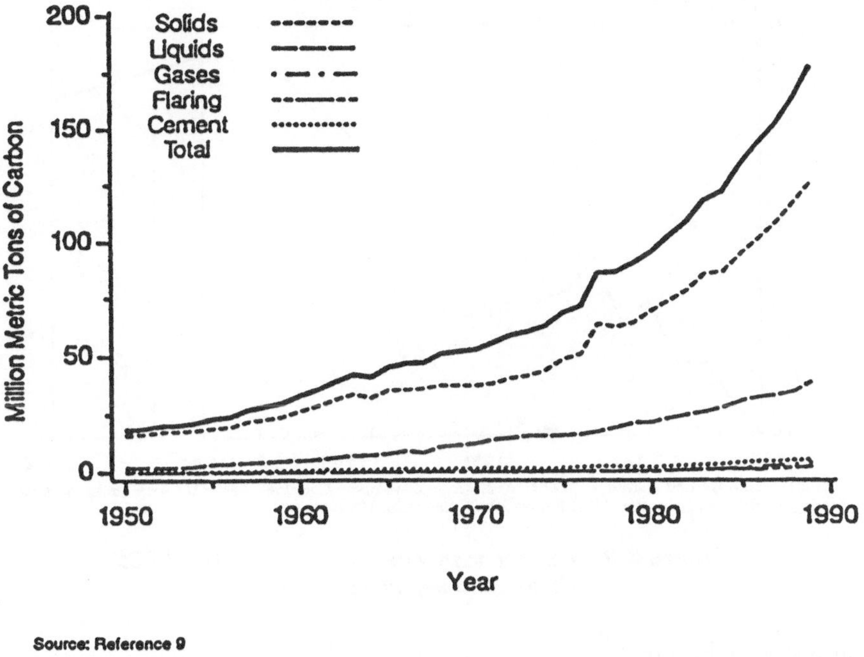

Figure 8.1 *CO_2 emissions from India, 1950–1989*

Table 8.2 *Scenarios for India of future carbon dioxide emissions (million tons of carbon)*

			Annual growth rate (%)
Modern energy sources[a]	*1985*	*2025*	
High	115	703	4.5
Low	115	615	4.2
Biomass sources[b]	*1986*	*2011*	
High	64	99	1.7
Low	64	70	0.4

Sources: a Pachauri, R K, Suri, V and Gupta, S (1991). *CO_2 Emissions from Developing Countries: Better Understanding the Role of Energy in the Long Term. Volume 3: China, India, Indonesia and South Korea*. July. LBL Report 30060.

b Ravindranath, N H, Somashekhar, B S and Gadgil, M (1992). *Forests: Case Studies from Seven Developing Countries, Volume 3: India and China*. August. LBL Report 32759.

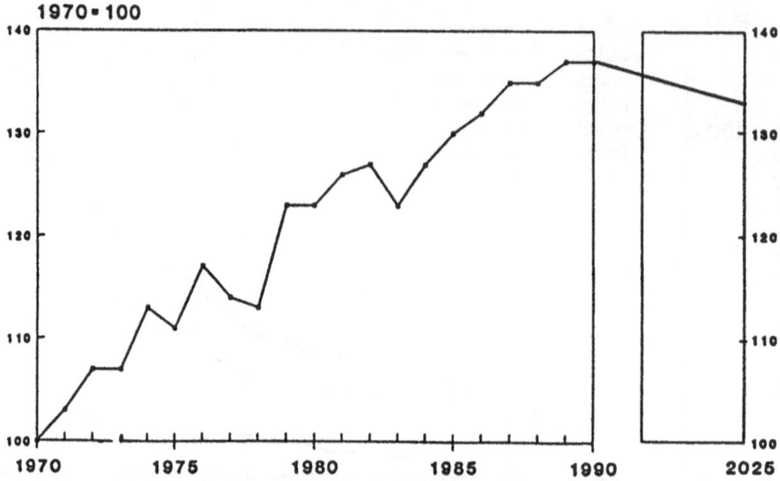

Source: J. Sathaye and A. Ketoff, *CO2 Emissions from Developing Countries: Better Understanding the Role of Energy in the Long Term, Vol. 1* (LBL Report 29607 Rev., Feb. 1991)

Figure 8.2 *Primary energy per GDP, India, 1970-2025 (HE scenario excluding biomass)*

tonnes of carbon in 1985, or 10 per cent of all carbon emissions emanating from the developing world. From a global perspective, India will account for 21 per cent of the *increase* in carbon emissions produced from energy use in developing countries between 1985 and 2025.[7]

Between 1970 and 1990, the intensity of India's primary modern energy use increased by 38 per cent (Figure 8.2). In various sectors of the Indian economy, the intensity of energy use per unit of value added exceeds that found in most industrialized and many developing nations. The increasing intensity in India reflects a replacement of human and animal draught power by mechanical and electrical devices. In particular, agricultural electricity use is increasing much faster (at almost 20 per cent annually) than the value added derived from this sector.[e] The high intensities also reflect the underlying inefficiency of energy use in industry, transport and power generation. For example, energy intensities of cement production for wet, dry and the wet-dry processes were as much as 20 per cent and, in some cases, 50 per cent above international norms.[8] Similar figures have been documented for steel, aluminum and other major energy intensive industries.

---

e Agricultural electricity consumption is reported to be increasing rapidly. However, there is little or no metered data regarding this consumption since most irrigation pumpsets do not have meters. Also no empirical studies of nationwide agricultural consumption have been conducted. It is conceivable that some of the reported consumption includes theft and transmission and distribution (T&D) loss. Reducing theft would reduce a utility's reported T&D loss and improve its financial soundness to donors.

A recent study has shown that direct and indirect carbon emissions may be higher from construction activity than any other component of India's final demand.[9] Energy-intensive materials such as glass, cement, bricks, steel, aluminum, and asphalt constitute the bulk of the components of a building or any other type of structure. Thus, emissions associated with construction are large. Reducing construction activity to decrease emissions is not a viable solution since a growing infrastructure is necessary to maintain the pace of economic development in India. Using materials more productively through improved designs of buildings and other infrastructure would be the better approach.

India's opportunities for curtailing emissions of carbon include rectifying its currently uneconomic allocation of fuels and inefficient energy-use patterns. Abundant but carbon-intensive coal resources satisfy almost 50 per cent of India's modern energy demand. In recent years, the costs of coal production and transportation have risen primarily because of changes in technology and low productivity. Arguably, more efficient fuel options (primarily oil, natural gas and renewables) could serve as economically viable substitutes for coal in the future.[10,11]

Determining which alternatives provide the most cost-effective means for India to restrain the growth of CO_2 entails a thorough economic evaluation of the available options. In the industrialized countries, economic evaluations of reducing carbon emissions have focused on taxation policies. However, in most developing countries where fiscal and technological resources are scarce, any effective emissions abatement strategy must go beyond evaluating the impact of changes in domestic taxes on levels of carbon emissions. Carbon conservation efforts must identify the types of energy-supply and energy-use technologies needed to restrain the growth of carbon. They must also assess the capital investment and foreign exchange requirements needed to acquire less carbon-intensive technologies and fuels.

Through the 1980s, imports of crude oil and petroleum products constituted the largest single commodity group in India's import bill. In Table 8.3, we show the share of oil imports in total imports in each year between 1980–81 and 1990–91. It also relates oil imports in this period to export earnings. Although we find no secular trend in Table 8.3, we can discern a clear pattern in which oil imports were linked to the growth in consumption of petroleum products and the trends in the domestic production of crude oil. The higher international oil price in 1980–81 boosted India's oil import bill. In response, India accelerated the exploitation of the offshore Bombay High Reserves. Thus, production of Indian crude tripled to 28.9 million tons in 1984–85.

The consumption rate accelerated in the second half of the 1980s. Consumption increased at an annual average rate of 5.5 per cent during the Sixth Plan (1980–85), but it increased at 6.8 per cent per year during the Seventh Five-Year Plan. As Bombay High was exploited fully in the mid-1980s and no new discoveries on the same scale as in the mid-1970s were

Table 8.3 *Petroleum imports (products and crude) as proportions of all imports and of export earnings*

Year	Oil imports (net) As % of all imports	As % of export earnings
1980-81	42	78
1981-82	37	64
1982-83	30	48
1983-84	20	33
1984-85	21	31
1985-86	22	40
1987-88	15	22
1988-89	14	19
1989-90	16	20
1990-91	23	30
Average	21	31

Source: Reddy, C R, D'Sa, A and Reddy, A K N (1992). 'The Debt-Energy Nexus: A Case Study of India', *Economic and Political Weekly*, July 4.

made, oil imports rose inexorably. Crude oil imports, therefore, increased in 1985–86, and petroleum product imports followed suit in 1987–88. As we show later, the nexus between oil demand and import payments has an important bearing on India's capacity to reduce emissions.

Generation, transmission and distribution of electricity are the most capital intensive of all energy activities. In the growing Indian economy, demand for electricity has increased faster than that for other forms of energy. Driven by this growth, demand for capital to finance the supply of electricity has also increased commensurately. Much of the capital to finance the energy sector is derived from the government budget. In the past, the government devoted an ever-increasing share of its budget to the energy sector. During the Sixth (1980–85) Five-Year Plan, for example, the government allocated 27.2 per cent of its plan outlay to the energy sector. This share increased to 30.6 per cent during the Seventh Plan. Even this increased share was insufficient to provide adequate power to the growing Indian economy. Power shortages are now common throughout the economy and 'brown-outs' in many cities during peak periods are the norm. The government is pursuing private sector power generation to alleviate power shortages. But the government has not yet set priorities in selection of technologies. Moreover, the management and operation of the power sector itself must be improved prior to seeking additional financing.

A recent analysis of the growth of carbon emissions and the economics of abating emissions from modern energy use in India highlights the relationship between abatement strategy on the one hand and capital investment and foreign exchange or hard currency requirements on the other (Table 8.4).[12]

Table 8.4 *Economic implications of reducing carbon emissions*

	1985	*2005*
GDP (billion US$)	193	512
GDP/capita (US$)	256	428
SCENARIO 1		
Emissions (million tons)	*110 mt*	*390 mt*
Cost	$38 billion	$107 billion
Investment	$8 billion	$29 billion
Investment/GDP (%)	4.1 %	7.0 %
Foreign exchange	$3.6 billion	$22.4 billion
FE/GDP (%)	1.9 %	4.4 %
SCENARIO 2		
Emissions (million tons)	*110 mt*	*340 mt*
Cost	$38 billion	$94 billion
Investment	$8 billion	$26 billion
Investment/GDP (%)	4.1 %	5.1 %
Foreign exchange (FE)	$3.6 billion	$22 billion
FE/GDP (%)	1.9 %	4.2 %
SCENARIO 3		
Case 1		
Emissions (million tons)	*110 mt*	*280 mt*
Cost	$38 billion	$95 billion
Investment	$8 billion	$23 billion
Investment/GDP (%)	4.1 %	4.5 %
Foreign exchange (FE)	$3.6 billion	$25 billion
FE/GDP (%)	1.9 %	4.8 %
Case 2		
Emissions (million tons)	*100 mt*	*280 mt*
Cost	$38 billion	$105 billion
Investment	$8 billion	$32 billion
Investment/GDP (%)	4.1 %	6.2 %
Foreign exchange (FE)	$3.6 billion	$23 billion
FE/GDP (%)	1.9 %	4.5 %

Scen. 1: Efficiency frozen at 1985 levels
Scen. 2: Includes efficiency improvements
Scen. 3: Lowest carbon emissions
Case 1: Lowest carbon emissions through greater fuel switching and fuel efficiency
Case 2: Lowest carbon emissions through increased reliance on renewables, e.g., solar, wind, hydro and biomass

Source: Mongia, N, et al. (1991), endnote 12.

The analysis assumes annual average rates for GDP and population growth. A linear programming model couples these rates to energy demand growth by sector and end-use. The model minimizes the cost of providing energy services to the Indian economy. Energy service may be provided by new energy supply or higher efficiency of supply and/or use. The model computes the investment and foreign exchange requirement for meeting the estimated demand for energy. Table 8.4 shows the foreign exchange requirement for fuel imports only.

In Scenario 1, energy intensity is frozen at 1985 levels. Energy sector investment as a proportion of GDP increases to 7.0 per cent by 2005. Foreign exchange requirements increase from 1.9 per cent in 1985 to 4.4 per cent by 2005. In each case, the sharp increase will require that financial resources be transferred to the energy sector from other sectors which will also demand more capital and foreign exchange.

Reducing the intensity of energy use as illustrated in Scenario 2, restrains the growth of carbon emissions to 340 million tonnes. Since this reduction is achieved primarily through cost-effective efficiency improvement, the cost of abatement is negative.[f] The Indian economy benefits from restraining carbon emissions growth. The scenario captures the many opportunities available to use electricity more efficiently, which reduces investment requirement.[13,14] However, the opportunities for reducing petroleum products demand are limited, and more difficult to implement, and this is reflected in the 4.2 per cent foreign exchange to GDP ratio, which changes little from Scenario 1.

Switching to less carbon-intensive fuels (Scenario 3) can reduce emissions further than in Scenario 2. This result is achieved through either the import of natural gas as illustrated in Case 1 or through the use of renewables as illustrated in Case 2. Renewables include the use of wind, solar, hydro and biomass resources. Natural gas imports increase the ratio of foreign exchange to GDP to 4.8 per cent while reducing the investment needs. Increased use of renewables drives up the investment requirement sharply to 6.2 per cent of GDP. Restraining emissions beyond what might be achieved through efficient fuel allocation and use would increase either capital investment or hard currency requirements.

The unit cost of conserved carbon rises from Scenario 1 to Scenario 3 as more expensive approaches are used to curtail carbon emissions (Table 8.5). By conducting alternate runs of Scenario 2 and placing progressively tighter constraints on carbon emissions, the cost of conserving carbon at levels between those in Scenarios 2 and 3 were determined. For example, the cost of conserved carbon is US$0.02 per kilogram when emissions are reduced from 340 million to 300 million tonnes.

f Net cost of abatement = Cost of providing given service through abatement strategy – Cost of providing same service by conventional means.

Table 8.5 *Unit cost of conserved carbon, India*

Carbon emissions (million tonnes)	*Unit cost of conserved carbon (1985 US$/kg)*
390-340	-0.27
340-300	0.02
300-280	0.05

The Scenario 3 costs of restraining carbon emissions are lower than those for Scenario 1. This implies that India can reduce carbon emissions at a net benefit to the economy, and the energy sector would therefore not require any new resources to reduce emissions. Scenario 3 costs compared to Scenario 2 are between 1 and 11 billion dollars higher. Many energy efficiency improvements are embedded in Scenario 2. Thus, if India's energy sector were to become more efficient, this analysis implies that it would need additional resources to reduce emissions. Which of the two paths (Scenarios 1 or 2) India takes to provide energy services will determine whether resources need to be transferred to or away from the energy sector, and consequently whether the nation would be justified in seeking resources from the world.

Restraining emissions from energy use in India will require that fuel allocation and energy efficiency be improved to their maximum potential. The scenario analysis shown in Table 8.4 includes many opportunities for improving efficiency of electricity use and supply. Additional measures to improve oil use efficiency would have further reduced the FE/GDP ratios shown in Table 8.4. What approaches might have yielded a more efficient use of oil in India? Four important shifts in strategy could be implemented:

1. a shift in long-haul freight movement from road to rail in order to reflect the economic cost of transportation for each mode;
2. a shift in cooking fuel from kerosene to LPG to encourage the use of more efficient stoves;
3. electrification of non-electrified households which would reduce kerosene consumed for lighting;
4. replacement of diesel pumpsets with electric pumpsets.[15]

These four strategies would have reduced demand for oil in the scenarios between 1980 and 1990 by 14.4 per cent. More importantly these steps would have reduced by 5 million tons the demand for kerosene and diesel, the two critical middle-distillates which are imported. The result of four strategies would be a reduction of the 1985 FE/GDP ratio of 1.9 per cent in Table 8.4 to 1.3 per cent. If we assume that similar improvement may be achieved by 2005, then the ratio for Scenario 2 would fall to 2.9 per cent from 4.2 per cent. A lower share of GDP allocated to importing oil would make the fuel import payment more manageable.

Strategies 3 and 4 would increase demand for electricity which would add to investment requirements shown in Table 8.4 for 2005. As we pointed out above, there are many options to improve efficiency of electricity use that are easily implementable. The options for improving efficiency of oil use involve a diverse set of actors which make them difficult to implement. For some end uses, such as lighting, electricity use is more efficient than oil use. Shifting to electricity in such selected end-uses would improve a nation's energy efficiency. Further, since improving system efficiency for electricity may be easier than for oil, increased electricity demand from strategies 3 and 4 could be better controlled.

In light of the cost, capital investment and foreign exchange parameters, to what extent can India restrain carbon emissions from modern energy use? Stabilization of emissions or limiting their increase to 20 per cent over a 20-year period has been discussed for the industrialized countries. Analyses show that this goal could be achieved without a net loss of GDP in some of the countries. The Energy Modelling Forum-12 in its deliberations on scenarios assumed that restraining emissions growth to a 50 per cent increase over a 20-year period was plausible for the developing countries.[g]

In contrast, Scenario 3 in Table 8.4 shows that, at best, emissions for 2005 could be held to 155 per cent (280 compared to 110 million tonnes) above the 1985 level. Further reduction in emissions would reduce annual GDP growth from the 4.9 per cent assumed in Table 8.4. Indeed, we estimate that to achieve the 50 per cent limit suggested by EMF-12 would require that the current pattern of unsatisfied energy demand continue in the future and that India's annual GDP growth be limited to 3 per cent.[16] Most probably, these requirements would be unacceptable to India.

Emissions and sequestration from forest biomass

Forests, defined as woodlands with more than 10 per cent crown cover, occupy about 20 per cent of the Indian land area. This proportion may be compared with 57 per cent for heavily forested Indonesia and Malaysia. Deforestation has led to a major decline in forest area in most countries; India is no exception. About one-fifth of the forests standing today are

g The Energy Modelling Forum was established in 1976 at Stanford University to provide a structural framework within which energy experts, analysts, and policy makers could meet to improve their understanding of critical energy problems. The twelfth EMF study focused on global warming and consisted of experts on economic analysis of climate change from the US and OECD countries. Results of the study were widely disseminated to policy makers in these countries.

extremely degraded. Half a million hectares or 0.8 per cent of total forest area was deforested in 1986. The primary conversion activity in India is agriculture followed by pasture and harvesting. The primary conversion activities vary by region. In the state of Karnataka, conversion to agricultural land accounts for 33 per cent, but submergence and resettlement due to power and irrigation projects account for 42 per cent, and mining for 21 per cent of the lost forest area between 1956 and 1984.[17]

Table 8.6 shows the carbon emissions from India associated with deforestation. These are divided into three categories: inherited, prompt and delayed. Inherited emissions are caused by past forest activities and occur in the base year. Prompt emissions are those that are generated immediately as a result of a forest conversion activity. Delayed are cumulative emissions that take place over time as decomposition of biomass occurs. Releases of carbon dioxide due to changes in soil organic carbon, both from forest conversion activities and areas under harvesting and afforestation programmes, are also included here.

The committed emissions displayed in Table 8.6 have been much debated. An earlier World Resources Institute estimate placed these emissions at 140 million tons.[18] More recent estimates from other sources place the committed emissions closer to the figure in Table 8.6.[19]

In the section above on energy, we discussed the potential and economics of restraining emissions from that sector. Growing biomass offers an opportunity to sequester carbon which would reduce net emissions from India. What is the potential for such offsets and to what extent might it be achieved while simultaneously pursuing or accelerating currently planned socio-economic development?

The development of agro-forestry and tree plantations on previously unforested lands can provide India with options to reduce its net emissions. The biomass density and carbon accumulation of new forests can exceed those of the initial natural vegetation depending on the silvicultural practices. In addition, afforestation projects can provide gainful employment to rural peasants who formerly earned their living through activities resulting in deforestation.

India has a strong, rapidly growing afforestation programme. India's afforestation process was accelerated by the enactment of the Forest Conservation Act of 1980 which aimed to stop forest clearing and degradation through a strict, centralized control of land-use rights. The afforestation activities resulted in a total of 11.5 million hectares as of 1986. Another 5.6 million more hectares were afforested between 1986 and 1989, raising the total planted area to 17.1 million hectares.

In Table 8.6, we show the carbon uptake, the annual carbon balance and the net committed emissions from India. The committed carbon uptake is almost twice as large as India's committed emissions. Over the years, the country would sequester 56 million tonnes of carbon from past afforestation activities. This figure is sufficiently large to make a dent in the emissions

Table 8.6 *India's forestry related carbon emissions and uptake, 1986*

Carbon emissions (MtC)				*Carbon uptake (MtC)*				*Annual carbon balance (MtC/year)*[a]	*Net committed emissions (MtC)*
Inherited[a] (1)	*Prompt* (2)	*Delayed* (3)	*Committed* (4) = (2) + (3)	*Inherited* (5)	*Prompt* (6)	*Delayed* (7)	*Committed* (8) = (6) + (7)	(9) = (1) + (2) - (5) - (6)	(10) = (4) - (8)
26	38	26	64	68.8	n.a.	120.0	120.0	-5	-56

a Inherited emissions for India were calculated using historic average deforestation rates for the past ten years.

Source: Makundi, W, Sathaye, J and Masera, O 1992. *Carbon Emissions and Sequestration in Forests: Case Studies from Seven Developing Countries, Volume 1: Summary*, August, LBL Draft Report LBL-32119.

from the energy sector. The annual carbon balance in Table 8.6 shows a net uptake of 5 million tonnes from forestry activities. Combining this figure with emissions from energy shown in Table 8.4 would reduce emissions from modern energy and forestry to 105 million tonnes of net carbon emissions.

These figures have important implications for future carbon emissions from India. The carbon sequestration occurred and is occurring through programmes with the main goal of promoting sustainable rural socio-economic development. Carbon sequestration is an unintended consequence of these actions. Given the potential for reducing net carbon emissions from India, the various factors that have contributed to this sequestration are worth noting. They include:

1 establishment of the Forest Conservation Act of 1980;
2 preparation of an environmental impact statement required with the beginning of the Fifth Five-Year Plan of 1975–80;
3 reduction of subsidies to forest-based industries beginning in the late 1970s;
4 increased industry-farmer links which have encouraged production of tree crops;
5 decentralized political decision making – village and district level authorities have been established in West Bengal and Karnataka, for example, that are far more motivated to ensure the prudent use of local resources;
6 growth of strong environmental movements in different parts of the country;
7 biomass fuel conservation programmes in all the states.

Strengthening these programmes would enhance carbon sequestration and accelerate rural development. As Saxena points out in a recent article, implementation of programmes will benefit some groups at the expense of others within India.[20] Most of the programmes proposed in the article, such as welfare forestry on forest lands, social security plantation, farm forestry for the poor, etc. will benefit the rural poor and the rich at the expense of lower level officials. However, not pursuing such programmes will make everyone lose in the long run.

Favourable scenarios have projected a net carbon uptake from forests in 2011 to be 121 million tons. Interpolating between the base year, 1986, and 2011 gives a net uptake of 57 million tons in 2005.[21] Thus, forests could offset India's modern energy related carbon emissions shown in Scenario 2, Table 8.4, by 17 per cent in 2005. India's net emissions from these two sources would be reduced to 283 million tons in 2005. This outcome would be achieved at no additional cost to those shown in Scenario 2. Further, given the exponentially higher sequestration potential for 2011, forests could offset as much as 25 per cent of the energy emissions in 2011.

Biomass use

Carbon is stored away or released when the biomass from a tree is utilized. The type of use and its duration determine the net carbon emissions. A tree burnt for the purpose of land clearing will release most of its biomass immediately. A tree providing lumber used to make buildings will store carbon away for decades. What is the best use of trees in order to sequester carbon?

Hall, Mynick and Williams point out that while sequestering carbon in forests is a relatively low-cost strategy for offsetting carbon dioxide emissions from fossil fuel combustion, substantially greater benefits can be obtained by displacing fossil fuel with biomass grown sustainably and transformed into useful energy using modern energy conversion technologies.[h,22] Biomass substituted for coal can be as effective as carbon sequestration, per ton of biomass, in reducing carbon emissions. However, fuel substitution can be carried out indefinitely, while carbon sequestration can be effective only until the forest reaches maturity. Also, greater biomass resources can be committed to fossil fuel substitution at any given time than to carbon sequestration because biomass (such as sugarcane bagasse) can be obtained from sources other than new forests. Thus, biomass can play a much larger role in reducing greenhouse warming by displacing fossil fuel than by sequestering carbon. Moreover, biomass energy is potentially less costly than the displaced fuel energy in a wide range of circumstances, so that the net cost of displacing carbon dioxide emissions would often be negative.

Conclusions

India's carbon emissions are likely to grow in the future because of the increasing energy and food consumption needed to support a growing economy. However, strengthening the ongoing afforestation programmes, increasing energy efficiency, and prudent use of renewable options in selected applications have the potential to offset a significant portion of the GHG emissions.

Implementing the three types of options will not be easy. Energy and forest products consumption and supply patterns, and forest land use are shaped by a large number of actors at various levels. Improving resource allocation and use patterns will require action at the national and interna-

h Use of forest biomass for lumber in buildings has the potential to store carbon for a period up to 100 years. In effect, the strategy to store biomass as lumber would store as much carbon as would be displaced by using the biomass as a substitute for coal. Economic benefit derived from each strategy may be quite different and should be considered in selection of the strategy.

tional levels. Reddy has outlined many barriers to energy efficiency improvement. Similar barriers exist to increased afforestation and renewable energy use. Energy consumers are often uninformed, first-cost sensitive, indifferent and helpless to improve efficiency.[23] National institutions are supply-biased, with little incentive to innovate. The government is uninterested, is short of capital and skills, has inadequate training facilities and limited access to hardware and software. Energy efficiency agencies are relatively powerless compared to their supply counterparts or they are part of the supply agency and therefore have no incentive to reduce demand for their product.

Further, bilateral and multilateral aid agencies target the supply aspects of energy systems with inadequate attention to demand-side measures. Other issues, such as an anti-innovation attitude, the large-is-convenient funder and the project-mode sponsor contribute to the lack of attention to the three options.

Many of the barriers listed above arise because there is no incentive for the various actors to behave differently. Concern about climate change can provide this incentive. The establishment of the GEF and the growing attention being paid to environmental issues at the World Bank is a positive sign which will alter future lending practices of multi-lateral institutions.[24] Increased attention to environmental issues holds out the hope that these and other similar institutions will begin to address the concerns of the poor, and not just those of the elite, in the developing countries. For example, dislocation of rural populations caused by building the Sardar Sarovar dam, coal mines and afforestation schemes are being discussed and addressed. Concern about climate change can improve on this dimension by explicitly developing projects which provide sustainable solutions to meet the energy, food, water and other needs of the poor. These projects will halt deforestation and/or lead to increased greening of rural areas in India.

Our analysis suggests that if India pursues basic-needs oriented development with emphasis on end-use efficiency, decentralized renewables and afforestation programmes, then its carbon emissions growth will slow and its economy will improve more rapidly. Simultaneously, it is in the interest of the developed countries to fund India's incremental costs of switching to less carbon-intensive technologies. Such technologies represent the most cost-effective path to economic development. For perhaps the first time in history, the interests of the developing world are aligned with those of the industrialized countries creating an unprecedented paradigm for future human development. More importantly, many of the measures to implement the three options have the potential to improve the condition of the poor in the developing countries. Efficient energy use and selected renewable options have been successfully demonstrated as necessary means to provide better water supply, lighting and fertilizer, which has fostered rural development.[25] Afforestation in India, through natural regeneration programmes, directly aids rural villagers.

Concern about the shared global problem of climate change offers a unique opportunity to align the interests of the developed and developing countries, rich and poor. While competition and dissimilar goals have often frustrated and defeated cooperative ventures, climate change offers a common incentive for collective action. Pursuing the socio-economic development goals of the South is consistent with the environmental goals of the North, and provides joint benefits to economy and ecology that are in the shared interests of all.

References

1 IPCC (1992). *1992 IPCC Supplement.* February

2 *Ibid*

3 Hall, D, Mynick, H and Williams, R (1991). 'Cooling the Greenhouse with Biomass Energy'. *Nature,* September 5; Hall, D, Mynick, H and Williams, R (1991). 'Alternative Roles for Biomass in Coping with Greenhouse Warming'. *Science and Global Security,* Volume 2, pp 1–39

4 Reddy, A (1991). 'Barriers to Improvements in Energy Efficiency'. *Energy Policy,* December, pp 953–961

5 Mitra A P (eds.) *Global Change: Greenhouse Gas Emissions in India, A Preliminary Report.,* Scientific Report No. 1. Prepared under the auspices of Council of Scientific and Industrial Research, New Delhi, June 1991

6 Boden, T, Sepanski, R and Stoss, F (Eds). *Trends '91: A Compendium of Data on Global Change.* ORNL Pub. No. ORNL/CDIAC-46, pp 442–445

7 Sathaye, J and Ketoff, A (1991). *CO_2 Emissions from Developing Countries: Better Understanding the Role of Energy in the Long Term. Volume 1: Summary.* February, LBL-29507

8 Bureau of Costs and Prices (1986). *Energy Audit of the Cement Industry.* Ministry of Industry, Government of India, New Delhi

9 Parikh, J, Gokarn, S and Barua, A (1992). *Climate Change and India's Energy Policy Options.* Indira Gandhi Institute of Development Research, February. Report prepared for the Rockefeller Foundation

10 Government of India, Bureau of Costs and Prices (1988). *Towards a New Energy Policy,* Delhi

11 Reddy, A, Sumithra, G Balachandra, P and D'Sa, A (1991). 'A Development-Focused End-Use-Oriented Electricity Scenario for Karnataka'. *Economic and Political Weekly,* April 6 and April 13

12 Mongia, N, Bhatia, R, Sathaye, J and Mongia, P (1991). Cost of Reducing CO_2 Emissions from India. *Energy Policy,* pp 978–986, December

13 Nadel, S, Kothari, V and Gopinath, S (1991). *Opportunities for Improving End-Use Electricity Efficiency in India.* Report prepared by American Council for an Energy Efficient Economy for the World Bank and US Agency for International Development, November

14 Reddy, et al. (1991), *op. cit.* (Reference 11)

15 Reddy, A (1981). 'A Strategy for Resolving India's Oil Crisis'. *Current Science,* Vol. 50, No. 2, pp 50–53

16 Sathaye, J (1992). *Carbon Emissions from Brazil, India and China*. Presentation to the EMF-12 meeting in Washington DC, May
17 Reddy, A (1987). *On the Loss and Degradation of Karnataka's Forests*. Paper presented at the International Conference on Tropical Forestry, 1–2 July, Bellagio
18 World Resources Institute 1990. *World Resources 1990–91*. New York, Oxford University Press
19 Kaul, O L (1991). *Forest Biomass Burning in India*. TERI Information Service on Global Warming, New Delhi, 2(2): 2–17
20 Saxena, N C (1989). 'Forestry and Rural Development'. *South Asia Journal*, Vol. 3, Nos. 1 and 2, pp 70–89
21 Ravindranath, N H, Somashekhar, B S and Gadgil, M (1992). *Forests: Case Studies from Seven Developing Countries, Volume 3: India and China*, August. LBL Report 32759
22 Hall, et al. (1991a,b) *op. cit.* (Reference 3)
23 Reddy (1991) *op. cit.* (Reference 4)
24 Reddy, A K N (1992). *Has the World Bank Greened?* Paper prepared for the Second Edition of the Green Globe Yearbook, Fridtjof Nansen Institute, Norway, September
25 Rajabapaiah, P, Jayakumar, S and Reddy, A (1993). 'Biogas Electricity – The Pura Village Case Study', Chapter 18, pp 787–816 in Johanson, T, Kelly, H, Reddy, A and Williams, R, (eds.) *Renewable Energy: Sources for Fuels and Electricity*, Island Press, Washington DC

9

Carbon abatement potential in West Africa

Ogunlade R Davidson[a]

Introduction

Africa's contribution to global greenhouse gas emissions is relatively small. Its share will grow, however, as poverty is eradicated by social and economic development. Also, all countries must cooperate and participate in carbon conservation strategies if the world is to avoid the possible adverse effects of climate change. The extent and magnitude of these effects on natural ecosystems are uncertain. But ecologically sensitive zones in Africa are among the world's most vulnerable areas. The population of Africa is relatively vulnerable to damages wrought by climate change due to its high dependence on natural systems for daily survival. Hence, African countries should participate actively in identifying potential for greenhouse gas abatement.

The climate convention negotiated at Rio in 1992 presents new financial opportunities to African countries. Their access to these funds will be enhanced if they understand fully greenhouse phenomena and related issues. Most African countries rely heavily on advice and funding from external donors in formulating and implementing their development strategies. These external agencies increasingly regard environmental concerns as important in their dealings with African aid recipients. To be successful, therefore, African development strategies must be more environmentally sensitive as well as economically sound.

In 1990, carbon dioxide accounted for 63 per cent of global greenhouse gas emissions. About three quarters of this total was emitted by developed countries.[1] However, as developing countries move up the development ladder and demand high quality energy, so too their share of the global total will grow. Furthermore, present land use practices such as intensive logging

a The author thanks the Catherine T and John D MacArthur Foundation for providing funds to undertake this study and the Lawrence Berkeley Laboratory and Center of Energy and Environment Studies of Princeton University for their support.

contribute to the destruction of the tropical rain forest, thereby reducing national carbon sinks. The search for global abatement strategies should involve all countries. The final approach, however, must also recognize their different responsibilities and capabilities in accordance with their social and economic circumstances.

The importance of the energy sector in abating carbon emissions is self-evident. It is the biggest single source of global carbon emissions. Although Africa contributes only about 3 per cent of total global carbon emissions (and Africans emit less than a quarter of the world average on a per capita basis)[2] its energy usage and greenhouse gas emissions will grow substantially. Thus, I focus on carbon abatement strategies within the energy sector of Africa, using West Africa as a case study.

The major challenge facing the region's energy sector is to substantially increase energy services delivered to consumers. In 1990, the annual per capita energy consumption in the region (excluding biomass energy) was less than 4 GJ. It ranged from less than 1 GJ per capita in Burkina Faso to over 7 GJ per capita in Ivory Coast. This figure is extremely low compared with the rest of the world. Indeed, due to declining economic performance and ecological degradation, energy usage stagnated in the 1980s (see Figure 9.1). West Africa is not short of exploitable energy resources *per se*,[3] but developing these resources will require innovative approaches. One such approach is to pursue carbon abatement strategies that also increase the availability of competitive, high quality energy services so vital to Africa's development.

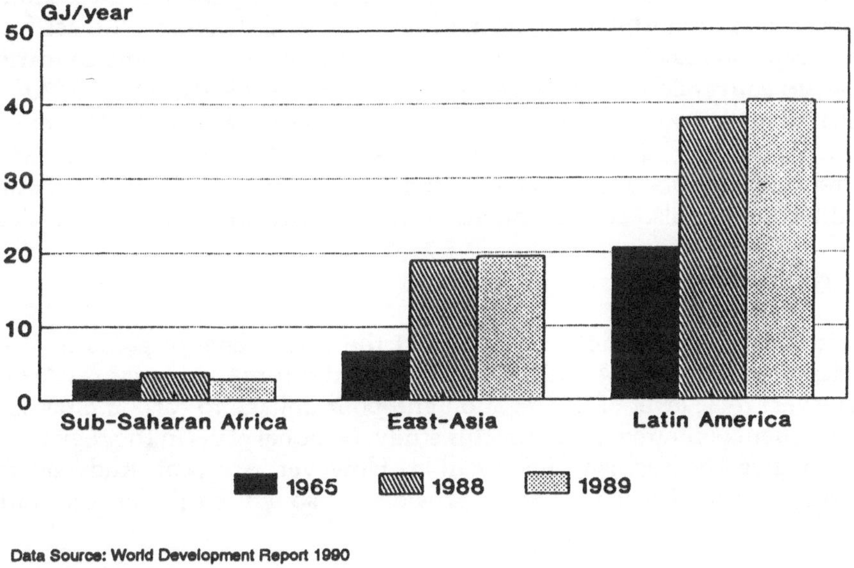

Figure 9.1 *Modern fuel use, per capita, comparison of developing regions*

In this chapter, therefore, I seek to identify carbon abatement strategies which also increase energy services for the West African region. While limited to West Africa, these results are relevant to the rest of the continent. I divide the chapter into four parts. In the first, I look at long term scenarios for energy consumption and carbon emissions. In the second, I discuss the technical options for reducing energy related carbon. In the third, I examine economic opportunities to implement some of these options. In the last section, I analyse assess policy issues which will facilitate the implementation of these technological and economic options.

Long-term energy and carbon emissions scenarios

West Africa forms about a fifth of Africa's land area, but has about a third of its population, a third of which in turn live in urban areas. The sixteen countries[b] of West Africa are socially and economically diverse, but their economies are dominated by a few sectors.

Energy production and use

Except for hydropower resources (which are found in most countries), only Nigeria has significant fossil energy resources. Energy consumption varies greatly among West African countries depending on their economic development. They remain highly dependent on biofuels, especially in the residential sector. Over 70 per cent of the total energy used in the region is from biomass,[4] mostly in the form of firewood and charcoal. Biomass fuels are cheap and easy to use and reduce the oil import bills for some countries. Biomass conversion technologies, however, are very inefficient (15–20 per cent).[5] The region also faces an acute and increasing scarcity of biomass fuels. About half of the non-desert areas in West Africa are projected to suffer from fuelwood scarcity by the end of the century.[6] Environmental and health concerns also pose problems with expanding biomass fuel supplies.[7]

Energy scenarios

Limited data make it difficult to project the future energy paths of West Africa. Studies by IIASA and IPCC[8,9] estimated future energy use for Africa. However, their level of aggregation and poor linkage to carbon emissions makes them of limited utility for this study. National plans in the region also do not give the requisite information. However, a recent study by the Lawrence Berkeley Laboratory provides a sound basis for an initial

b Benin, Burkina Faso, Cape Verde, Gambia, Ghana, Guinea, Guinea Bissau, Ivory Coast, Liberia, Mali, Mauritania, Niger, Nigeria, Senegal, Sierra Leone, and Togo.

investigation of long-term energy and carbon emissions scenarios for the region.[10] Among the 17 countries studied, there were three West African countries: Ghana, Nigeria and Sierra Leone. In total, they contain about 20 per cent of the land area of the region and 69 per cent of the population (due to Nigeria's large size). In addition, they consume almost 60 per cent of the region's total energy.

I created long-term scenarios by using an end-use methodology that estimates energy consumption as a function of expected changes in national economic and demographic structures and fuel intensities. These changes were assumed to be strongly affected by growth in GDP, population and fuel prices. Primary supply and delivered energy were estimated for each end-use sector. The scenarios then aggregated energy supplies by fuel type required to meet the sectoral end-use demands, relying on real GDP as the indicator of overall economic activity.

Two energy and associated carbon emissions scenarios were constructed for each country. The scenarios focused on the situation in 2025 rather than a quantitative path over time. The base year was set as 1985 (except for Ghana where 1987 was used). The first scenario assumed 'business as usual' and was called the High Emissions Scenario (HES). The second scenario introduced measures to increase efficiency and to restrain carbon emissions and is referred to as the Low Emissions Scenario (LES). Both scenarios incorporated the same basic economic and demographic assumptions (see Table 9.1).

Demographic and economic assumptions

The population of Africa grows at more than 3 per cent annually, the highest rate in the world. Population in West Africa grows at a rate only slightly lower than that of the whole continent. However, the scenarios[11] assumed a slower growth rate in recognition of regional family planning activities, improved education, and the recent adoption of a population policy in

Table 9.1 *Demographic and economic data, Ghana, Nigeria, Sierra Leone*

	Ghana		*Nigeria*		*Sierra Leone*	
	1987	*2025*	*1985*	*2025*	*1985*	*2025*
Population (millions)	13.4	42.7	96	317	4	9
Population AAGR (%)	-	2.9	-	3.0	-	2.3
GDP (US$ billion)	4.59	22.76	75.0	373.0	0.87	2.88
GDP/capita (US$)	343.0	533.0	788.0	1176.0	247.0	327.0
GDP AAGR (%)	-	4.0	-	4.0	-	3.0
GDP/capita AAGR (%)	-	1.2	-	1.0	-	0.7

AAGR: annual average growth rate

Nigeria,[12] the most populous country in Africa. The scenarios assume that populations in the region will increase between two-and-a-half and three times between 1985 and 2025. For comparison, the World Bank has used an even slower growth rate, projecting 34 million by the year 2025 for Ghana versus 43 million in this study; and 255 million for Nigeria in 2025 versus 317 in this study. This author and the World Bank use almost the same figure for Sierra Leone.[13]

In contrast to rising population, economic growth in the region has fallen in the last two decades. High oil imports and low export commodity prices combined with poor local response strategies were responsible for this poor economic performance. In addition, Ghana suffered from serious drought in the early 1980s, which crippled agricultural production. The drought also lowered the water level of the Volta dam that generates over 97 per cent of Ghana's electricity. In Nigeria, the oil price boom since the mid-1970s led to a spurt of unprecedented GDP growth. The subsequent decline in oil prices and weak internal policies left the economy floundering. In Sierra Leone, excessive public spending between 1979 and 1981 and reduced foreign exchange earnings since the mid 1970s have reduced its ability to import oil. The resultant energy shortages have been disastrous.

All these countries instituted economic reforms recently. Ghana's policies revived GDP growth which reached 6 per cent in the latter part of the 1980s. Nigeria's reforms improved its GDP growth rate to 4 per cent annually over the same period. Sierra Leone is emulating some of Ghana's more successful measures.

In light of these developments, the scenarios assumed GDP growth rates of 3–4 per cent up to the year 2025. For comparison, the International Panel on Climate Change assumed slightly more than 4 per cent for their high growth case; and less than 3 per cent for their lower growth case. In the scenarios, these growth rates result in a five-fold increase in the GDPs of Ghana and Nigeria by 2025, and a three-fold increase for Sierra Leone.

Sectoral analysis of carbon emissions

The scenario applies carbon emission coefficients to the resulting sectoral patterns of energy transformation and end use. The residential, transportation and industrial sectors combined account for over 80 per cent of the carbon emissions (from fossil fuel and biomass usage) in the base year and in 2025 in both the HES and LES in Ghana, and over 90 per cent in Nigeria and Sierra Leone (see Figure 9.2).

In the base year, the transport sector generates most carbon – 59 per cent in Ghana, 53 per cent in Nigeria and 65 per cent in Sierra Leone. By 2025, the anticipated increase in the use of liquefied petroleum gas (LPG) and kerosene in households makes the residential sectoral emissions surpass those of the transport sector in Ghana and Nigeria. In addition, Nigeria is building a

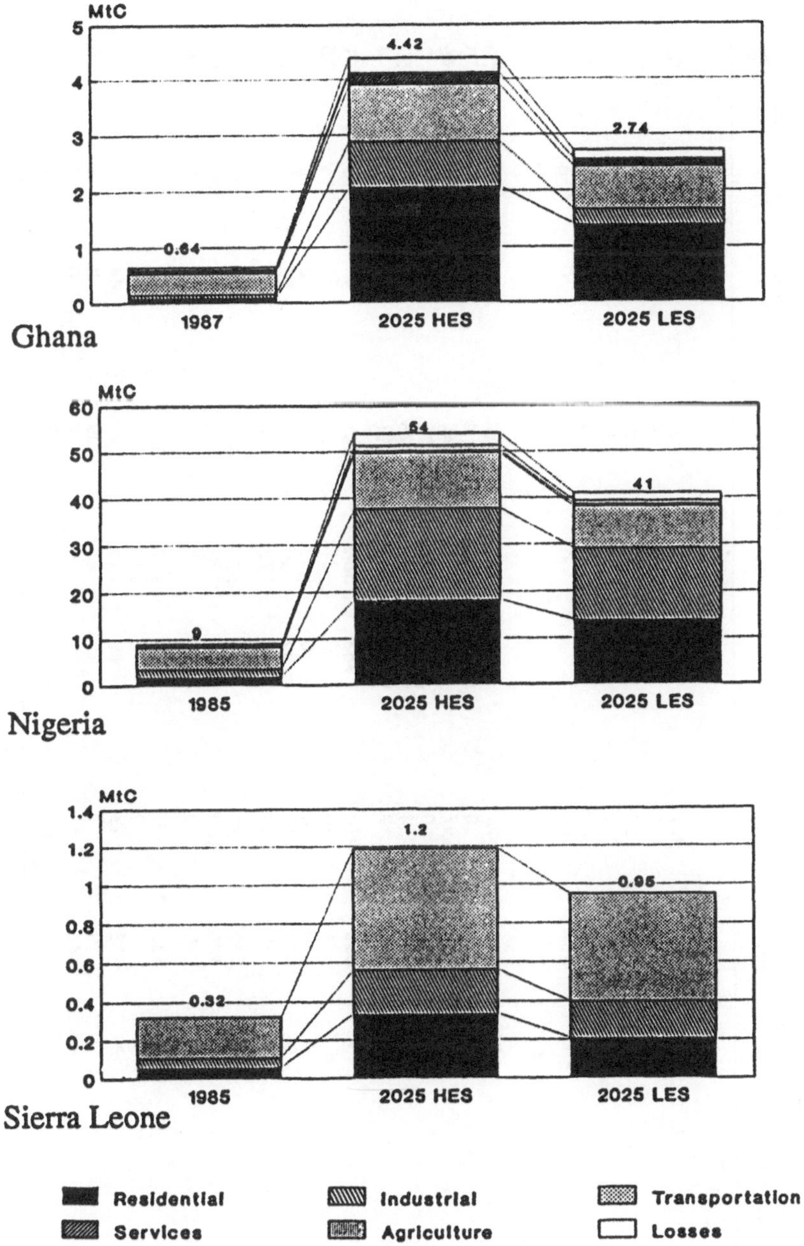

Figure 9.2 *Carbon emissions from energy use, Ghana, Nigeria and Sierra Leone*

Table 9.2 *Primary energy supply and delivered energy demand, Ghana, Nigeria, Sierra Leone*

	Ghana			*Nigeria*			*Sierra Leone*		
	1987	*2025 HES*	*2025 LES*	*1985*	*2025 HES*	*2025 LES*	*1985*	*2025 HES*	*2025 LES*
Primary Energy Supply									
Total	232	884	707	1482	4117	3205	56	130	108
Coal (%)	0	0	0	0	6	5	0	0	0
Oil (%)	14	22	20	27	45	39	30	48	46
Natural gas (%)	0	5	0	8	22	30	0	0	0
Biomass (%)	64	39	36	62	23	21	70	46	49
Hydro and renewables (%)	22	34	44	3	4	5	0	6	5
Delivered Energy Demand									
Total	192	616	474	1345	3292	2617	53	118	100
Coal (%)	0	0	0	1	4	4	0	0	0
Oil (%)	16	27	27	28	50	44	25	45	46
Natural gas (%)	0	0	0	2	7	15	0	0	0
Biomass (%)	78	56	54	69	29	26	74	51	50
Electricity (%)	7	17	19	2	10	11	1	4	4
Residential	130	358	270	983	1584	1231	41	75	61
Oil (%)	4	24	26	7	31	35	4	15	15
Natural gas (%)	0	0	0	0	0	1	0	0	0
Biomass (%)	95	63	61	92	56	50	95	80	81
Electricity (%)	1	13	13	1	12	14	1	5	5
Transport	19	52	38	259	691	484	10	32	29
Oil	100	100	100	100	100	100	100	100	97
Biomass (%)	0	0	0	0	0	0	0	0	3
Industrial	24	135	109	92	939	832	2	11	10
Coal (%)	0	0	0	2	12	12	0	0	0
Oil (%)	11	13	13	40	45	29	84	91	91
Natural gas (%)	0	0	0	25	23	39	0	0	0
Biomass (%)	47	45	41	22	6	7	0	0	0
Electricity (%)	41	42	46	11	13	13	15	9	9

HES: high emissions scenario
LES: low emissions scenario
All energy quantities shown are in exaJoules (J)

natural gas plant that will increase emissions from the domestic and industrial sectors in Nigeria by 2025.

Residential carbon conservation

As is evident in Table 9.2, the residential sector dominates total energy use in African countries.[14] In the scenario's base year, households in all the countries consume around 70 per cent or more of total delivered energy. This dominance stems from the widespread use of inefficient fuelwood and charcoal cooking devices. However, the scenarios assume that the intensity of biomass use (that is, GJ per household) declines by 2025 in all these countries. This outcome follows from two trends that can be discerned, namely, the introduction of improved woodfuel stoves and charcoal kilns, and fuel substitution.

With respect to the latter, the proportion of electrified households in the scenarios is estimated to rise from about 30 per cent today to more than 60 per cent by 2025 in all the countries. In Ghana, for example, the current programme aims at electrifying 90 per cent of households by 2025.[15] The scenarios assume that efficiency improves at point of end use because rising incomes result in the replacement of old, inefficient equipment with new, efficient household appliances and lights. However, most of the old stock will be used second-hand by lower-income households. This transfer will limit the reduction of household electricity demand from efficiency measures. Similarly, substituting oil-based fuels for biomass will improve the quality of life. However, this shift will not reduce carbon emissions.

Innovative consumer financing offers the possibility of introducing much more efficient electrical appliances and lighting systems. If these are incorporated, as in the LES, then household use of electricity could be reduced by about 40 per cent in Nigeria and Sierra Leone, and 30 per cent in Ghana (see Table 9.2). These improvements could be dramatically increased if the devices were designed and made locally and if fiscal measures were employed to reduce the high front-end costs to consumers.

Transport carbon conservation

The large share in carbon emissions of the transport sector is a result of its dependence on oil-based fuels. The rise in incomes postulated in the scenarios will increase the number of vehicles in all countries by 2025. The car fleet will grow even faster than all vehicles as the number of two-car families and imported and locally available second-hand cars will also increase.

Trucks and buses will play a significant role by 2025 in transporting freight and for mass transit, because of the poor quality of railway systems and increased demand for public transport. Transport fuel consumption will grow substantially, but this growth will be moderated by the high cost of imported vehicles and fuel, as has occurred already in Ghana and Sierra

Leone. Trip distances are expected to fall in all countries over the next forty years.

Fuel intensities (litres/km) fall in both the HES and LES for all the countries, but this trend is limited by the expected large number of old and inefficient vehicles that are likely to remain in the fleet by 2025. Declines in this ratio of 20 per cent relative to the base year were estimated in the HES for Ghana and Sierra Leone; and 30 per cent for LES. Traffic congestion is assumed to fall in Nigeria thereby lowering its fuel intensity in 2025 to 50 per cent in the HES, and 65 per cent in the LES, of that in 1985. The scenarios for Sierra Leone also introduced some ethanol in light of its success in other African countries. However, ethanol offers great promise to save energy and conserve carbon in the transport sector which may be understated in this study.[16] The scenarios also assume that enacting performance standards for imported vehicles, improving driving practices, and upgrading maintenance facilities will reduce energy demand and carbon emissions.

Industrial carbon conservation

Industrial activities, mainly mining and light and intermediate manufacturing play a limited role in the regional economy. The oil boom in Nigeria stimulated its industrial sector over the last two decades during which industrial production doubled. Industrial activities are expected to expand greatly throughout the region over the timespan of the scenarios. Growth areas are likely to include agro-based industries, petrochemicals, and processing of mining products. By 2025, Nigeria's iron, steel and fertilizer plants are expected to be operating. These activities will boost industrial energy intensive activities in the region.

Gains from energy conservation efforts will be determined largely by international developments due to reliance on imported machinery. As a developmental latecomer, the region can benefit from efficient technologies developed elsewhere, provided that it avoids dumping of old technology. Biomass fuels may also supply a major amount of industrial energy by 2025, thereby reducing the carbon intensity of the manufacturing sector. The adaptation and local manufacture of spare parts for industrial machinery and equipment could further improve industrial energy intensities.

In the LES scenario, Ghana and Sierra Leone reduced carbon emissions due to increased use of agricultural wastes for electricity production in their agro-based industries. In Nigeria, carbon is conserved in industry by increasing usage of natural gas. The LES also incorporated more housekeeping measures resulting from energy auditing and the setting of energy efficiency guidelines for new industries.

Electrical sector carbon conservation

Electricity generation comprises a small but growing share of the total energy demand in the region. Also, due to supply constraints, the level of unfulfilled demand is increasing. These factors explain the very high growth in electricity demand by 2025. In Ghana, for example, it increased eight- and seven-fold in the HES and LES respectively. In Nigeria, it increased ten- and eight-fold; and in Sierra Leone, fivefold in both scenarios.

Hydropower will play a major role in expanding electricity supplies in Ghana and Sierra Leone, while Nigeria will rely more on natural gas. Today's high transmission and distribution (T&D) losses will fall from 25–32 per cent to 15–20 per cent by 2025, mainly as result of improving technical efficiencies and less theft of power.

These countries have huge quantities of unexploited hydro resources. Ghana and Nigeria have tapped less than 50 per cent and Sierra Leone only 10 per cent of estimated potential hydropower. In the HES, 80 per cent of Ghana's electricity is generated by hydropower in 2025. In the LES, 100 per cent is supplied by hydropower. Nigeria develops about 75 per cent of its hydro potential in the HES and 95 per cent in the LES. Coal provides only 10 per cent of Nigeria's electricity by 2025 in the HES, and 5 per cent in the LES.

In Sierra Leone, hydropower rises from 3 to 50 per cent of electrical energy in both scenarios. The HES relies on oil to generate the remaining 50 per cent of electricity, and biomass supplies 30 per cent of the shortfall in the LES.

The major constraint on exploiting this hydropower potential is the shortage of capital in the region. The scenarios indicate that electrical demand will increase two to three times faster than GDP by 2025. This trend implies that power projects will require investment capital at the expense of other development sectors – a serious challenge for these countries.

Aggregate analysis of carbon conservation

According to the HES, by 2025 Ghana, Nigeria, and Sierra Leone will emit 4.4, 54 and 1.2 million tonnes of carbon respectively. These projected emissions are a seven- six- and four-fold increase over current emissions for each of these countries. These emissions in 2025 can be reduced by 36 per cent, 25 per cent and 13 per cent respectively if measures to conserve carbon are introduced as assumed in the LES (see Table 9.3).

In all the countries, carbon emissions increase much faster than energy consumption during the four decades because of the shift from biomass to petroleum fuels. In the HES for Ghana and Nigeria, energy demand increases by factors of 2.2 and 2.5 respectively, while carbon emissions increase 6 and 7 times. This shift is less dramatic in Sierra Leone where energy use slightly more than doubles but carbon emissions increase almost fourfold.

In the LES, energy efficiency and carbon conservation measures lower the ratios of energy consumption to carbon emissions in all countries. Growth in carbon emissions surpasses even the high population growth rates of these countries, resulting in increases in carbon emissions per capita between 1985 and 2025 (see Table 9.3). This outcome varies among the

Table 9.3 *Carbon emissions from Ghana, Nigeria and Sierra Leone*

	Ghana			Nigeria			Sierra Leone		
	1987	2025 HES	2025 LES	1985	2025 HES	2025 LES	1985	2025 HES	2025 LES
Carbon emissions (mt)	0.6	4.4	2.7	9	54	41	0.3	1.2	1.0
Residential (%)	15	47	51	21	34	34	19	28	22
Industrial (%)	8	18	10	18	36	37	15	19	20
Transport (%)	59	23	27	53	22	22	65	52	58
Services (%)	7	5	5	2	1	1	1	*	*
Agriculture (%)	4	*	*	1	2	2	*	*	*
Losses	6	6	6	5	5	4	*	*	*
CO_2/capita (kg)	48	104	64	99	171	129	93	140	111
CO_2/commercial energy (kg/GJ)	3	5	4	6	13	13	6	10	9

HES: high emissions scenario
LES: low emissions scenario

* less than 1 per cent

countries. It is most pronounced in Nigeria where dependence on fossil fuels increases per capita annual carbon emissions to 171 kg. Ghana's greater reliance on hydropower keeps its per capita carbon emissions down to 104 kg.

Carbon emitted per unit of commercial energy used also increases between 1985 and 2025 due to the increased use of petroleum based fuels. This feature is well illustrated by Nigeria which will use more and more carbon intensive fuels.

Extrapolated analysis for the region

This three-country analysis provides a basis on which to estimate the future energy use and carbon emissions for the whole West African region. As mentioned above, these three countries comprise most of the region's population and about 60 per cent of its energy use. Both scenarios assumed a region-wide annual population growth rate of 2.9 per cent and a GDP annual growth rate of 4 per cent.

In the HES, carbon emissions increase almost sixfold. In the LES, in which conservation measures are taken, carbon emissions fall by almost a quarter (see Table 9.4). On a per capita basis, carbon emissions increase by 75 per cent in the HES but only 30 per cent in the LES. Expanded use of fossil fuels and the dominant role of Nigeria accounts for the more than twofold increase in carbon emitted per unit of commercial energy used (see Table 9.4).

In short, total energy consumption will increase substantially in the region as will fossil fuel use. But the energy system will also become be more efficient. The transport and industrial sectors observed will become more important throughout the region.

Table 9.4 *Extrapolated data for the West African region*

	1985	*2025 (HES)*	*2025 (LES)*
Population (millions)	169	537	537
Population AAGR (%)	-	2.9	2.9
GDP/capita ($US)	551	987	587
Carbon emissions (MtC)	13	77	58.6
CO_2/capita (kg)	92	161	120
CO_2/commercial energy (kg/GJ)	5.6	12	11.8

HES: high emissions scenario
LES: low emissions scenario
AAGR: annual average growth rate

Options for rational energy use and carbon conservation

In general terms, the policy options available to reduce energy use and carbon emissions are: low-carbon fuels and fuel substitution; renewable energy technologies; large-scale biomass development; energy efficiency technologies; energy pricing; and carbon taxes. This section will review each of these policy options as they pertain to the region.

Development of fossil fuel resources

West Africa has abundant exploitable energy reserves, including more than 2.3 billion tonnes of crude petroleum, about 90 billion cubic metres of natural gas; some high and poor quality coal deposits; vast hydropower resources; uranium; and large quantities of renewable sources of energy.[17]

It will be easier to attract the requisite external development capital for oil and gas development than for coal as these sources are exportable and emit one third to a half of the carbon per unit energy output as does coal. These fossil fuels, however, are distributed unevenly with more than 80 per cent located in Nigeria. Moreover, more than 80 per cent of oil produced in the region is exported, even as some countries within the region suffer from an acute shortfall of oil. Additional prospecting may uncover more fossil fuel resources, although exploration is risky and expensive. The positive signs reported in the Gulf of Guinea[18] suggest that joint ventures with foreign partners may be productive.

Presently, Nigeria flares more than 70 per cent of its natural gas production. The completion of its liquefaction plant now under construction[19] will supply local electrical power and industrial energy needs and allow exports to nearby countries. The un-exploited hydropower resources of West Africa are estimated to be at least 40,000 MWe.[20] Nearly all countries in the region contain hydropower potential. Hydropower may be constrained both financially and environmentally, however. The climatic impacts of large-scale hydro development must be studied before large investments are sunk into this option.

Fuel substitution

Fuel substitution is possible in households and in the electrical supply system. In the Sahelian countries, LPG substitutes for woodfuels as a cooking fuel in urban households. Countries with such programmes include the Gambia, Senegal, Burkina Faso, Mali and Niger. Ghana promotes widespread use of LPG in households and institutional users.

Replacing woodfuels with petroleum based fuels may increase carbon emissions unless the biomass is produced on a renewable basis.[c] In any case, shifting to gas will likely reduce the pressure on forests that have not been cut by woodfuel suppliers. Natural gas offers a more attractive substitute than coal. Not only is natural gas less carbon-intensive, but gas-fuelled combined gas turbines can be built at lower total cost than other fossil fuel power plants, and even large hydropower plants.[21]

Renewable energy sources

Modern renewable energy sources which emit little or no carbon have great potential in the region, especially for small and decentralized applications and to meet the needs of dispersed, vulnerable groups. The high front end costs, however, must be made more affordable for the intended users.

At present, some mature renewable energy technologies have proven their viability in the region.[22] Solar energy devices for household use, and stand-alone power systems especially for communication and water pumping, are very promising. Technical and pricing problems remain, however. Recent studies show that wind speeds equivalent to those exploited in Denmark exist along the coast of Senegal and Mauritania.[23] The falling cost and increasing technological maturity of wind generators offer the possibility of integrating windpower into a national distribution grid.

Biomass energy

Biomass energy utilization in the region is limited to small-scale energy applications except for a few agro-based industries. Recent work has shown that this energy source can be used in large-scale end uses and can compete with other modern fuels.[24] If these systems (referred to as biomass integrated gasifiers/gas turbines or BIG/GT) are well planned, then their net carbon emissions can be minimized by replanting programmes. The configuration of these systems can be varied. The most promising application is in the sugar industry, a major part of the regional economy.

These technologies are likely to be commercialized by the mid-1990s, after which time they could play a major role. A pilot plant is already under construction in Brazil and another is contemplated for the Ivory Coast. The successful introduction of ethanol as transport fuel in Zimbabwe and Malawi can be emulated in West Africa, especially in land-locked countries.

---

c Woodfuel, however, may produce more greenhouse warming due to its greater release of other greenhouse gases such as methane. The overall relative greenhouse impact of the transition from woodfuel to fossil fuels depends on this issue, as well as the country-specific relative fuel cycle conversion device efficiencies.

Energy efficiency technologies

Energy efficiency technologies offer the region many opportunities to increase energy services from existing and new energy transforming capital stocks. Energy efficiency, therefore, should be treated as equivalent to supply options at the margin. The potential for using energy efficiency technologies in developing countries is great, not least because their energy sectors are so wasteful.[25]

The most attractive sectors for implementing these technologies are the most carbon-intensive in the long-term scenarios of this study, namely, the residential, transport and industrial sectors. Three sets of efficiency technologies have promise: capital-intensive measures which will require external financial assistance; housekeeping and retrofitting at little or no cost; and improved household devices. The first requires external assistance from the highly industrialized countries which will develop most of these technologies. The other two can be undertaken largely by local efforts.

Relatedly, transmission and distribution (T&D) losses offer additional potential to conserve carbon. Up to 15 per cent improvement in overall T&D efficiency is easily achievable in Nigeria and Ghana by the year 2025. Most of the technical measures needed to reduce T&D losses are not expensive and are quickly recovered by utilities, as occurred in Sudan where the benefits were 12 times the installed cost of efficiency measures.[26] Improving the thermal efficiency of the whole power system, which can drop as low as 22 per cent, as in Benin, can also slow the growth of emissions in the regional power sector.

Improved woodfuel stoves and charcoal kilns have great technical potential to reduce energy consumption and to conserve carbon, especially in urban areas. These devices, however, have limited acceptability in rural areas because of social and financial constraints. Nonetheless, designs have been tested successfully and disseminated in Kenya, Sudan, Burkina Faso and Niger.[27]

Energy pricing

Energy pricing is a useful instrument for controlling the quantity and type of energy used. However, prices are also used to promote welfare goals in many poor countries by inter-fuel and cross-consumer subsidies. Nonetheless, cost recovery rather than social equity guides energy pricing policy in most of the region today. Reforming electricity prices is linked to changing billing and revenue collection practices. The failure to bill and collect revenue has left many utilities unable to realize revenues based on nominal high tariffs.

Biomass fuels are largely unregulated and suppliers often use their market power to exploit consumers. Prices for energy services in the region badly need to be adjusted to reflect real economic and environmental costs. The analysis that should underlie a price reform strategy should consider several innovative measures including carbon taxes.

Carbon taxes

Events at the global level affect the macroeconomic environment of countries in West Africa. Carbon taxes are one recent method proposed to fund collaboration between developed and developing countries for carbon conservation. Revenues from carbon taxes could fund businesses with minimal adverse environmental effects in developing countries that would use equipment and expertise supplied by the developed countries.

Obstacles and strategies

A significant number of financial, institutional and technical obstacles exist which block the region from implementing these options.

Lack of investment finance is a major obstacle to energy development in the region. This scarcity arises from the inability of local economies to generate domestic capital and diminishing access to foreign capital. Establishing mechanisms to mobilize local capital as occurred in Ghana can improve the situation.[28] However, the region's inability to repay its existing energy loans casts doubts on its ability to take on new debt to increase its energy efficiency and to reduce carbon emissions. The continuous fluctuation of exchange rates further erodes the region's capacity to meet its debt obligations. However, stimulating local investments, minimizing foreign inputs, and establishing better planning mechanisms can reduce these difficulties.

'Institutional inertia' also affects the implementation of new ideas. The power sector is especially susceptible to institutional conservatism. Also, many energy institutions are rated in terms of their technical efficiency rather than on their energy or financial performance. The generally poor management of energy and related institutions also hinders the effective implementation of these measures.

Institutional reform, therefore, is an essential ingredient of a carbon abatement strategy. Separating generation from distribution, as in Nigeria, and privatization of utilities, as in the Ivory Coast, exemplify such reforms. New institutional mechanisms to produce and use biomass energy are also needed urgently due to its potential importance in the energy sector of the region. Adequate information must also be disseminated to equipment producers, government personnel, local R&D personnel, and technology users if these technologies are to diffuse rapidly. An institutional framework for the collection, organization, storage and retrieval of this information is another prerequisite for success. Computers can provide cheap access to international information flows. Harmonizing equipment standards will also ensure increased compatibility between equipment, a crucial issue for developing countries at the receiving end of donor programmes.

Finally, improving training facilities and programmes and improving labour markets will increase the supply of technical competence when and where it is needed, as well as reducing dependence on foreign assistance. A rational allocation of local human resources, developing well-planned follow-up programmes, enhancing a well-articulated local R&D system, and establishing collaborative programmes with external agencies can further ameliorate these problems. Increasing public awareness to promote these options will also facilitate their use.

Economic opportunities for implementation

The region's poor economic prospects accentuate the need to take advantage of least cost energy strategies and foreign funds available to support such programmes. The technology choice model used by energy planners should be selected carefully to ensure that the least cost investment strategy that minimizes foreign exchange requirements is selected. It is essential that this model places efficiency improvements at end use on an equal footing with additional supply options. Unfortunately, inadequate information on the region restricts the level of analysis now possible to a largely qualitative exploration of the least cost strategy to provide energy services and conserve carbon. Electricity generation will be singled out here because of its capital intensity and developmental significance.

Electricity generation

The HES and LES scenarios indicate that as countries in the region develop economically, so the ratio of energy to GDP will fall. This outcome is the result of substituting more efficient energy forms for inefficient biomass fuels. In the scenarios, the primary energy per unit GDP declines between the base year and 2025 by 24 per cent (HES) and 38 per cent (LES) in Ghana; 44 per cent (HES) and 56 per cent (LES) in Nigeria; and 34 per cent (HES) and 46 per cent (LES) in Sierra Leone. Also, as Table 9.2 reveals, the primary energy supply more than triples in the HES and falls only by 20 per cent on average in the LES in all the countries. Moreover, the scenarios show that power sector investment grows faster than GDP. This result implies that financial resources will have to be diverted from other non-energy sectors.

The scenarios indicate that electrical supply can be increased by greater energy efficiency at end use and by reducing T&D losses. Nonetheless, a major increase in power generation will be required to meet demand. Hydropower is the most available supply option but also needs the biggest front end investment, much of which is in foreign exchange. Technological maturity makes it difficult to reduce historical costs of hydropower.

The development of large-scale hydropower is necessary in the region.

However, international donors are reluctant to invest in hydropower because of environmental problems associated with large dams. Nonetheless, countries that can obtain the requisite resources may do well to exploit hydropower because of its extensive backwards and forwards developmental linkages.

Natural gas development using combined cycle plants is more attractive as the investment cost is only about US$600/kWe-installed.[29] This option is attractive for Nigeria which has large domestic gas reserves. It remains unclear, however, if its gas-powered electricity can be exported to neighbouring countries at competitive prices. Much will depend on whether local technical inputs are used to minimize the costs of developing this power supply.

Three options should be explored in the region to maximize carbon abatement. First, gas-fuelled combined cycles for electricity generation cost less than oil or coal fuelled plants (which cost US$800–1500/kWe). Also, the gas cycle has a higher thermal efficiency and emits less carbon and almost no sulphur dioxide or particulates. It also takes less time to construct and has lower maintenance requirements, being of modular design.

Second, agricultural wastes should be used as fuel in steam-generating plants and later in large-scale energy applications. These systems are expected to be commercially competitive with conventional systems within a few years. Third, many low cost steps should be taken at many end use sites to improve the overall efficiency of the energy system. This option is important because it requires little or no foreign help.

Carbon conservation in other sectors

This section is necessarily qualitative, as carbon emissions disaggregated by individual abatement measures could not be counted in the scenarios. The results are shown in Table 9.5. The aggregate energy and carbon savings from adopting least cost technologies were evaluated by subtracting the LES from the HES, sector-by-sector.

Residential sector

The residential sector offers great potential for energy and carbon savings. The LES assumed a widespread deployment of improved biofuel stoves and charcoal kilns; improved lighting and electrical appliances; and fuel substitution. Large-scale dissemination of improved woodfuel stoves and charcoal kilns will produce significant energy savings.

Most of the options in this sector do not require much financial investment or intervention by government, as the case of Kenya has illustrated[30]. The primary role of government is to facilitate technological R&D. However, the supporting infrastructure needed for this measure may require foreign assistance.

Table 9.5 *Energy and carbon savings (HES – LES)*

	PJ saved	% total	MtC saved	% total
GHANA	142	100	1.68	100
Residential	88	62	0.68	41
Industrial	26	18	0.53	32
Transport	14	10	0.28	20
NIGERIA	675	100	13.00	100
Residential	353	52	4.42	34
Industrial	107	16	4.27	33
Transport	207	31	2.86	22
SIERRA LEONE	18	100	0.25	100
Residential	14	78	0.13	51
Industrial	1	5	0.04	15
Transport	3	17	0.07	29
WEST AFRICA (*)	969	100	16.71	100
Residential	517	53	6.45	39
Industrial	156	17	6.15	37
Transport	296	30	4.11	24

* Extrapolated values

Transport sector
In the transport sector, three sets of important measures can produce energy and carbon conservation. These are: improved traffic management and vehicle maintenance; fuel substitution and modal shift; and increased vehicular efficiency.

The first set will give only moderate energy and carbon savings. Conversely, these measures require relatively little investment, and some can be locally financed and managed. The second set involves capital intensive measures such as improved urban transport, ethanol, and more railways. These measures will give substantial carbon savings but need large investments with significant foreign exchange requirements. Third, more efficient vehicles rely largely on foreign technological innovation by car manufacturers. Nonetheless, it is important that governments regulate imports and local assembly of cars to increase vehicular efficiency at low cost.

Industrial sector
Industrial activities promise large carbon reductions. The emerging iron and steel industry in Nigeria, which will draw on raw materials from Sierra Leone, Guinea and Liberia, can be built to high standards of energy efficiency and implement standard housekeeping measures to obtain moderate energy and carbon savings at low cost.

Despite the lack of disaggregated data, these options were assessed to give an indicative representation of their cost and emission reduction potential. Table 9.6 summarizes the results.

Policy issues for the region

Policies to foster the energy and carbon conservation potential of the region fall into two categories: regional and national.

Regional policies

The region can only fully benefit from the highly competitive external funds if regional mechanisms are created to analyse and develop well-articulated projects. Such options include: strengthening existing institutions such as the Economic Community of West African States (ECOWAS); creating linkages between national organizations; and developing nodes of analytical skills. The Climate Change Convention recommends that all these steps be taken.[31] The commitments contained in the convention to promote technology transfer, education, and training, etc. can only be realized in West Africa if there is a strong regional framework. The increasing difficulties of donors in meeting aid commitments and competition between recipients imply that a minimum technological capability must be established in the

Table 9.6 *Cost assessment of carbon abatement options*

	Savings potential		*Ranking*	*Cost*
	Energy	*Carbon*		
Residential				
Improved stoves and kilns	High	Low	Negative	Low
Efficient electric lights and appliances	Medium	High	Middle	Medium
Fuel substitution (LPG and kerosene)	Low	Low	Positive	High
Transport				
Fuel substitution (ethanol)	Low	High	Positive	High
Traffic management	Medium	High	Middle	Low
Improved maintenance	Low	Medium	Middle	Medium
Urban transport	High	High	Positive	High
Industry				
Housekeeping measures	Medium	Medium	Negative	Low
Standards	Medium	Medium	Negative	Low
New technology	High	High	Positive	High
Electricity				
T&D losses	Medium	Medium	Negative	Low
Increased supply	High	High	Positive	High

Rankings refer to the possible point in a cost-emission reduction curve.

region if it is to tap the climate change related aid programmes. The operation of the Global Environment Facility has already demonstrated the need for such a regional structure. Fragmented regions will be disadvantaged in global fora.

National policies

At a national level, specific areas that require policy attention are: fuel pricing for all fuels; an R&D structure with strong links to policy-makers, industries and end-users; the promotion of education and awareness on energy and environmental issues; the development of improved information systems; and the establishment of enforceable and feasible standards.

Governments must also create an environment conducive to the full participation of the private sector, while regulating its activities. Financial mechanisms such as loan guarantees, insurance schemes, and fiscal incentives should be explored also to encourage private sector involvement. An overriding task is to establish an appropriate mechanism to ensure that governmental decisions and goals on energy development and climate change are spelled out clearly and remain flexible in the face of rapidly changing global events.

Conclusions

I have demonstrated that although West Africa's contribution to global carbon emissions is small, its role in a global carbon abatement strategy is important. It is inevitable that energy consumption and carbon emissions will increase as the region develops. However, many of the measures discussed in this study are available to countries in the region at low cost and with external donor support.

To exploit these opportunities, the countries of West Africa must develop strong regional initiatives. There are many barriers to realizing the technological potential to increase energy efficiency and to reduce carbon emissions. But these are susceptible to a variety of policy initiatives. A regional approach is an important prerequisite for successful mobilization of local capital and technological capabilities.

References

1 IPCC, 1990, *Climate Change, The IPCC Response Strategies*, Working group 111, WMO/UNEP, Cambridge University Press, Cambridge, Massachusetts

2 IPCC, 1990, Energy and Industry Subgroup Report, WMO/UNEP, Cambridge University Press, Cambridge, Massachusetts

3 British Petroleum, *1991 BP Statistical Review of World Energy*, BP, June 1991, London

4 Ryan, P, 1991, 'Some Policy and Economic Realities of Biomass Development and Management in Africa and Asia,' paper to the conference on technologies for a greenhouse-constrained society, Oak Ridge National Laboratory, June 11–13 1991

5 Dutt, G S, *et al*, 1991, 'Bio-energy Alternatives for Cooking,' *Fuels and Electricity from Renewable Sources of Energy*, T Johansson *et al* (eds), forthcoming, Island Press, Washington DC

6 Biswas A K, 1986, 'Renewable Energy and Environmental Policies in Africa,' *Energy Policy*, June 1986

7 Smith, K, 1991, 'Household Air Pollution in Developing Countries,' paper to the conference on energy and environment issues in the residential sector, Mexico, December 2, 1991

8 IIASA, 1981, *Energy in a Finite World: Paths to a Sustainable Future*, Ballinger, Cambridge, Massachusetts

9 IPCC, 1990, *Climate Change: The IPCC Scientific Assessment*, WMO/UNEP, Cambridge University Press, Cambridge University Press, Cambridge, Massachusetts

10 Sathaye, J and Ketoff, A, 1991, *CO_2 emissions from developing countries: better understanding the role of energy in the long term*, Volume 1, Summary, LBL-29507 UC-350, Lawrence Berkeley Laboratory, February 1991

11 Davidson, O R, *et al*, 1991, *CO_2 emissions from developing countries: Better understanding the role of energy in the long term*, Volume IV, Ghana, Sierra Leone, Nigeria and the Gulf Cooperation Council Countries (eds), Sathaye, J and Goldman, N, Lawrence Berkeley Laboratory, LBL-30061 UC-350, July 1991

12 Adegbulugbe, A, 1991, 'Energy Demand and CO_2 Emissions Reduction in Nigeria,' *Energy Policy*, Volume 19, No. 10, December, 1991

13 World Bank, 1992, *World Development Report*, Washington DC

14 Davidson *et al* (1991), *op cit* (reference 11)

15 Wreko-Brobby *et al*, 1988, *Population and Energy Resources in Ghana*, Ministry of Energy, Ghana

16 Davidson, O R, 1992, *Transport Energy in Sub-Saharan Africa: Options for a Low-emissions Future*, Princeton University, Center for Energy and Environmental Studies research report 267, February, 1992, Princeton University, New Jersey

17 UN Economic Commission for Africa, 1984, *Special Memorandum to the Ministers on Conference on Africa's Economic and Social Crisis*, ECA, Addis Ababa

18 World Bank, 1989, *Sub-Saharan Africa: from Crisis to Sustainable Growth*, World Bank, Washington DC

19 Sharma, V C, and Sharma, A, 1991, 'Nigeria's Primary Energy Reserves: Domestic Consumption and Future Outlook,' *Energy*, Volume 16, No. 5, May 1991, pp 817–822

20 WEC, 1989, *1989 Survey of Energy Resources*, World Energy Conference, London

21 Moore, E, and Crousillat, E, 1991, *Prospects for Gas-Fueled Combined-Cycle Power Generation in the Developing Countries*, World Bank Industry and Energy Department Working Paper, Energy Series no. 35, March 1991

22 AFREPREN, 1992, *Renewable Energy in Africa*, O R Davidson and S Karakezi (eds), Zed Books, London, England

23 World Meteorological Organization, 1981, *Meteorological Aspects of the Utilisation of Wind as an Energy Source*, WMO technical note 157, Geneva

24 Larson, E, 1991, *A Developing Country-Oriented Overview of Technologies and Costs for Converting Biomass Feedstocks into Gases, Liquids and Electricity*, Princeton University Center for Energy and Environmental Studies report no. 266, Princeton University, New Jersey

25 Levine, M, and Meyers, S, 1992, 'The Contribution of Energy Efficiency to Sustainable Development in Developing Countries,' *Natural Resources Forum*, February, 1992

26 UNDP/World Bank, 1984, *Sudan Power System Efficiency Study*, June 1984, Washington DC

27 Davidson, O R, and Karakezi, S, 1992, *A New, Environmentally-Sound Energy Strategy for the Development of sub-Saharan Africa*, AFREPREN, MacScan Publications, Nairobi, Kenya, January, 1992

28 Wreko-Brobby, C, 1992, 'Energy Institutions in Africa: A case of Ghana National Energy Board,' AFREPREN/UNEP/RISO, May 1992

29 Moore and Cousillat (1991), *op cit* (reference 21)

30 Karakezi, S, 1989, 'Responding to a global need for local action in domestic energy,' in *Nordic Seminar on Domestic Energy in Developing Countries*, Lund University

31 INC, 1992, 'Report of INC for a Framework Convention on Climate Change,' United Nations, May 1992

10

Abatement of carbon dioxide emissions in Brazil

Jose Moreira and Alan Poole[a]

This chapter concentrates on energy-related carbon dioxide (CO_2) emissions, the principal greenhouse gas, of which Brazil is an important contributor at a global level. The clearing of forest land for non-energy uses is its largest source of CO_2 emissions. Quantitative estimates of this source are controversial, however. It is not possible to be precise as to the amount and costs of carbon conservation associated with land use changes in Brazil. For this reason, this chapter will focus on the abatement of fossil CO_2 emissions released from the energy sector. But we will also discuss policies to decelerate deforestation and outline some crucial interrelationships between land use, energy policy, and carbon conservation strategies.

Brazil energy economy

Contemporary energy related emissions of CO_2

In 1990, Brazil's total energy consumption reached 183.7 million tonnes of oil equivalent (MTOE). In that year, 37 per cent of Brazil's primary energy came from fossil fuels, 26 per cent from biomass fuels and 37 per cent from hydroelectricity (see Table 10.1). Important features of Brazil's energy balance are: the preponderant role of hydropower in electricity generation (94.4 per cent); the relatively large amount of biomass that is used at an industrial scale (alcohol, charcoal and wood – see Table 10.2); and the small penetration of natural gas.

The energy system resulted in 1990 in the emission of approximately 51.6

a We would like to acknowledge the collaboration of Gilberto de M Januzzi, Department of Energy, University of Campinas, David Zylbersztajn and Juliana Fries, Institute of Electrotechnology and Energy, University of Sao Paulo, for the provision of part of the data about energy efficient technologies.

Table 10.1 *Energy supply and consumption, Brazil 1990 (MTOE)*

	Coal	Natural gas	Petroleum	Subtotal fossil	Biomass	Hydro	Nuclear	Other primary[b]	Electricity
Gross internal supply	9.21	3.75	55.06	68.02	47.33	67.75	0.58	183.68	n.a.
Electricity generation[a]	1.15	0	1.15	2.30	0.72	67.75	0.58	71.35	71.74
Other energy sector	0.82	1.16	1.93	3.91	13.78	-	-	17.69	10.88
Non-energy use	0.06	0.59	8.94	9.59	0.33	-	-	9.92	0
Final energy demand	7.18	2.00	43.04	52.55	32.50	-	-	85.05	60.86
Residential	0	0.16	5.09	5.25	8.49	-	-	13.74	13.85
Commercial/services	0	0.07	0.65	0.72	0.18	-	-	0.90	12.09
Agriculture	0	0	3.16	3.16	2.15	-	-	5.31	1.83
Transport	0.01	0	26.77	26.78	5.65	-	-	32.43	0.34
Industry	7.17	1.76	7.28	16.21	15.92	-	-	32.13	32.85

Source: *Boletim do Balanco Energetico Nacional*, 1991.
Data for fuel include derivatives of the primary energy source (for example, coke from coal, alcohol from sugarcane). Electricity (last column) is calculated assuming that 1 MWhe = 0.29 TOE.

a Includes transformation and other losses and energy use (for example, in refineries).
b For gross internal supply and electricity generation includes all energy consumption, thereafter electricity is excluded. The actual inputs for electricity generation are included in this column, since these are different than the coefficient of 0.29 TOE used in last column. There is a slight discrepancy.

Lessons from climate change in Brazil

Each year, much human suffering has been caused by climate hazards in Brazil and billions of dollars have been lost . . . The Brazil study concludes that human actions have inadvertently led to increased societal vulnerability to climate variations. Deforestation in the Northeast has made the semi-arid region more vulnerable to droughts. Inadequate urban planning in Rio de Janeiro has made the city much more vulnerable to floods. Deforestation in the Upper Paraguay River Basin may be altering the pattern of floods and droughts in the Pantanal (Great Swamp) region, thus contributing to ecological imbalance.

On the other hand, some action has been taken to increase resilience to climate variations. The modernization of the salt industry in the state of Rio Grande has made that industry more resistant to heavy rains. Agricultural research has led to the development of several new crop varieties more resistant to climate variability. Relief action in Northeast Brazil has [reduced] the very heavy impacts of droughts on the poor rural population.

It is clear that the same climatic event may have different impacts according to local socioeconomic and environmental characteristics. Rainfall that brings terrible floods to Rio de Janeiro is a beneficial event in the Pantanal area. Drought may cause huge losses to agriculture in the Northeast but can also bring economic benefits for the salt industry in the same region. Moreover, while the majority of the poor population will suffer from droughts, a small group of large land owners and businessmen may indeed profit from them. Climatic variations can thus have differing effects on different regions, ecosystems, economies and social classes.

Government policy needs to integrate short-term relief actions during extreme climatic events with long-term actions aimed at increasing societal resilience to climate variability and change. It also needs to pursue a goal of sustainable development, by seeking to increase the technological capacity of people to face climatic extremes, reduce the social impacts of the weather, reduce poverty (since the poor are the most vulnerable) and improve our knowledge and thus increase our capacity to accommodate adverse variations of climate. A policy of sustainable development would pursue both a reduction in greenhouse gases emissions and an improvement of the capacity of the environment to adapt to possible future climate changes. To achieve this will require international cooperation to enable the transfer of resources and technology that will allow developing countries to use the most environmentally appropriate available technologies to their development process.

Excerpt from M Parry, A Magalhaes, Nguyen H Ninh, *The Potential Socio-Economic Effects of Climate Change, A Summary of Three Regional Assessments*, UN Environment Programme, Nairobi, Kenya, 1991, p. 12.

million tonnes of carbon (TC) as CO_2[b] from the combustion of fossil fuels (Table 10.3).

To simplify our analysis, we assumed that hydroelectricity has zero CO_2

b In this paper we are only considering carbon emission due to CO_2 production. Carbon equivalent emissions due to other greenhouse gases are not quoted in our model or in any other scenarios referred too. Carbon is carbon as carbon dioxide throughout this chapter.

Table 10.2 *Biomass supply and consumption, 1990 (MTOE)*

	Wood for charcoal	*Other wood*	*Sugarcane*	*Other*[b] *biomass*	*Total*
Gross internal supply	12.31	15.14	18.14	1.74	47.33
Electricity generation	0	0.12	0.23[c]	0.37	0.72
Other energy sector[a]	6.43	0	7.41[c]	0.06	13.78
Non-energy use	0	0	0.33[d]	0	0.33
Final energy demand	5.88[e]	15.14	10.17	1.31	32.50
Residential	0.63	7.86	0	0	8.49
Commercial	0.06	0.12	0	0	0.18
Agriculture	0.01	2.14	0	0	2.15
Transport	0	0	5.65	0	5.65
Industry	5.19	4.90	4.52[c]	1.31	15.92

Source: *Boletim do Balanco Energetico Nacional*, 1991.

a Includes transformation losses and fuel use; specifically losses in producing charcoal from wood and alcohol from cane; b Basically pulp mill liquor; c Sugarcane bagasse; d Alcohol; e Charcoal.

emissions.[c] Similarly, we assumed that sugarcane supply (alcohol and bagasse) and 'other renewables' (pulp industry liquors) also produce zero CO_2 emissions. In the case of alcohol, however, we included fossil fuel inputs to alcohol production.

Estimates of carbon emissions from wood uses are problematic and vary widely according to end use sector. Most wood used in the residential and agricultural sectors does not appear to contribute to deforestation. The relatively dispersed use in rural areas does not exceed natural regeneration. In this regard, Brazil's fuelwood problem is unlike that in many other countries. Little information is available to quantify deforestation in the areas that it occurs. In this study, we simply assumed that 20 per cent of the fuelwood used in the residential and agricultural sectors results in net CO_2 emissions.

Conversely, fuelwood used in the industrial sector is more geographically concentrated. Demand from this subsector of fuelwood consumption often destroys natural forests in Brazil. Again, precise information as to rates and magnitudes of this phenomenon are scarce. We have assumed that 60 per cent of industrial fuelwood contributes to CO_2 emissions. Charcoal, however, presents a special case requiring the analyst to adopt additional assumptions. The impact on the forests of converting wood to charcoal (the

c An estimate for US hydropower carbon emissions is 3TC/GWhe (US Department of Energy 1990); but any simple extrapolation, especially for reservoir filling, to Brazil would be artibrary. The release of methane from reservoirs complicates this issue (Moreira and Poole 1992). It is believed that the emissions remain small in most cases. However, further study of this issue is warranted.

Table 10.3 *CO_2 emissions from energy consumption (MTC)*

	Fossil fuels[a]		*Biomass*[a,b]	*Total*	
Gross internal supply	51.64	(59.77)[e]	11.44	63.08	(71.21)[e]
Electricity generation	2.24		0	2.24	
Other energy sector	3.29		3.31[c]	6.60	
Non-energy use		(8.13)[d]	0		(8.13)[d]
Final energy demand	46.11		8.13	54.24	
Residential	4.48		1.94	6.42	
Commercial/services	0.60		0.05	0.65	
Agriculture	2.72		0.44	3.16	
Transport	23.03		0	23.03	
Industrial	15.21		5.70	20.91	

a For conversion factors see [5].
b Sugarcane and 'other biomass' assumed to have no CO_2 emissions. For wood it is assumed that deforestation results from 20% of residential, commercial and agricultural use; from 60% of industrial fuelwood use; and from 50% of charcoal use. These shares are somewhat arbitrary as discussed in the text.
c All due to conversion of fuelwood to charcoal.
d CO_2 equivalent of fossil fuel used, if they were burnt. See text.
e Includes non-energy CO_2 equivalent in parenthesis. (See note d above).

largest use of fuelwood) is highly controversial. An estimated one third only of Brazil's charcoal is supplied from planted forest. The remainder is made with fuelwood obtained by clearing the natural forest, mostly the drier forests of the *cerrado*. Most of this land is cleared for grazing and/or agriculture of which charcoal is a mere by-product. We assume, therefore, that 75 per cent of the charcoal from natural forest results in deforestation. Thus, we estimate that roughly 50 per cent of fuelwood used to make charcoal results in net CO_2 emissions.[d]

On this basis, about 40 per cent of total fuelwood use generates annual CO_2 emissions of 11.4 millions of tonnes of carbon as carbon dioxide (TC) – or 22 per cent of that resulting from fossil fuel use. Evidently, fuelwood is not yet a renewable resource in Brazil. In spite of its importance, fuelwood-related carbon emissions constitute only 3–4 per cent of a total deforestation-related emission each year of about 300 million tonnes of carbon (MTC).

Total (fossil fuel plus biofuel) energy-related emissions of CO_2 in 1990 are estimated to be about 63 MTC. This figure increases to about 71 MTC if the non-energy use of fossil fuels is valued at its combustion equivalent. This figure is roughly 470 kgC/capita per year.

d That is, we assume that 33 per cent of charcoal is sourced from CO_2-neutral regenerative planted forests; and 75 per cent of the other 66 per cent of the charcoal therefore contributes to CO_2 emissions. Thus, 50 per cent (0.75 × 0.066) of total charcoal contributes to carbon emissions.

Existing energy scenarios and their impacts

An official study was recently performed in Brazil as a basis for policy recommendations (SNE 1991a). This study analysed two basic policy postures. The first, a reference scenario called 'tendencies', assumed that current political and economic constraints in the energy sector (including those on pricing policy) will persist. The second scenario, called the 'alternative', postulated substantial changes in the *status quo*. This latter scenario sought to actively pursue greater efficiency in the overall use of energy resources; to promote competition and private sector investment in energy supply; and to stimulate the use of certain energy forms in some applications – principally biomass, but also coal and natural gas.[e]

Relative to 'tendencies', the 'alternative' scenario contains moderate reductions in energy use. In both scenarios, lower and higher economic growth cases were studied. In the alternative scenario, primary energy use falls approximately 9 per cent by 2000 and approximately 18 per cent by 2010 in both the lower and the higher economic growth cases (see Tables 10.4 and 10.5). The fact that the reduction is proportionately the same in both the lower and higher growth cases is somewhat odd. One might expect a proportionately greater reduction in energy use in the higher economic growth case, since faster growth permits the more rapid penetration of new technology in the marketplace. Other key differences between the alternative and tendencies scenarios are that in the former, hydroelectricity and petroleum products are projected to grow less, and biomass use to grow more in the latter.

Total fossil fuel emissions are 12 per cent less in the low growth alternative case relative to the 'tendencies' in 2000 and 21 per cent less in 2010. However, the fossil fuel emissions in the lower growth version of the 'alternative' scenario are still 42 per cent and 103 per cent higher than in 1990. Transport offers the biggest abatement in fossil fuel CO_2 emissions in the National Energy Secretariat (SNE) scenarios. Even so, this sector remains the biggest single source of carbon (see Table 10.6).

Fossil carbon emissions from electricity generation do not fall in the alternative scenario, despite substantial reduction of total electricity consumption. Indeed, thermal generation increases substantially, though this growth is mostly the result of increased use of biomass (mainly from sugarcane). Thus, the 'alternative' scenario retains the significant increase in

e SNE used a macroeconomic model to estimate the growth of each subsector and obtained energy demand by multiplying an energy intensity for each subsector by subsectoral activity levels. SNE assumed a $25/bbl price for oil in 2000, and analysed each subsector of the 'tendencies' scenario to identify possible savings that are reflected in the 'alternative' scenario. Much of the reduced energy demand in the 'alternative' scenario arises from SNE's shift in the composition of the industrial subsector economy toward less energy intensive activities, especially in the iron and steel industry.

Table 10.4 *Official energy scenarios – basic characteristics*

				Low economic growth				*High economic growth*			
				Tendencies		*Alternative*		*Tendencies*		*Alternative*	
	1970	*1980*	*1990*	*2000*	*2010*	*2000*	*2010*	*2000*	*2010*	*2000*	*2010*
Gross domestic product (1990=100)	37.3	85.3	100	146.3	238.3	146.3	238.3	163.5	292.5	163.5	292.8
GDP per capita (1990=100)	58.5	105.7	100	122.6	172.7	122.6	172.7	137.0	212.2	137.0	212.2
Population (millions)	95.9	121.3	150.4	179.5	207.5	179.5	207.5	179.5	207.5	179.5	207.5
Primary energy consumption (MTOE)	74.7	137.1	183.7	266.5	407.3	241.8	330.3	288.1	473.9	262.9	386.6
Fossil fuel CO_2 emissions (MTC)[a]	24.3	52.0	59.8	95.4	153.9	84.6	121.5	104.1	181.8	92.6	143.9
Fossil fuel emissions (1990=100)	40.6	87.0	100	159.5	257.4	141.5	203.2	174.1	304.1	154.8	240.6
Fossil fuel emissions per capita (kg C/cap.)	253	429	397	532	742	471	586	580	876	516	693
Fossil fuel CO_2 emissions per unit of GDP (1990=100)	109.1	102.0	100	109.2	108.0	96.8	85.8	106.5	103.9	94.8	82.2

Source: projection from *Reexame de Matriz Energética Nacional*, Ministerio da Infraestrutura, Brasilia, 1991.

a Based on gross internal supply of fossil fuels and thus includes a non-energy component.

Table 10.5 *Sources of energy supply – official scenarios (%)*

			Tendencies		*Alternative*	
	1980	*1990*	*2000*	*2010*	*2000*	*2010*
Oil	39.1	30.0	31.7	33.1	30.2	30.5
Natural gas	0.8	2.0	4.1	4.6	4.7	6.0
Coal	4.2	5.0	5.8	6.2	5.8	6.3
Other fossil	0	0	0.3	0.2	0.4	0.2
Subtotal fossil	44.1	37.0	41.9	44.1	41.1	43.0
Nuclear	0	0.3	1.1	1.9	1.2	1.4
Hydro	26.8	36.9	35.8	36.0	32.8	32.8
Sugarcane	6.5	9.9	8.2	7.7	10.0	10.0
Fuelwood	22.0	14.9	11.6	9.0	13.5	11.4
Other biomass/renewable	0.7	1.0	1.4	1.3	1.4	1.4
Subtotal biomass	29.2	25.8	21.2	18.0	24.9	22.8
Total	100	100	100	100	100	100
Gross internal supply (MTOE)	139.2	183.7	288.1	473.9	262.9	386.6

Sources: *Reexame da Matriz Energética Nacional*; Balanco Energética Nacional.
High economic growth case.

fossil fuel emissions found in the 'tendencies' scenario. From 9 TC/GWhe today (Moreira 1991a), the carbon intensity of electricity increases to 23 TC/GWhe in 2000. Even so, this ratio is still very low by international standards.[f]

The 'alternative' scenario projects large economic benefits. Energy sector investments, for example, fall by US$26 billion between 1991 and 2000 and $58 billion between 2001 and 2010 (in the high economic growth case) (SNE 1991a). Moreover, almost all of the carbon abatement in the alternative scenario result from 'no regrets' policy measures. That is, these steps are justified on non-greenhouse grounds (Ayres 1991), and accounting for CO_2 emissions only makes already economically viable changes even more desirable. Thus, these steps should be taken regardless of the scientific and other uncertainties surrounding greenhouse phenomena.

We have briefly described these scenarios to delineate the CO_2 emissions implied by official energy planning. The SNE study is also the only recent example of an integrated energy analysis in Brazil. This planning did not emphasize CO_2 abatement, however. New scenarios are needed badly to evaluate the costs of carbon abatement. The following end-use analysis is a first step in that direction.

---

f The average developing country carbon intensity in 1987 for electricity was 170 TC/GWhe.

Table 10.6 *Fossil CO_2 emissions by activity*

	Million tonnes of carbon			*Percentage*		
	1990	*2000 Tend.*	*2000 Alt.*	*1990*	*2000 Tend.*	*2000 Alt.*
Thermal electricity generation	2.24	7.73	7.72	4.3	9.2	10.4
Other energy sector use and losses	3.97	7.39	6.51	7.6	8.8	8.8
Subtotal energy sector	6.21	15.12	14.23	11.9	18.0	19.2
Subtotal final fuel demand	46.11	68.81	59.65	88.1	82.0	80.8
Residential	4.48	6.53	6.08	8.6	7.7	8.2
Commercial/public	0.60	0.61	0.57	1.1	0.7	0.8
Agriculture	2.72	4.49	4.18	5.2	5.3	5.7
Transport	23.03	32.22	27.38	44.0	38.4	37.1
Industry	15.21	24.94	21.47	29.1	29.7	29.1
Total fossil fuel	52.32	83.93	73.88	100	100	100

Source: based on Secretaria Nacional de Energia, *Reexame da Matriz Energética Nacional*, 1991. Low economic growth scenario.

Methodology of subsector analysis

An important input for CO_2 emissions abatement policy formulation is the analysis of sets of energy end-uses and transformations that constitute the various fuel cycles. This approach is not widely used in Brazil, and is most advanced in the electrical sector. In sectors where quantitative analysis is rudimentary, we will make qualitative observations. The time horizon for quantitative analysis is the year 2000. While politically realistic, this horizon is quite short for energy planning or carbon abatement analysis.

To construct a CO_2 emission abatement cost curve entails that we address a series of methodological issues in the Brazilian context. These include: the reference scenarios; and the parameters and scope of cost-benefit analysis. Choosing a reference scenario is not trivial – especially in Brazil where key economic and political factors are highly uncertain. The reference scenario greatly influences the estimate of abatement potential. Here, we take the socio-economic assumptions and the energy projections of the low growth, 'tendencies' scenario (described above) as our reference scenario.

Cost-benefit parameters used in the analysis must be comparable to those used in other countries, or subject to sensitivity analysis. A key parameter is the cost of capital, which is higher in Brazil than in the industrialized countries. We assume a 12 per cent discount rate. This discount rate is quite high but it reflects Brazil's high debt burden and the scarcity of internal investment capital.

The scope of the cost-benefit analysis can greatly affect the evaluation of net costs and benefits. This issue is relevant to quite specific technologies (such as vehicles) as well as to broader system changes (such as shifting transport modes). Analysis of the latter type of changes, however, is particularly susceptible to changes in analytical scope. The narrowest definition of benefits is the energy saved or substituted. We always incorporate such benefits in this study. In cases such as electrical equipment, this definition of benefits is adequate. But in others, it gives an erroneous estimate of net costs. Governments, for example, undertake many energy-related investments to achieve multiple objectives, including stimulating productivity, improving the quality of products, responding to environmental constraints, even improving social welfare (as with much rural electrification or public transport). In such cases, much broader definitions of benefits enter the picture which may outweigh the values attributed to energy savings or carbon abatement.

Finally, we used current 'frozen' costs in 2000. That is, our analysis of energy savings (and associated carbon reduction) in 2000 is based on current new technology costs and current energy prices. We selected only more efficient technologies that are already economic compared with existing technologies. We also assumed no increase in the real price of oil, oil derivatives, and electricity in 2000. This approach is conservative as the cost of new technologies often falls as it enters service; and the price of fossil fuels may be expected to increase in real terms as it becomes increasingly scarce.

Energy subsector analyses

Energy sector use and transformation

Electricity generation

Electricity generation in Brazil is dominated by hydropower (94 per cent). As already noted, we assume that hydropower generates zero net CO_2 emissions. We expect that this share will fall by 2000. In the official SNE scenarios, the share falls from 94 to 85–88 per cent (ELETROBRAS 1990).[g] Here we assume a moderate shift to thermal generation such that hydropower falls to 90 per cent of total generation by the year 2000. By then, 20 per cent of newly available generating capacity coming on-line would be thermal (of which 10 per cent would be coal-fired; 40 per cent would be natural-gas-fired; 30 per cent would be fuel-oil- and refinery residues-fired; and 20 per cent would be biomass-fired). We estimate that generating 1 GWhe would result in 37.2 TC emissions in Brazil in 2000, an increase from today's 9 TC/GWhe. This coefficient would be very low by international standards. The analysis also assumes that 16 per cent of total generated

g These projections are not consistent with the current ten-year plan for the electrical sector, but we will ignore this fact in this study.

electricity is lost in transmission and distribution losses incurred in delivering power to the residential, commercial and public illumination sectors.

The shift to greater thermal generation is a response to a series of problems faced by the power sector. These include: a profound financial crisis (hydropower is capital intensive); the large market uncertainty confronting supply planning (larger hydropower has long lead times); and the need to attract more private sector investment in generation, which tends to favour thermal power plants. A more hydropower-intensive scenario is conceivable but we have not reviewed this possibility as it would require extensive, system-level analysis beyond the scope of this chapter. It is noteworthy, however, that Brazil's thermal electrical sector would not be very carbon-intensive.[h]

Other energy transformation

This end use includes all energy inputs to extract and process fuel oil rigs, refineries, sugar plantations, etc. Half of the sectoral carbon emissions are due to the conversion of fuelwood to charcoal. As is generally the case with biomass fuels used in industry, the production of charcoal is inefficient. This inefficiency can be greatly reduced; this must be done in any case to ensure that fuelwood from sustainably managed forests for charcoal manufacture is economically viable. Reducing conversion waste in the charcoal fuel cycle will deliver the biggest abatement of carbon emission in the subsector. Although less dramatic, incremental improvements in efficiency are also available in most industries. These small but pervasive gains would accumulate into significant reductions of fossil fuel usage and carbon emissions.

Electricity final demand

Electricity generation will remain a relatively small source of CO_2 emissions for many years in Brazil. Yet it is still desirable to reduce these emissions by demand side management. Many of these opportunities to increase efficiency have a large 'negative cost.' Exploiting these opportunities will also relieve the financial pressures felt by the electrical sector in Brazil.

Residential sector

The residential sector consumes about 22 per cent of total electricity and contributes 30–37 per cent of the electricity evening peak. We have considered a range of technical options that could be implemented in this sector up to 2000. For each of the following final uses, we estimated the

h Natural gas in cogeneration and high efficiency central stations have low emissions. The use of sugarcane bagasse results in no increased emissions from alcohol production inputs. The heavy refinery residues used are a by-product of other petroleum derivative demand.

Table 10.7 *Residential sector technologies to limit CO_2 emissions implementable by 2000*

	Old technology (1)	New technology (2)	Avoided C (kg/y/unit) (3)	Levelized annual cost ($/kg C) (4)	Annual cost of saved elec. ($/kg C) (5)	Net annual cost ($/kg C) (6) = (4–5)	Total C avoided (MTC/y)	Total net cost (million $/y)
	Incandescent bulb (standard)	*Efficient incandescent*	0.260	0.192	1.989	-1.797	0.0385	-69.26
Lifetime (h)	1000	1000						
Power (W)	60	54						
Usage (h/y/lamp)	1000	1000						
Cost (US$/lamp)	0.50	0.55						
	Incandescent bulb (standard)	*Fluorescent (standard)*	1.560	0.485	1.989	-1.504	0.1386	-208.40
Lifetime (h)	1000	8000						
Power (W)	60	24						
Usage (h/y/lamp)	1000	1000						
Cost (US$/lamp)	0.50	7.00						
	Incandescent bulb (standard)	*Compact fluorescent*	1.910	1.338	1.989	-0.651	0.1129	-224.50
Lifetime (h)	1000	8000						
Power (W)	60	16						
Usage (h/y/lamp)	1000	1000						
Cost (US$/lamp)	0.50	17.00						
	Electric shower	*Solar shower*	17.350	4.146	1.989	2.157	0.2563	553.00
Lifetime (y)	15	15						
Consumption (kWh/y)	500	100						
Cost (US$/unit)	10.00	500.00						
	Refrigerator (standard)	*Efficient refrigerator*	5.210	2.255	1.989	0.266	0.0857	22.80
Lifetime (y)	15	15						
Consumption (kWh/y)	600	480						
Cost (US$/unit)	200	280.00						
	Air conditioner (standard)	*Efficient air conditioner*	7.460	1.297	1.989	-0.692	0.0170	-11.80
Lifetime (y)	12	12						
Consumption (kWh/y)	860	688						
Cost (US$/unit)	490.00	550.00						

carbon abatement that can be obtained by improving electricity end-use efficiency: lighting, water heating, refrigeration and air conditioning. In Table 10.7, we show the technical and economic specifications for the new technologies and those being replaced; the amount of carbon abated per year per unit of equipment; the levellized annual cost of the new technology; and, in the last column, the net costs when the annual value of the electricity saved is deducted from the levelized costs.[i] A negative figure means that the new technology brings economic savings, even before considering carbon credits.

Lighting

Based on the present stock of residential lamps (220 million incandescent lamps and eight million fluorescents), we estimate the stock in 2000. We assume that 50 per cent of old, inefficient incandescents will be replaced by efficient incandescents; 30 per cent by fluorescents; and 20 per cent by compact fluorescents. Such a programme would reduce carbon emissions by 0.29/MTC in 2000, providing a total net annual levelized benefit of US$502 million.

Water heating

In 1987, 75 per cent of Brazil's 26.3 million electrified households had electric showers. Assuming that the number of showers increases by 2.1 per cent per year up to 2000 and further, that solar water heaters achieve a 50 per cent penetration level in that year, we estimate that another 0.256 MTC can be conserved at an annual net cost of US$553 million.

Refrigeration

Refrigeration accounts for about 33 per cent of total residential electricity use and is very wasteful (Jannuzzi and Schipper 1991; Geller 1991). We projected the stock of refrigerators in the year 2000 based on today's stock of 27 million. Assuming that 50 per cent of the stock in 2000 is efficient, we

---

i Levelized cost refers to a technique used by engineering economists to present value and then annuallize a stream of costs and benefits over time that is associated with a project life cycle. It is conducted to enable a comparison between different methods of delivering the same product or service, given that each one has a differently timed flow of front end costs, annual operating and maintenance costs, benefits, and final salvage value. For example, if compact fluorescent lamps are compared with incandescent lamps, how can these options be compared given that the first option costs US$17 and lasts an average of 8000 hours, whereas the second costs only US$0.5 and lasts an average of 1000 hours. Given a real discount rate, a levelized cost can be computed for both options that converts each into an annual, mathematically equivalent cost per unit output – in this case, US$4.44 and US$5.68 respectively. Thus, the first option saves the consumer US$1.24 per year. As it also emits 0.69 versus 2.6 kg of carbon per year, it also reduces carbon emissions by 1.91 kg/year. The cost of conserved carbon, therefore, is US$1.24/1.94 kgC per year, or $0.65/kgC conserved by replacing incandescent lamps with compact fluorescents.

Table 10.8 *Commercial and public sector technologies to limit CO_2 emissions, implementable by 2000*

	Old technology (1)	New technology (2)	Avoided C (kg/y/unit) (3)	Levelized annual cost ($/kg C) (4)	Annual cost of saved elec. ($/kg C) (5)	Net annual cost ($/kg C) (6) = (4–5)	Total C avoided (MTC/y)	Total net cost (million $/y)
	Incandescent bulb (standard)	*Efficient incandescent*	0.830	0.229	2.690	-2.461	0.0053	-12.95
Lifetime (h)	1000	1000						
Power (W)	100	48						
Usage (h/y/lamp)	1920	1920						
Cost (US$/lamp)	0.80	0.90						
	Incandescent bulb (standard)	*Fluorescent (standard)*	4.330	1.309	2.690	-1.381	0.0618	-111.16
Lifetime (h)	1000	8000						
Power (W)	100	48						
Usage (h/y/lamp)	1920	1920						
Cost (US$/lamp)	0.50	22.50[a]						
	Incandescent bulb (standard)	*Compact fluorescent*	2.290	1.345	2.690	-1.345	0.0485	-96.26
Lifetime (h)	1000	8000						
Power (W)	60	16						
Usage (h/y/lamp)	1200	1200						
Cost (US$/lamp)	0.50	17.00						
	Fluorescent (standard)	*Efficient fluorescent*	1.000	0.990	2.690	-1.700	0.0645	-109.68
Lifetime (h)	8000	8000						
Power (W)	48	36						
Usage (h/y/lamp)	1920	1920						
Cost (US$/lamp)	7.00	10.00						
	Incandescent bulg (standard)	*Mercury/sodium*	15.050	1.928	1.548	0.380	0.015	5.74
Lifetime (h)	1000	12000						
Power (W)	175	80						
Usage (h/y/lamp)	3650	3650						
Cost (US$/lamp)	1.00	80.00[b]						

a Includes lamp (8000h), ballast (12000h) and light fixture (20000h).
b Average cost of mercury and sodium package with 80% mercury lamps.

estimated that 0.086 MTC can be abated at a total net cost in that year of US$22.8 million.

Air conditioning

We assumed that 2.85 million air conditioners will be in use in residences in the year 2000. Of this total, 80 per cent will be installed after 1990. We further assumed that all will be energy saving units. On this basis, 0.017 MTC can be conserved in 2000 at a total net benefit of US$11.8 million per year.

Commercial, service sector and public lighting

Total energy consumption of the commercial and services sectors has increased threefold during the past fifteen years, and electricity accounts for more than 90 per cent of the increase. We considered only opportunities to increase the efficiency of lighting. In Table 10.8. we present the technical and economic data for the new and old technologies, and quantities and costs of carbon abatement. We assumed the current tariff for commercial electricity of US$0.10/kWhe. (In the future, we would prefer to use an estimate of the economic cost of electricity). We also assumed that 26 million incandescent and 48 million fluorescent lamps were used in these two sectors in 1990 (Jannuzzi *et al* 1991). We postulated a 2 per cent annual growth rate for incandescent sales and 3 per cent for fluorescents (taken as 40W lamps). We estimate that all the incandescent lamps could be replaced in the year 2000 as follows: 20 per cent by krypton-filled bulbs; 45 per cent by standard fluorescents; and 35 per cent by compact fluorescents. This substitution would conserve 0.116 MTC of emissions and increase total net benefits that year by US$220 million. Replacing all fluorescents of 40W with more efficient 32W units with electronic ballasts would conserve another 0.064 MTC, and increase total net benefits by US$110 million.

In Table 10.8, we also present results for improvements in public illumination. If 800,000 incandescent lamps are replaced by mercury lamps and 200,000 by sodium lamps, then 0.015 MTC could be conserved at an annual cost of US$5.7 million.

Industrial sector

To calculate industrial electricity demand in 2000, we assumed an average industrial economic growth of 3.5 per cent per year (as suggested by the low growth 'tendencies' scenario) and the same energy intensity (total energy/GNP) as in 1988. We analysed electricity consumption for each industrial subsector and factored in technological improvements believed to be currently economic. We included the following measures: housekeeping; efficient lighting; and more efficient electric motors, variable speed electric motors, electric ovens, and electrolytic processes. We priced industrial electricity at an average US$0.058/kWhe. In Table 10.9, we list technical and economic data for these improvements.

Table 10.9 *Industrial sector technologies to limit CO_2 emissions, implementable by 2000*

	(1) Avoided carbon (MTC/y)	(2) Levelized annual cost (US$/TC)	(3) Annual cost of saved electricity (US$/TC)	(4)=(2)-(3) Net annual cost (US$/TC)
Housekeeping measures	0.32	409	1560	-1151
Lighting	0.02	988	1560	-572
High efficiency motor	0.115	383	1560	-1177
Variable speed drivers	0.197	357	1560	-1203
Electric ovens and boilers	0.152	197	1560	-1363
Electrolytic processes	0.075	389	672	-283

Housekeeping measures

From many evaluations performed by electric utilities (CEMIG 1989; CESP 1990) it is clear that electricity demand can be reduced by 10 per cent with better end-use management. These measures should include:

- a better choice of electric motor size, especially by avoiding oversized motors which are common (Latone *et al* 1990);
- appropriate design of the factory's internal electric distribution grid;
- installation of small size transformers in parallel with the main one, to be used during idle factory periods;
- correction of the load factor;
- avoidance of short term peak demand (through the use of demand controllers); and
- better mechanical coupling between electric motors and the equipment driven by them.

These actions would conserve 0.32 MTC for a total annual net benefit of US$368 million.

Lighting

The industrial sector contains about 14 million fluorescent 40 watt lamps. We assumed that this stock grows by 3.5 per cent per year and that all new lights will be the 32 watt efficient type (with new ballast). These steps would conserve 0.020 MTC/year yielding a total annual net benefit of US$11.3 million.

High efficiency electric motors

These devices are available in Brazil and offer an improved average efficiency of 7 per cent. We assumed that 60 per cent of motors will be replaced by high efficiency models up to the year 2000. This step would save 3.1 TWhe in that year (Geller 1991), saving Brazil US$135 million and avoiding 0.115 MTC emissions.

Variable speed electric motors

Variable speed motors can be run partly loaded without decreasing energy efficiency. More important, in applications such as refrigeration and air circulation, such motors can operate partly loaded with less energy consumption than do fixed frequency motors that provide only on/off cycles. About 30 per cent of the total motor market (measured in kWhe/year) is used to drive variable loads. If half of this load is met by variable speed motors, Brazil would reap a total annual net benefit of US$237 millions and conserve 0.197 MTC.

Electric ovens and boilers

At least 10 per cent of the electricity used in electric ovens and electric boilers can be avoided by recycling the exhaust heat or by installing more efficient equipment (Geller 1991). We expect that the retrofit rate will be low. Nonetheless, one third of the potential saving could be achieved by the year 2000. We estimate that the total annual net savings that year would be US$207 million, thereby conserving 0.152 MTC.

Electrolytic processes

Studies on electricity intensive industries (CNE 1989) have shown that improvements in electrolytic processes can reduce electricity consumption by 7 per cent in metallurgical industries and by 10 per cent in chemical industries. Brazil could avoid generating 2 TWhe by this means in 2000, saving a total annual net benefit of US$21.2 million and reducing carbon emissions by 0.075 MTC.

Final electricity demand summary

Modern technologies can improve end-use electricity efficiency, although their cumulative impact on carbon emissions is small in a country like Brazil where most of the electricity is provided by hydroelectric power plants. We estimate that the potential to conserve carbon in the residential, commercial, service, public illumination and industrial sectors in 2000 is about 1.7 MTC – or only about 3.2 per cent of the total fossil fuel carbon emissions of 1990. This result follows from the predominance of hydroelectricity supply in Brazil. Nonetheless, these technologies should be promoted in any case because most bring net economic benefits. The few which involve net costs today should become cheaper as the technology is accepted more widely and economies of scale are achieved. The only high cost technology where this trend may not hold is solar heating to replace electric showers.

Final fuel demand

The final consumption of fuels is the predominant source of CO_2 emissions in Brazil (see Table 10.3). We focus on the transport sector in this section because it is the largest source of fossil CO_2 emissions.

Transport

Transport is central to the achievement of carbon emission abatement. In Brazil, this sector is responsible for 44 per cent of fossil carbon dioxide emissions. Most (82 per cent) of these emissions come from road transport (SNE 1991b). Transport-related carbon emissions result from three broad factors: the carbon emission coefficient of the fuel used; the efficiency with which different energy forms are used in different modes and markets; and the demand for transport services in different markets and different modes that serve these markets.

In this section, we review the measures which influence these three determining factors in the transport sector. Fuel substitution can modify the first factor, the carbon emission coefficient. Vehicle efficiency affects the second. 'Structural changes' influence both the demand for different modes of transport; and the performance of the vehicles operating within them, that is, the second and the third factors listed above. These first two classes of measures are dominated by energy sector objectives and priorities. The economics of the third, the category of 'structural changes,' is determined by non-energy societal benefits. Consequently, economic analysis of the third factor is more complex than for the first two.

Fuel substitution

The emergence of CO_2 emissions as an issue will have a profound effect on the development of substitutes for petroleum derivatives in the long run. The only large scale commercial substitutes in the world which reduce rather than increase emissions are ethanol from sugarcane (the production of which is concentrated in Brazil) and compressed natural gas.

Alcohol

Brazil's well-known alcohol programme (PROALCOOL) now fuels nearly five million Brazilian cars with pure (hydrated) ethanol. The rest use a gasoline-alcohol mixture. PROALCOOL is now in a difficult situation due to the collapse of oil prices in 1986. Some have suggested that alcohol output should be reduced gradually (World Bank 1990). We have not calculated the societal cost of maintaining current output, nor the cost of increasing the output level. Such an estimate should include the impact on sugar prices of reducing alcohol output (Borrel 1991) and on employment – especially in the Northeast where the sugarcane industry is in crisis. However, the net cost of merely maintaining the output of alcohol should be lower than for expanding it (see Table 10.10).

The 'alternative' SNE scenario projected moderate growth of 2.1 million of tonnes of oil equivalent (MTOE) or 37 per cent by 2000, thereby increasing the market share of ethanol from 17 per cent to 19 per cent of total transport fuel. We adopted this estimate in our own scenario. This substitution would conserve about 1.5 MTC (see Table 10.10), even allowing for fossil fuel inputs for alcohol production. The net cost of this expansion is heavily

Table 10.10 *Transport sector opportunities to limit CO_2 emissions by 2000, preliminary examples*

	Avoided CO_2 (MTC)[a]	Net cost/TC (US$)[a]
FUEL SUBSTITUTION		
Alcohol: maintain existing output	4.2	not available[b]
Alcohol: expand output		
Current technology at $25/b gasoline	1.5[e]	230[c]
Current technology at $35/b gasoline	1.5[e]	115[c]
New technology at $25/b gasoline	very small	75[d]
New technology at $35/b gasoline	very small	-35[d]
Natural gas	0.2[f]	near zero (+,-)
VEHICLE EFFICIENCY		
Improvement in automobiles	1.9-2.5[k]	-135[j]
Diesel engine	0.9[h]	-30/-40[i]
STRUCTURAL CHANGE		
Highway system recovery	2.4[g]	800[g]
Improved urban transportation	0.4-0.8[l]	near zero (+,-)

a Includes rough estimates of fossil fuel inputs for alcohol production (15% of alcohol output) and of refinery efficiency for gasoline (95%). Elsewhere in column a 95% refinery efficiency is assumed.

b Estimated to be lower than increasing alcohol output. Baseline scenario assumes maintenance of existing output.

c Assumes a litre of hydrated alcohol is equivalent to 0.7 litres of gasoline (small efficiency credit). Cost of good exiting distillery is $0.20 per litre. Allows for refinery efficiency of 95% in gasoline production and fossil fuel inputs equivalent to 15% of alcohol production.

d Assumes alcohol production cost of $0.14 litre, only available by end of the decade.

e Based on 'alternative' scenario relative to 1990 (see text).

f Based on 'alternative' relative to 'tendencies' scenario.

g Assumes baseline of 14 MTOE diesel consumption and 3.7 MTOE gasoline in 2000 compatible with 'tendencies' scenario assuming same proportion of total diesel and gasoline transport use as today. Assumes an average 15% improvement for all vehicles. While up to 40% improvement is possible from worst to best conditions, not all of the roads needing improvement, about 50% of the roads are in the 'worst' category.

h Assumes 25% of market of 21.4 MTOE in 'tendencies' scenario shifted to this engine type.

i Assumes conservatively average vehicle use at 45,000 km per year with 3-year engine lifetime, engine 25% more expensive and average efficiency improved by 15%.

j Assumes existing gasoline price ($0.26 per litre). Average cost of measures is $0.17 per litre, adjusting cost estimates of Ledbetter and Ross (1990), for 12% discount rate.

k Ledbetter and Ross (1990) estimate US average fleet fuel economy could increase 25% by 2000. Assume here that in 'tendencies' baseline average Brazilian fuel economy increases to existing US level (estimated at 9.2 km/litre).

l See text. This is the least defined case. It helps to illustrate the impact of a relatively small effort to improve urban transportation.

influenced by assumptions as to gasoline prices. At a gasoline price of US$25 per barrel, we estimated the net cost of expansion to be US$230/TC. Changes in alcohol production technology may also substantially reduce costs by improving the utilization of the residues of sugarcane processing in the cogeneration of electricity. In Brazil, conventional steam turbine technology does not offer much hope of reducing alcohol production costs. But the new gasification/gas turbine (BIG/GT) technology could reduce costs significantly (Ogden *et al* 1990). Table 10.10 illustrates this possibility. Such major reductions are likely to be commercially proven only by the end of the decade.

Natural gas

Natural gas is promoted as a substitute for diesel, primarily in public transport. The main goal is to reduce atmospheric pollution (NO_x, SO_x, particulates) in metropolitan areas. Several cities aim to replace all diesel in vehicle fleets by around 2000. The 'alternative' scenario estimates that 0.9 MTOE of natural gas may be used in this fashion which would conserve O.2 MTC. Some of this gas may displace gasoline instead of diesel, since this is more lucrative at current relative prices. The cost of the measure (excluding environmental benefits) is near zero. That is, the natural gas option roughly breaks even with diesel at today's prices.

Road transport vehicle efficiency

Significant improvements in vehicular fuel economy are possible. The rate of improvement will be determined primarily by technological innovation in the automotive industry in the industrialized countries and secondarily, by the pace that these changes penetrate the Brazilian market.

Light vehicle efficiency

The current Brazilian automobile averages about 7.5 km/l (gasoline equivalent). The SNE 'tendencies' scenario projected an increase of 30 per cent in light vehicle (basically automobile) fuel consumption to 17 MTOE, incorporating modest improvements in fuel economy. We extrapolated from US data on trends in automobile fuel efficiency (Ledbetter and Ross 1990), and considered only measures that cost less when operated at the consumer retail price of gasoline today. On this basis, we estimated that the average fuel use of the automobile fleet could be decreased by 20–25 per cent relative to the 'tendencies' scenario, saving 3.4–4.3 MTOE of fuel per year.[j] The associated carbon abatement depends on whether these savings cut gasoline

j To obtain this estimate, we assumed 10 per cent of the vehicle stock are replaced each year. But in Brazil, surveys show that new cars are driven twice as far as old cars – typically 24,000 km/year versus 12,000 km/year. We assumed that two-thirds of the new cars run on gasoline; that the present fleet has 11 million cars; and that five million of the latter run on pure ethanol, and six million use a blend of gasoline with ethanol and methanol. Seven per cent of the fleet retire each year. The vehicle stock increases, therefore, by 3 per cent per year. On this basis, automobiles will increase by 30 per cent by 2000 – the same as assumed by the SNE 'tendencies' scenario. We assumed that new vehicles are 23 per cent more efficient than those that they replace. Since our estimate of the number of cars in 2000 is the same as SNE, but the new vehicles are more efficient, our scenario saves an additional 1.9–2.5 MTC in 2000.

rather than alcohol usage. If we assume that two-thirds of the fuel saving is gasoline, then carbon emissions would fall by 1.9–2.5 MTC. Based on US costs, the average cost of these measures would be negative (see Table 10.10) – even if we ignore the likely benefits of reducing the emissions of other local pollutants. Our projected large fuel saving contrasts strikingly with that of the SNE official 'alternative' scenario, which projected that only 0.4 MTOE could be saved by increasing automotive fuel efficiency.

Heavy vehicle efficiency

We estimated heavy vehicular fuel use from figures available for diesel consumed by Brazilian road vehicles. On this basis, this end use in 1990 was 15.5 MTOE. The SNE 'tendencies' scenario projected that it would increase to 21.5 MTOE. More efficient diesel engines offer substantial fuel savings at a zero or slightly negative net cost (Cummins 1991). We assumed that 15 per cent of the fuel can be saved in this sector. As the initial cost of the engine is about 25 per cent greater than less efficient motors, the net cost would be US$30–40/TC conserved. If the more efficient motors achieved a 25 per cent additional market share than in the SNE 'tendencies' scenario, then another 0.9 MTC would be conserved. Other measures can improve heavy vehicle fuel economy, including improved maintenance and motor regulation and more appropriate sizing of vehicles to their tasks.

Structural changes

In this section, we outline structural changes that foster energy efficiency. These are: the repaving and rehabilitation of the existing highway system; and the reform of urban transport to improve public transport and the overall productivity of urban transport infrastructure. Both approaches entail large public sector investments in which non-energy costs and benefits usually determine policy decisions. Broad rather than narrow economic analysis must be used to estimate the economic feasibility of these changes.

Highway systems recovery

Brazil has an extensive, but badly deteriorated, intercity highway system. Roughly half of its 130,000 km system requires intense reconstruction. The economic cost of this decayed infrastructure is high in terms of trip time and reliability, vehicle maintenance and lifetime, and hazard. The poor system also reduces energy efficiency. For inter-urban trucks and buses the loss may be as high as 40 per cent on poor quality paved roads relative to well-maintained roads (GEIPOT 1989). Assuming an average fuel economy gain of 15 per cent and a baseline interurban vehicle fuel consumption consistent with the 'tendencies' scenario, CO_2 emissions in 2000 could be reduced by about 2.4 MTC. The investments required are large and the lifetime of the assets is as short as five years (Lee 1991). The cost (if fully charged to carbon abatement) could exceed US$800/TC conserved (including only fuel savings). Much – perhaps most – of this reconstruction is economically justified without reference to carbon emissions. In that case, the energy savings and carbon abatement can be treated as a by-product obtained at zero marginal cost.

Improved urban transportation

Approximately one third of transport fuel is consumed in the capital cities and larger metropolitan areas. Changes in the structure and operation of the urban transport systems can influence the evolution of fuel demand, though the evaluation of this potential is still in its infancy. Measures are diverse and include land use control, disciplining the automobile's use of road space, coordinating traffic flow and strengthening collective transport (Poole *et al* 1992).

The city of Curitiba has already addressed the problem of urban transport in a comprehensive manner. The city has coordinated urban land use, roadspace and collective transport policy for more than fifteen years, and has improved bus systems and traffic controls. Fuel consumption per car in Curitiba is about 30 per cent less than the average for other cities of its size in Brazil (Lerner 1989). This differential may indicate the potential impact of such measures if adopted widely in Brazil.

An aggressive programme to address urban transport imperatives might result in CO_2 emissions savings of 5–10 per cent relative to the SNE 'tendencies' scenario for 2000. This potential could be realized in spite of the inertia of a decentralized system involving thousands, even millions of actors. If lower cost solutions are emphasized in the next decade, then this energy and carbon savings could be achieved at zero net cost. Such measures would include creating and integrating public transport systems, and controls on traffic flow and parking.

Changing land-use trends

By far the largest source of anthropogenic carbon dioxide emission in Brazil is deforestation, principally in Amazonia. The carbon stock of the seasonal and rainforest vegetation in Amazonia is estimated to range from 140 to 200 TC/ha; that of pasture is 10 TC/ha; and of cropland, 5 TC/ha. The forest carbon stock may be adjusted as new information becomes available on subsurface biomass of the vegetation. Changing land use also reduces soil carbon content. In pasture soil, for example, the carbon content may be about 10 per cent of the approximately 100 TC/hectare of forest soil, or about 90 TC/ha less than in forests. (Houghton *et al* 1991).

Thus, assuming a deforestation rate in Amazonia of 1.8 million hectares per year, gross CO_2 emissions would be 250–360 MTC (though not all appears immediately in the atmosphere). To this figure should be added emissions from deforestation in other regions of Brazil. Unfortunately, we have no estimates for this source. Although substantially smaller, these are not insignificant.

Biological processes also continually accumulate carbon from the atmosphere, as is the case with regrowth of natural vegetation on deforested areas, abandoned land, or forest plantations. The rate of natural regrowth can vary by a factor of twenty in humid tropical areas depending on the local land-use situation (Nepstad *et al* 1990). The scale of this countervailing sequestration is poorly understood.

Despite these uncertainties, it is clear that Brazil's annual emissions from deforestation (250–360 MTC) dwarf those of fossil fuel use (60 MTC) as well as from biomass use for energy (about 11 MTC). This fact is consistent with the observation that fuelwood use is not a major factor in overall deforestation, though it may be significant in some regions (for example, charcoal from *cerrado* and mangroves). The primary direct causes of deforestation are clearing for pasture and cropland, with logging often opening up the occupation process.

Focusing on Amazonia, any substantial decrease in the rate of deforestation is likely to be associated with decreased economic growth. Macroeconomic modelling suggests that for every 1 per cent reduction in deforestation regional GDP would have to fall by roughly 1.7 per cent (Reis 1991).[k] While pessimistic, the model suggests a first approximation of the cost of CO_2 abatement by halting deforestation, roughly US$4/TC according to the model's author. This low cost (equivalent to a tax of $0.50/barrel of oil) is probably an upper limit, since the model assumes historical relationships. A strategy to change these relationships should be both cheaper and allow a less drastic trade-off between economic growth and deforestation. Such a strategy must go beyond police enforcement or reducing/eliminating legal and financial incentives to deforestation, though these are important (for example, Binswanger 1991). New or modified economic activities must be developed or strengthened both in forested and deforested areas (Sawyer 1990) based on land-use zoning. Settlement and economic activity, for example, should be stabilized, consolidated, and (in many areas) intensified in the largely deforested areas along the frontier and the 'pre-frontier'. While complex, restructuring Amazonia's economy is likely to be a large, 'no regrets' source of CO_2 abatement.

The relationship between land-use trends and energy policy has been little explored in Brazil. The most important such interaction is fuelwood for industry and charcoal. This nexus is the most important direct energy-related source of deforestation. A key issue is whether a decisive move to put

k We use this estimate as a first approximation of the cost of halting deforestation. Further study may show that the economic gains that accounted fully for societal values preserved by halting deforestation are greater than the financial costs to Amazonia itself. This issue requires much more study and is beyond the scope of this paper.

these uses on a sustainable basis is justified or whether they should be phased out.

Another important land-use issue in relation to energy arises from hydroelectricity development in Amazonia. The relative priority, rate of development, and ultimate potential may all be influenced by a strategy to minimize deforestation. The infrastructure and migrations occasioned by hydro are the key concern. Some projects may provoke deforestation. Others help to decrease it as, for example, on the Tocantins river (Moreira *et al* 1990). This indirect effect on carbon emissions is likely to be larger than differences in direct electricity CO_2 emissions resulting from alternative scenarios of hydropower/thermal generations (as discussed above).

Two subsidiary issues also connect land-use and energy policy issues. The unavailability of electrical power to isolated communities (most especially in Amazonia) constrains economic development. Poverty, in turn, fosters more carbon-emitting and intensive resource exploitation (Poole *et al* 1990). Relatedly, fuels such as diesel sold for use in Amazonia are subsidized (Reis 1991). The common denominator of CO_2 emissions reinforces the need to consider energy, land use and regional development together.

Conclusion

The analysis of CO_2 evolution and abatement measures is still incipient in Brazil both in regard to energy and to land-use change. The creation of credible, systematic and internationally comparable 'abatement cost' curves for Brazil is still not possible. As a consequence there is as yet little basis on which to agree on specific CO_2 limitation targets. The problem is exacerbated by the wide range of uncertainty surrounding the prospects for economic growth. The work briefly described here is part of an effort to better understand the potential and economics of CO_2 abatement. In Figure 10.1, we summarize the tentative results of this study for the costs and abatement potential of measures to reduce fossil energy emissions. We identified about 13.7 MTC[1] of technologically feasible and economically justified carbon abatement, relative to the SNE 'tendencies' scenario in the year 2000. This reduction potential amounts to about 16 per cent of the SNE reference projection for carbon emissions in that year. We did not allow for the more pervasive energy and carbon reducing effects of technological innovation in all end-using sectors; nor did we include the impact of shifts in the sectoral composition of the economy on our estimate of energy and carbon conservation. Moreover, as our bottom-up calculation is extended to other sectors, we expect to increase substantially the size of the carbon reduction potential above 13.7 MTC.

1 Being 0.659 MTC from Table 10.7; 0.1951 MTC from Table 10.8; 0.8799 MTC from Table 19.9; and 11.8 MTC from Table 10.10.

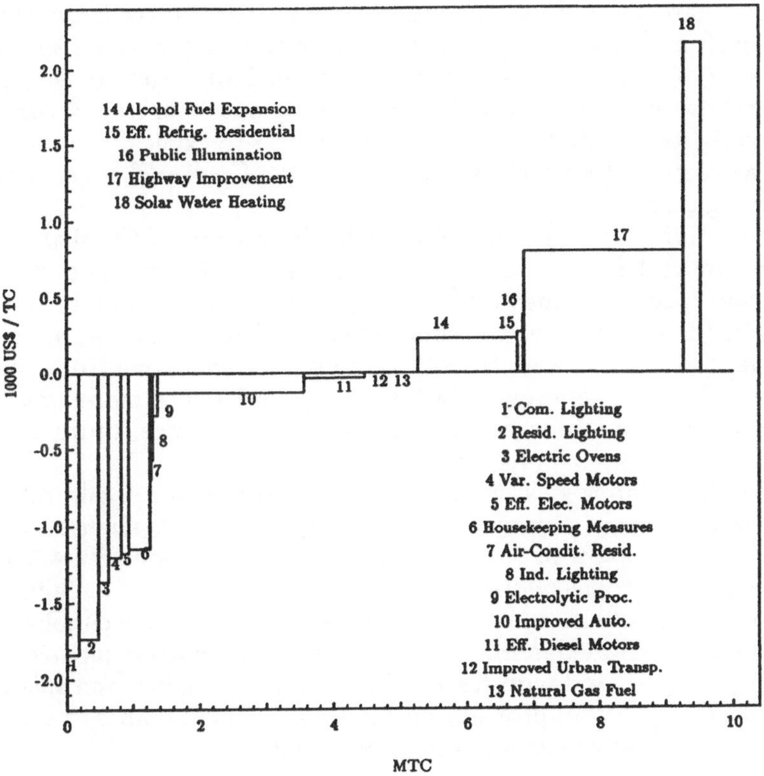

Figure 10.1 *Cumulative annual carbon emissions avoided by 2000 for technology improvements in Brazil*

Although the quantitative analysis is incomplete and preliminary, it permits some observations which are relevant for policy. It appears that substantial savings in CO_2 emissions can be achieved at 'negative cost' or very low cost (say, less than US$10/TC), both in energy and land-use change. These savings should be substantially cheaper than those of many measures being considered by the industrialized countries. However, the fact that these are 'no regrets' savings does not mean that they are easy to achieve. This fact is relevant for a possible policy of international resource transfers, which should be very attentive to 'no regrets' opportunities in developing countries.

In Brazil's energy sector, the major 'no regrets' savings involve increasing energy efficiency in all consuming sectors. As explained earlier, conditions and policies favouring greater energy efficiency are also likely to favour, and are associated with, higher medium term economic growth (low inflation, correct price signals, investment in modernizing processes and products, competition, etc). Energy efficiency itself should directly contribute to improved overall productivity and thus to economic growth.

With regard to land-use change, we argue that the trade-off between Amazonia's economic growth and reducing deforestation is not so acute as some suggest. However, a strategy is needed urgently to change the economic dynamics of Amazonia's frontier regions. The case of Amazonia also highlights the need for cost-benefit analysis to explicitly consider who pays the costs and who receives the benefits of policy changes aimed at conserving carbon.

The medium term potential for relatively low cost CO_2 abatement is probably much larger for land-use change than in the energy sector. This judgement does not mean that energy efficiency should be ignored, however. Many measures to reduce energy-related emissions are as cost effective, or even more so, than those to reduce emissions from land-use change. Moreover, energy is fast becoming relatively more important. Finally, as we have observed, energy policy can influence future land-use, especially in the Amazon region.

Using the technologies discussed in this chapter, the total amount of carbon abatement is approximately 13.5 million TC. In Figure 10.1, we plotted a total amount of 9.3 million TC, since we excluded the 4.2 million TC due to the use of alcohol at the existing output (as shown in Table 10.10) because we did not know the net cost associated with this technology. The opportunities identified in this chapter are not the complete picture. Other technologies exist, and with their full inclusion on the demand side of the Brazilian energy matrix, probably more CO_2 abatement than is forecasted in the official alternative scenario can be achieved.

References

Ayres, R U (1991) 'No regrets options for greenhouse gas abatement,' in *A Comprehensive Approach to Climate Change*, ed. T Hanisch, Center for International Climate and Energy Research, Oslo

Binswanger, H (1991) 'Brazilian Policies that Encourage Deforestation in the Amazon', *World Development*, 19(7)

Borrel, P (1991) 'How a Change in Brazil Sugar Policies Would Affect the World Sugar Market,' PRE, Working Paper, World Bank, Washington DC

CEMIG (1989) Companhia Energética de Minas Gerais, 'Estudo de Economia de Energia', Belo Horizonte-MG, Brazil

CESP (1990) Companhia Energética de São Paulo, Personal communication, São Paulo, Brazil

CNE (1989) Comissão Nacional de Energia, 'Estudo Sobre Energo Intensivos, Assessoria Técnica', April, 1989, Brasilia, Brazil

Cummins Newsletter (1991) published by Cummins Engine Co. Inc., Columbus, USA

ELETROBRAS (1990) Programa Decenal de Geração 1991/2000, Sistemas Interligados SE/CO/S, N/NE. Planning of Grupo Coordenador do Planejamento dos Sistemas Elétricos, Rio de Janeiro, Brazil

GEIPOT (1989) Empresa Brasileira de Planejamento de Transportes. *Manual para Cálculo de Custos Operacionais de Veículos Rodoviários*, Ministério dos Transportes, Brasilia, Brazil
Geller, H (1991) *Efficient Electricity Uses: a Development Strategy for Brazil*, American Council for an Energy Efficient Economy, Washington, DC, USA
Houghton, R A, Skole, D L and Letkowitz, D S (1991) 'Changes in the landscape of Latin America between 1850 and 1985: II Net release of CO_2 to the atmosphere,' *Forest Ecology and Management*, 38, pp 173–199
Jannuzzi, G M, Gadgil, A, Geller, H, Sastry, A (1991) 'Energy efficient lighting in Brazil and India', *Proceedings of the 1st European Conference on Energy-Efficient Lighting*, Stockholm, 28–30 May
Jannuzzi, G M and Schipper, L (1991) 'The structure of electricity demand in the Brazilian household sector', *Energy Policy*, 19(11)
Latone, C O F, Nobre, E C and Burgoa, J A (1990) 'Diagnóstico do Potencial de Conservação de Energia na Indústria'. *Proceeding of the 5th Congresso Brasileiro de Energia 'desafio dos anos 90'*, November 5–9, 1990. Rio de Janeiro, Brazil
Ledbetter, M, and Ross, M (1990) 'Supply Curves of Conserved Energy for Automobiles'. Lawrence Berkeley Laboratory, Berkeley, USA
Lee, S H (1991) Planning Manager, Secretaria de Transportes e Obras, Santa Catarina, Brazil, Oral Communication
Lerner, J (1989) 'Curitiba Mass Transit Systems', Workshop on New Energy Technologies, Transportation and Development, International Institute for Energy Conservation, Washington DC, September
Moreira, J R and Poole, A D (June 1991) 'Energia e Melo Ambiente', Secretaria de Ciencia e Tecnologia, Federal Government, Brasilia, Brazil
Moreira, J R and Poole, A D (1992) 'Hydroelectricity and its Constraints' draft for publication, ed. Thomas Johansson, 'Fuels and Electricity from Renewable Sources of Energy', Island Press.
Moreira, J R, Poole, A D and Serrasqueiro, M A (1990) 'Alternativas Energéticas da Amazonia'. Part IV of Projeto/Análise da Implantação de Grandes Projetos Energéticos - O Caso do Setor Elétrico do Brazil. Institute of Electrotechnology and Energy. University of São Paulo
Nepstad, D, Uhl, C and Serro, E A (1990) 'Surmounting Barriers to Forest Regeneration in Abandoned Highly Degraded Pastures, A Case Study from Paragominas, Para, Brazil', in *Alternatives to Deforestation*, ed. Anthony Anderson, Columbia University Press, New York
Ogden, J M, Williams, R H and Fulmer, M E (1990) 'Cogeneration Applications of Biomass Gasifier/Gas Turbined Technologies in the Cane Sugar and Alcohol Industries', Center for Energy and Environmental Studies, Princeton University, Princeton
Poole, A D, Ortega, O and Moreira, J R (1990) 'Energia para o Desenvolvimento da Amaznia', Instituto de Eletrotécnica e Energia for UNDP/SUDAM, Macrocenários Amazonia 2010, Belém, Brazil
Poole, A D, Pacheco, R S and Melo, M A C (1992) *Collective Transport Policy in Developing Countries: A Case Study of Brazil*, International Development Research Institute, Ottawa
Reis, E (1991) 'A Amazonia e o efeito Estufa, Perspectivas da Economia Brasileira - 1992, IPEA - Instituto de Pesquisas e Economia Aplicada, Brasilia, Brazil
Sawyer. D (1990) 'The Future of Deforestation in Amazonia: A Socioeconomic and

Political Analysis', in *Alternatives to Deforestation*, ed. Anderson, Columbia University Press, New York

SNE (1991a) 'Reexame da Matriz Energetica Nacional', Ministerio da Infraestrutura, Federal Government, Brasilia, Brazil

SNE (1991b) 'Balanco Energetico Nacional', Secretaria Nacional de Energia, Ministerio da Infraestrutura, Federal Government, Brasilia, Brazil

USDOE (1990) *The Potential of Renewable Energy*, Interlaboratory White Paper, Office of Policy, Planning and Analysis, Washington DC

World Bank (1990) *Brazil Energy Strategies and Issues Study Pricing and Investment Policy*, Washington DC

11

Thailand's demand side management initiative: a practical response to global warming

Peter du Pont, Mark Cherniack, Michael Philips, and Somthawin Patanavanich[a]

Introduction

Today, Thailand's power sector accounts for nearly 33 per cent of the country's total carbon dioxide (CO_2) emissions. During the next decade, the power sector will surpass transportation as the major source of CO_2 emissions in Thailand. By the year 2011, we project that the sector will account for 43 per cent of Thailand's CO_2 emissions.

To keep up with the rapidly rising demand for electricity, the Electricity Generating Authority of Thailand (EGAT) is planning to more than triple its capacity over the next fifteen years, from 10,000 to more than 30,000 megawatts (MW) by 2006. Most of the new plants being built will burn lignite and coal. This trend will lead to greatly increased emissions of carbon dioxide and other greenhouse gases.

Here we focus on the potential for reducing CO_2 emissions in the power sector, since this is one of the most rapidly growing sources of such emissions in Thailand. It is also the sector for which there are the best data on costs and associated CO_2 reductions. The cost of conserved carbon (CCC) for residential electricity conservation measures ranges from *negative* US$155 to *positive* $41 per tonne of carbon as CO_2 equivalent in the year 2001. The average CCC for a comprehensive electricity conservation effort covering all subsectors (industrial, commercial, and residential) may be even lower, around *negative* $190 per tonne of carbon.

At present, there are inadequate data to enable similar analyses for the transportation and industrial sectors. In the future, work needs to done to apply the principles used here to perform analyses of the technical potential and cost feasibility of CO_2 reduction measures in the transportation and industrial sectors as well.

We also describe the crucial role that the multilateral development banks can play in providing the financial, technical, and policy support necessary to seriously address the problem of growing CO_2 emissions in rapidly industrializing countries.

Even if current predictions of global temperature rise are overstated, the economic impacts of continued high rates of growth in energy use worldwide will be untenable. The World Bank estimates that developing countries will require an average of US$100 billion annually just for capital expenses in their power sectors alone during the next decade. The amount needed over the next three decades is an estimated $4 trillion. Foreign exchange currently pays for about 38 per cent of these capital expenses. Yet only $10–12 billion per year is expected to be available from multilateral and bilateral agencies, the main providers of foreign exchange for electricity supply projects (Philips 1991). A shift to cleaner, cheaper fuels and improved energy efficiency will reduce the debt burden of developing countries, and thus yield additional societal benefits besides just reducing emissions of greenhouse gases.

Shift in Thai CO_2 emissions

Thailand is not currently a major contributor to global warming, in terms of emissions of carbon dioxide and other greenhouse gases. The World Resources Institute has estimated that Thailand emits 1.2 per cent of the world's total of greenhouses gases (125 million of a total of 10.6 billion tonnes of C annually as CO_2 equivalent (SEI 1992)).[a] In 1987, Thailand emitted 1.3 tonnes of carbon per capita, matching the world average (SEI 1992). By 1990, Thailand's CO_2 emissions had declined to about 0.98 tonnes of carbon per capita (TTCGE 1991).

Increases in CO_2 come from two major sources: deforestation, which increases CO_2 emissions directly and also reduces the uptake of CO_2 from the atmosphere; and fuel combustion, which emits CO_2 directly into the air. In the past, deforestation was by far the largest cause of Thailand's CO_2 emissions. As Thai forests are depleted and energy use increases, however, fossil fuel combustion will rapidly become the dominant source. In the near future, as Thailand's economy expands and its burning of fossil fuels grows, so too its contribution to global warming will increase. The U-shaped curve of total CO_2 emissions implied by Figure 11.1 tells the story.

a In 1990, Thailand had the following characteristics: population of 56.7 million; GDP/capita of US$1,661; exchange rate of 25 baht/US$; total modern energy demand of 30,284 KTOE; renewable energy demand of 10,969 KTOE; fossil fuel fraction of primary energy demand of 71 per cent; estimated total CO_2 emissions of 54 million tonnes, being 30 million tonnes from deforestation, and 24 million tonnes from fuel combustion.

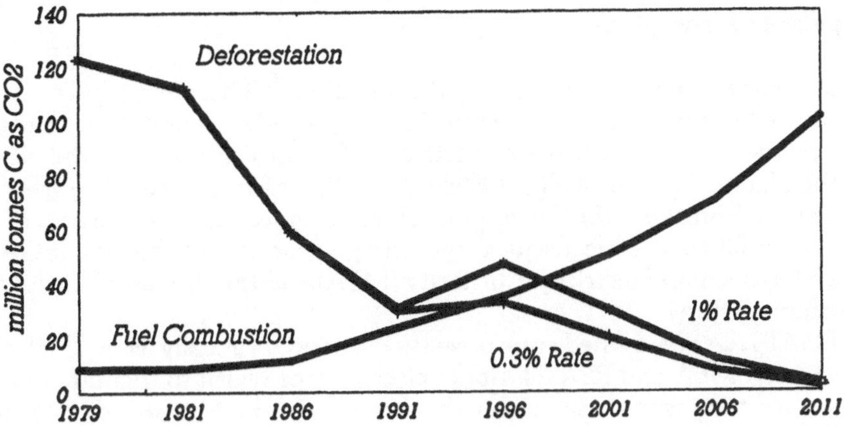

Figure 11.1 *Thailand's net CO_2 emissions from deforestation and fuel consumption*

In Thailand, the ratio of CO_2 emissions from deforestation to emissions from fuel consumption has declined dramatically from 13.7 in 1979 to nearly 1 in 1991. CO_2 emissions from fossil fuels are expected to rise fourfold over the next 20 years, to 100 million tonnes by 2011.

The transition from deforestation to fuel consumption as the main source of CO_2 emissions is a possible pattern that will occur in other countries – with formerly abundant forest reserves – that are on the road to industrialization. The drop in the amount of CO_2 released due to deforestation in Thailand's case has two primary causes: reduction in the amount of remaining forests and the nationwide logging ban instituted by the government in January 1989. The large increase in fuel consumption will come from two main sectors: transportation and power. This chapter focuses on the potential to reduce greenhouse gas emissions from the rapidly growing power sector.

Thailand's policy responses to global warming

During the preparations for the United Nations Conference on the Environment and Development (UNCED) in Brazil in June 1992, Thailand's government showed significant interest in international environmental issues, and particularly in addressing global warming. In addition to hosting a series of high-level international conferences on the environment, the government attempted to identify the most effective response strategies at its disposal for dealing with climate change. Thailand's national report to UNCED concluded that the country's most significant contribution will be a comprehensive set of energy conservation programmes that, if aggressively pursued, could reduce projected increases in Thai CO_2 emissions from the power sector by more than 2.5 million tonnes of carbon annually over the next decade (TTCGE 1991).

Thailand's energy picture

The agency responsible for setting the direction of Thai energy policy is the National Energy Policy Council (NEPC), a cabinet-level committee that sets the policies governing fuel and electricity. The operating arm of the NEPC is the National Energy Policy Office (NEPO), which is under the Office of the Prime Minister. NEPC is responsible for overseeing the kingdom's three electric utilities, the Electricity Generating Authority of Thailand (EGAT), the Metropolitan Electricity Authority (MEA) and the Provincial Electricity Authority (PEA).

EGAT is a state-owned enterprise that produces virtually all of Thailand's electricity. MEA and PEA distribute electricity provided to them by EGAT. They are state enterprises under the direction of the Ministry of Interior.

The agency charged with taking the lead on energy conservation and renewable energy activities is the Department of Energy Development and Promotion (DEDP), which is under the Ministry of Science, Technology and Energy. In summary, NEPC and NEPO develop energy policies and the electric utilities; DEDP, and other institutions (for example, the Petroleum Authority of Thailand) are responsible for implementing those policies.

Thailand's primary energy use grew at an average annual rate of 13.4 per cent between 1985 and 1990. The rapid, sustained growth is due to the overall pace of growth of the economy and expansion of industrial, construction, and transport activities. The Seventh Economic Plan projects an economic growth rate of 8.2 per cent from 1992 to 1996. It is expected that energy demand will grow at an even faster rate, for example 10.3 per cent in the power sector (NEPO 1991).

Transportation accounts for the largest share of primary fossil fuel demand at about 30 per cent, followed closely by the power sector (29 per cent) and the industrial sector (20 per cent) (see Figure 11.2). Energy use in

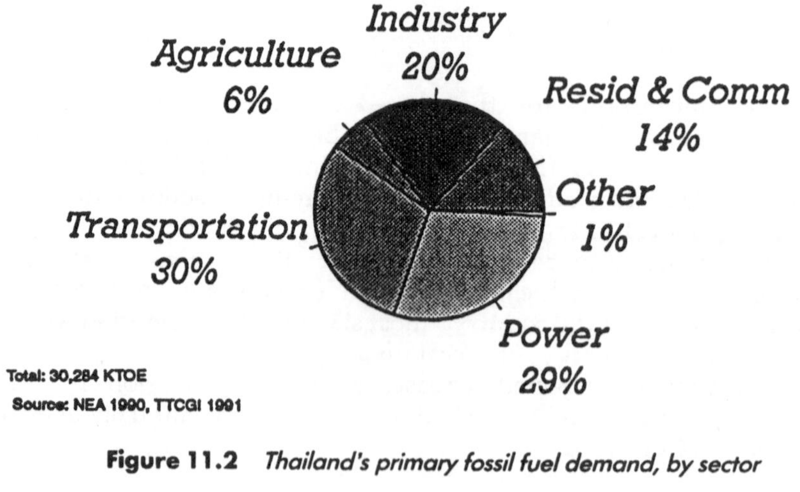

Figure 11.2 *Thailand's primary fossil fuel demand, by sector*

the power sector is rising faster than energy use for transportation, however. By 2006, power generation will account for some 37 per cent of Thailand's primary fossil fuel demand, compared to 35 per cent from the transport sector (TDRI 1991). To keep up with the rapidly rising demand for electricity, EGAT projected in 1990 that its installed capacity in the year 2006 would have to reach 24,900 (EGAT 1990). The estimate was recently revised upward to more than 30,000 MW (EGAT 1992).

Sources of CO_2 emissions

Petroleum is the largest fuel source in Thailand, accounting for 56 per cent of primary energy demand. Most of this fuel is used in the transportation sector, which contributed 40 per cent of Thailand's CO_2 emissions in 1991. The next largest sources of CO_2 emissions are the power sector (33 per cent) and industry (15 per cent). During the next decade, the power sector will surpass transportation as the major source of CO_2 emissions in Thailand. By the year 2006, we estimate that the power sector will account for 43 per cent of Thailand's CO_2 emissions from fuel consumption (see Figure 11.3).

Thailand's policy is to promote the use of domestic energy resources in order to reduce the burden on the country's balance of payments. Domestic energy resources comprise gas, lignite, biomass (including fuelwood) and hydropower. The use of fuelwood and hydropower is unlikely to grow significantly, due to the government's efforts to encourage reforestation and to growing public opposition to the building of dams.

The implications of this strategy of relying on domestic and inexpensive energy resources can be seen most clearly in the power sector. In 1991, oil and natural gas together accounted for 66 per cent of electricity generation. By 2006, a complete shift will have taken place. Oil and gas together will account for just 36 per cent of primary energy supply to electricity

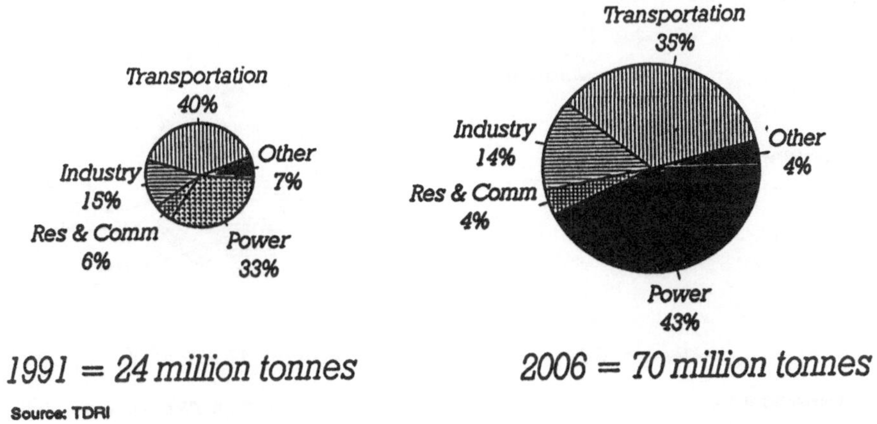

Figure 11.3 *Thailand's CO_2 emissions by sector, 1991 and 2006*

generation, while coal and lignite will make up 55 per cent. Lignite is readily and cheaply available in Thailand. Coal, while not mined locally, is available in the region (from Australia and Indonesia) and its price is expected to remain stable because of the large available resources.

In the power sector, where most of the lignite and coal is used, consumption will rise more than fourfold - from 11.6 million tonnes in 1991 to 49 million tonnes annually in 2006 (EGAT 1990). In Figure 11.4, we show the dramatic shift in fuel mix that will occur in the energy and power sectors. Two 1,000-MW nuclear plants are scheduled to come on line in 2006, with four more units following soon thereafter.

End-use energy efficiency policies

Considerable scope exists to improve end use efficiency in Thailand's energy consumption. Recent studies have estimated that, in the power sector alone, efficiency improvements could reduce the projected growth in demand by 25–35 per cent over the next decade (Monenco 1991, IIEC 1990). Similar improvements can be made in thermal (non-electric) energy use in factories, but this area has not been adequately studied.

Thailand is currently pursuing two main initiatives to improve end-use energy efficiency: an energy conservation law and a utility-run demand side management (DSM) programme.

The Energy Conservation Promotion Act

The National Assembly passed Thailand's first-ever energy conservation law in February 1992. The Energy Conservation Promotion Act gives the Department of Energy Affairs (DEA, formerly the National Energy

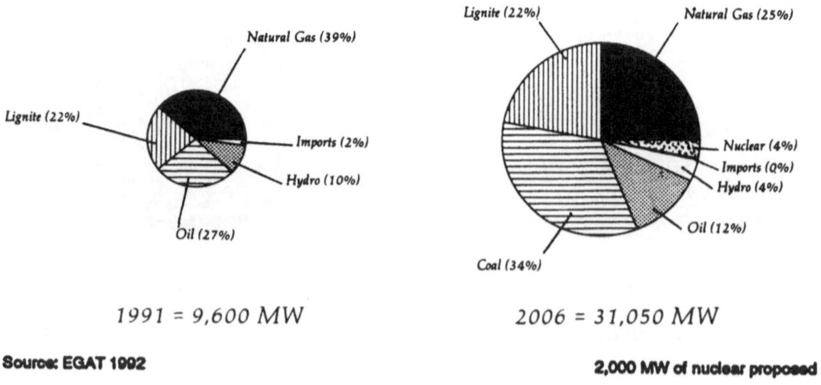

Figure 11.4 *Thailand's electricity production by fuel type, 1991 and 2006*

Administration), the power to issue an energy code for new buildings. Initially, the code will be voluntary, and developers of new buildings will be encouraged to abide by it. The code will establish minimum performance levels for the insulating properties of building materials and glazing, and will recommend levels of lighting and energy intensity.

At the same time, the law will require the owners of large buildings and factories (with power demand of more than 2,000 kW) to appoint an energy conservation manager and to submit a comprehensive energy management plan to DEA. The owners will have three years to submit the plan; if they fail to do so, they will receive a surcharge on their electric bill. In its initial stage the requirement to submit an energy conservation plan will apply to 220 large industrial facilities and 60 buildings.

The Energy Conservation Promotion Act also empowers DEA to establish minimum efficiency standards for electric appliances and energy-consuming equipment. It is likely that these standards will be coordinated with the efforts of the Thailand DSM Office to set minimum energy efficiency requirements for its various programmes.

Finally, the law also establishes the Energy Conservation Promotion Fund, which will be available for energy conservation and renewable energy projects in all sectors. The cabinet has recommended that the initial amount in the fund be US$60 million. Thus there will be two major sources of funds for end-use efficiency projects in Thailand: the Energy Conservation Promotion Fund (for all sectors and fuel types) and the DSM programme, described below (for electricity measures).

The demand side management programme

In November 1991, the Thai government approved a five-year Demand Side Management Master Plan that will allocate US$188.5 million to the purchase of energy-efficient equipment in the commercial, industrial, and residential sectors. The plan calls for the three utilities (the Electricity Generating Authority of Thailand – EGAT, the Provincial Electricity Authority – PEA, and the Metropolitan Electricity Authority – MEA) to establish jointly a Demand Side Management Office, which will operate a comprehensive set of energy conservation programmes (Cherniack and du Pont, 1991).

Both the Energy Conservation Promotion Act and the DSM Master Plan will bring about efficiency increases in all sectors and help to reduce Thailand's CO_2 emissions substantially. In the case of the DSM, no central government monies need to be allocated for the CO_2 reduction measures, since the cost will be borne by utilities, just as they currently bear the cost of building power plants. The following section describes the principles of DSM, how it is being applied in Thailand, and the potential for reducing CO_2 emissions from the power sector.

When it passed its DSM Master Plan in November 1991, Thailand became the first Asian country to incorporate energy efficiency formally into its power planning process. Top management at the Thai electric utilities, with some prodding from the National Energy Policy Council, decided to spend money to produce energy efficiency as a future resource for the electric power system. The Thai DSM initiative was inspired by a decade of experience with DSM at more than 500 North American utilities. These utilities now spend US$1.2 billion annually, and have sponsored 1,400 different DSM programmes (see box).

The role of demand side management

International commitment is growing to sustainable development at the local and national levels in a wide variety of social and economic conditions. Three important criteria for establishing sustainable development in the power sector are efficiency in electricity generation, transmission and distribution; efficient use of electricity; and renewable energy resources, with emphasis on solar (thermal and photovoltaic), wind (mechanical and electrical) and biomass (small and large scale).

The major components of end-use efficiency in the power sector include the principles of integrated resource planning and demand side management. These principles are the fundamental basis for sustainable energy planning in the power sector.

Integrated resource planning (IRP) is an effort to fully integrate both supply- and demand-side options into a utility system expansion plan that provides reliability, lowest total system cost, and acceptable levels of risk. It is a rational enhancement to the least-cost generation planning framework that has governed utility expansion planning worldwide for decades and been promoted by the Bank and other multilateral lenders. What distinguishes IRP from current power system supply planning is that conservation – the efficient end-use of electricity – competes directly with generating resources for consideration in meeting a utility's future load growth requirements. Both supply and demand are considered formally in the utility planning and resource acquisition process. IRP can also be called least-cost utility planning (LCUP) since it describes the truly lowest cost system for meeting all electricity service needs in society.

Demand side management (DSM) can be defined as the systematic effort by the electric utility to influence the timing and magnitude of customer electricity use in order to optimize power system operation and planning. DSM includes tariff pricing mechanisms, load management techniques and increased efficiency in all end-uses of electricity. An integrated resource plan typically includes a comprehensive demand side management plan that is implemented by the utility to create the least-cost operation and expansion of the power system.

Beginning in the United States in the early 1980s, electric utilities began the most advanced work in the world on developing demand side resource options for their electric power systems. In North America, where most DSM work to date has taken place, more than 500 utilities have sponsored DSM programmes; these included 1,000 programmes for residences and 400 programmes for commercial and industrial buildings.

Carbon dioxide emissions from electricity generation are expected to rise dramatically in the US during the next 20 years, increasing by 68 per cent between 1990 and 2010. As a share of emissions from fuel combustion, CO_2 emissions from the power sector will rise from 31.5 per cent to 39.4 per cent during this period

(Faruqui and Haites 1991). DSM programmes have the potential to reduce electricity use in the year 2010 by 20 per cent (Hirst 1991). They also have the potential to reduce projected emissions of CO_2 from the power sector by up to 18 per cent (Faruqui and Haites 1991).

Because the cost of most energy conservation measures is typically less than the cost of building new power plants, the adoption of DSM measures by a utility actually saves money (that is, produces a net benefit). This means that the net cost of conserved carbon for DSM measures is negative in most cases. One US study estimated that about 23 per cent of US carbon emissions in 2010 could be eliminated at a negative cost of conserved carbon, and that up to 39 per cent of emissions could be eliminated if all the conservation measures were adopted. The most expensive of the conservation measures would have a net cost of conserved carbon of just US$43 per tonne of carbon (Atkinson et al. 1991).

In Table 11.1, we list the estimated programme costs and savings associated with the Thai DSM effort. These programmes will provide financial incentives for customers to purchase energy-efficient equipment. The utilities compare the cost of purchasing electricity savings to the cost of building new power plants. Only measures that cost less than the cost of building new generation capacity are included in the programmes. The average long-term cost of savings from all of the DSM programme measures is US$0.017/kWh (see Figure 11.5). When these costs are compared to EGAT's adjusted long-term cost of US$0.062/kWh (see below) to produce new electricity supply, it is clear that the least-cost investment for the utilities is in energy efficiency.

Thailand's five-year plan aims to save an estimated 238 megawatts (see Figure 11.5). Although this is a pioneering effort for an Asian utility, the US$188.5 million allocation represents just a small portion of the estimated US$36 billion that the Electricity Generating Authority of Thailand will need for its capacity expansion programme over the next decade (EGAT,

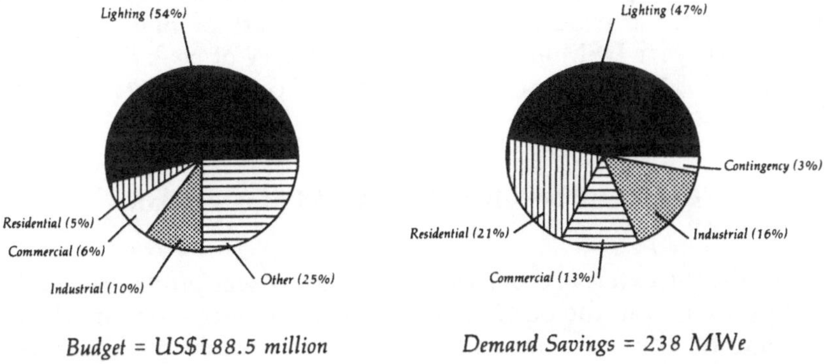

Figure 11.5 *Thailand's power sector, 5-year DSM budget and demand savings*

Table 11.1 *Thailand's proposed demand side management programmes*

Programme	*Peak savings (MWp)*	*Energy saved (GWh/yr)*	*Cost ($m)*	*CSE (cts/kWh)*	*CAP ($/kW)*
Lighting	133	677	101.0	2.2	759
Residential refrigerators	28	186	6.0	0.4	214
Residential air conditioning	23	117	3.0	0.3	130
Commercial programmes	15	180	12.0	0.8	800
Industrial motors	30	225	19.0	1.1	633
Contingency	9	42	11.5	3.6	1,278
Whole programme	238	1,427	188.5	1.7	792

Notes

EGAT estimate combining New Commercial Building Design Programme with Peak Shaving.

Whole programme budget includes laboratory and testing, consulting, training, programme administration, and public relations.

CSE (cost of saved energy = (CRF * DSM Cost)/(GWh/yr saved) — EGAT LRMC = 4.2 cts/kWh

CAP (cost of avoided peak) = DSM cost/kWp saved — EGAT new capacity = $1,500/kW

CRF = capital recovery factor

0.147 CRF lighting — 0.117 CRF commercial

0.131 CRF refrig — 0.131 CRF motors

0.131 CRF A/C — 0.131 CRF contingency

Source: International Institute for Energy Conservation, 'Thailand: Promotion of Electricity Energy Efficiency. Final Report of Pre-Investment Appraisal,' report to World Bank/Global Environment Facility & United Nations Development Programme, Bangkok, 1993.

1992). In fact, capital constraints are one reason that EGAT stands to benefit from an aggressive DSM programme.

On average, the DSM options outlined in the five-year Master Plan will provide MW savings at a cost of less than half the cost of building new capacity. The 238 MW is just the tip of Thailand's iceberg of efficiency potential, however. Studies conducted by various Thai agencies and outside consultants of Thailand's electricity end uses have identified an achievable DSM potential of at least 2,000 MW over the next decade (Monenco 1991, IIEC 1990). This represents nearly 20 per cent of EGAT's planned system expansion, which can be avoided at half the cost of new capacity. An aggressive, 10-year DSM effort to save 2,000 MW of peak demand would yield EGAT US$2.9 billion net savings in capital costs for system expansion (see Table 11.2).

Costs and benefits of the DSM Master Plan

Traditionally, the external costs associated with power production have not been accounted for in the tariff structure. These externalities are the cause of damage to the environment and human health in Thailand and surrounding regions from the normal operation of power plants that burn fossil fuels such as oil, gas, coal, and lignite. By not including these externalities in the cost of electricity, power system planners are assigning a

Table 11.2 *Thailand's capital investment choices*

	10-year achievable DSM potential	=	2,000 MW
	2,000 MW of power plants[a]	=	US$4.5 billion
minus	2,000 MW of DSM	=	US$1.6 billion
equals	10-year DSM net cost savings	=	US$2.9 billion
Other major infrastructure projects			
	Second stage expressway	=	US$1.2 billion
	Nationwide sewage treatment	=	US$2 billion

a The cost of 2,000 MW of delivered power is calculated based on a 15% reserve margin and 14% power transmission and distribution losses. It is equivalent to 2,817 MW of installed capacity.

This table contains the cost savings accrued by EGAT, government-run utility, for investing in an aggressive 10-year DSM plan. These cost savings are compared to similarly-sized investments that the Thai government is planning for other large infrastructure projects. EGAT plans to add 13,100 MW over the next decade at an approximate cost of US$19.7 billion. The cost of new capacity is US$1,600/kW, while DSM measures for Thailand cost US$800/kW on average (source: du Pont and Biyaem 1992).

value of zero to the environment. Many state and national governments have recognized this fact and have begun to add an externality surcharge to the tariff. As of mid-1992, the Thai government had not yet established an environmental resource accounting method that could be applied to the power sector. The DSM Master Plan assigns the following credits to DSM measures in comparison to supply side power generation options: 14 per cent for transmission and distribution losses, 15 per cent for the reserve margin, and a 15 per cent environmental credit.

EGAT's long-term avoided cost for new power plants is US$0.043/kWh. After factoring in the recommended credits, EGAT's adjusted avoided cost is US$0.062/kWh. DSM measures will be judged cost-effective if they cost less than this amount. At present, Thailand has no formal environmental regulations that can influence the choices made by utility planners. Clearly, the establishment of an environmental accounting system could allow Thai planners to assess the environmental benefits of demand side management.

CO_2 reductions from the DSM Plan

In order to assess the CO_2 benefits of DSM, it is necessary to use a detailed model that relates the energy savings from each DSM measure to the actual operation and fuel mix of the generating plants in the power system. To date, this analysis has only been carried out for Thailand's residential sector.

Utility system planners from the Tellus Institute in Boston conducted a workshop in May 1991 with the System Planning Department of EGAT and staff from Thailand's National Environment Board. Using power system data supplied by EGAT, and cost and savings data provided by a residential DSM assessment for Thailand, the workshop team quantified the environmental benefits of the combined DSM measures for the residential sector (EGAT 1991, Bartels 1991, Parker 1991). The benefits occur when the utility burns less diesel for peaking plants and less coal and lignite for baseload plants. The savings occur when the utility defers or avoids the need to build future power plants.

The system planners used the Energy Conservation Model (ECO), which is a computer software package that calculates the costs, resource impacts, and environmental externalities of demand side management programmes. It also determines the revenue and rate impacts of the programmes. Data on residential conservation measures were input from the Parker (1991) study of residential electricity use in Thailand, the most complete analysis of energy conservation potential in this sector. The simulation assumed eight different conservation programmes were run over a ten-year period (one rural and one for Bangkok) for each of four major electricity uses: cooling, lighting, refrigerators, and cooking (rice cookers).

The costs of residential efficiency improvements were based on data collected in Thailand during 1991 (Parker 1991). These costs were used to calculate the cost of saved energy for each technology or package of measures (Parker 1991, Cherniack and du Pont 1991). The costs of conserved carbon (CCCs) were then calculated based on the costs of saved energy for the measure, the utility's adjusted long-run marginal cost of production, and the CO_2 reductions derived using the ECO model. This method for calculating CCCs is explained in Akbari and Rosenfeld (1990).

Figure 11.6 shows a supply curve of conserved carbon for the Tellus/EGAT workshop. The cost of conserved carbon (CCC) ranges from *negative* US$155 to *positive* $41 per tonne of carbon as CO_2 equivalent in the year 2001. A study of the CO_2 emission reduction potential from conservation measures in US homes found a range of CCCs of *negative* US$92 to *positive* US$43 per tonne of carbon (Atkinson et al. 1991). Table 11.3 compares the two studies.

As explained in several other sources, negative costs of conserved carbon are common for most energy conservation measures. In fact, they benefit, rather than burden the economy (Akbari and Rosenfeld 1990, Atkinson et al. 1991). The Tellus numbers indicate an energy savings potential of more than 3,200 gigawatt hours (GWh) in the residential sector by 2001. The associated CO_2 reductions from the residential subsector could total 1 million tonnes of carbon (as CO_2) by 2001. This figure is 5 per cent of projected CO_2 emissions from the overall power sector in 2001.

Since the residential sector accounts for only 25 per cent of Thailand's electricity demand, it stands to reason that additional CO_2 reductions (beyond the 1 million tonnes for the residential sector) from commercial and industrial conservation measures will be substantial. These data have not

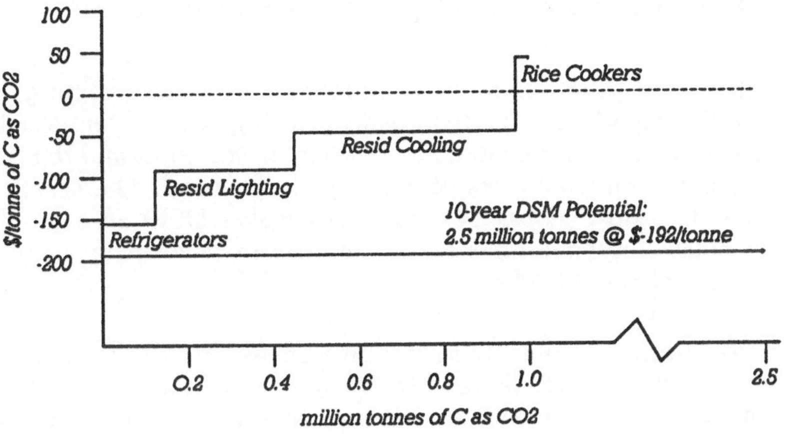

Figure 11.6 *Carbon abatement curve for Thai residential sector, 2000*

been analysed yet in the Tellus (or any other) model, but a rough estimate can be made by assuming that Thailand's achievable 2001 DSM potential of 2,000 MW (11,500 GWh) is spread evenly across the EGAT fuel generation mix through the year 2001.[b]

This 'back-of-the-envelope' analysis yields a CO_2 reduction of about 2.5 million tonnes of carbon annually, or 13 per cent of CO_2 emissions from the overall power sector in 2001 (Monenco 1991, IIEC 1990, Busch 1990, Parker 1991). The average cost of conserved carbon for DSM measures in all sectors is about *negative* US$190/tonne of carbon.

Table 11.3 *Comparison of CCCs for residential electricity conservation measures (US$/tonne carbon)*

End use	*United States*	*Thailand*
Lighting	-92.2	-96.2
Refrigerators	-30.1	-155
Space conditioning	+9.9	-
Air conditioning	-	-44.5
Water heating	-	-
Rice cookers	-	+40.7
Other	-36.9	-

Source: Atkinson *et al* 1991

b 2,000 MW was cited as the low end of a range of achievable potentials for efficiency improvements by the year 2001 (Monenco 1991, Cherniack 1991).

Why should other developing countries adopt DSM?

The benefits of DSM programmes and integrated resource planning (IRP) are substantial, especially given the rapidly growing demand for electricity by all sectors of the economy in many developing countries and the rising capital and environmental costs of meeting the demand. The search for mechanisms to reduce global CO_2 emissions makes DSM an even more attractive policy for governments and utilities in developing countries. Briefly, the benefits of DSM are:

- It enables utilities to operate at the lowest possible cost of production. This cost minimization keeps tariffs stable and under control. Net system benefits will increase because investments in needed new power plants are avoided for as long as possible or avoided completely. Also deferred or avoided are transmission system expansion, operation, maintenance and fuel costs.
- End-use efficiency reduces customers' bills. This leaves everyone with more money which can be spent in other parts of the economy. Foreign debt obligations for power plant construction can be reduced also leaving additional funds for other infrastructure development priorities.
- Demand side management reduces the uncertainty in projecting future electricity demand because a portion of the expected load growth is managed through efficiency programmes.
- Investments in efficiency can be made in smaller amounts and adjusted more quickly to meet changes in demand. Therefore, the risk of underbuilding or overbuilding new power plants is reduced. The use of expensive capital is optimized.
- Integrated resource planning reduces social conflict over natural resource utilization because fewer power plants of any type have to be sited.

Other Asian developing countries are also exploring the feasibility of implementing DSM programmes on a large scale. *Tenaga Nasional Berhad* (TNB), the private power utility that serves Malaysia, has about 3,650 MW peak demand which is expected to grow to around 5,500 MW in 1995. The DSM project proposed by TNB, which began in 1992, is to conduct a small-scale test of commercial building retrofits. Several buildings will undergo 'super retrofits.' After the retrofits, the project team will monitor the end uses of electricity in these buildings and compare these results to monitored use in a group of control buildings.

The purpose of the test will be to document the capacity and energy savings acquired by replacing standard ventilation and air conditioning equipment, inefficient lighting systems, and other end uses with more efficient technology. A second objective is to expose the TNB staff to the analysis, development, and monitoring of DSM projects. The TNB staff will be trained in the fundamentals of DSM project development and operation.

The role of the multilateral development banks

Developing countries must seriously address the issue of climate change. To do so, they will require monetary as well as technical assistance from the developed countries, as promised in the Climate Change Convention. The most appropriate vehicle for funnelling funds into worthy CO_2 reduction schemes is the multilateral development banks (MDBs), which largely shape investments in the power and transportation infrastructures in developing nations.

With energy investments averaging more than US$5 billion per year (over $3 billion for The World Bank alone), MDBs are the world's most important sources of capital for energy investments in developing countries and Eastern Europe. Through their economic advice, loan conditions, and investments in both energy-producing and energy-consuming sectors, these banks shape energy development and consumption in these countries. They also strongly influence the investment priorities of bilateral and multilateral assistance agencies, commercial banks, and other investors. Thus, energy investment in many countries turns to a great extent on the decisions of these banks.

The MDBs cannot singlehandedly effect a transition in developing countries to a greater emphasis on energy efficiency; government, industry, and financial institutions in these countries must be full partners. The banks, however, because of their vast influence on the countries' investment policies, can be leaders in this transition.

Energy service needs should be met by evaluating both new supplies and demand-side efficiency improvements, and selecting from among the options according to cost.[c] This is called an integrated, least-cost planning approach. For the electric power sector, loans should be based on the lowest cost methods of delivering electricity services, either by improving the efficiency with which these services are provided or by increasing the supply itself. The same integrated least-cost approach can be applied to oil, gas and water resources.

To accelerate their learning process in this area, the multilateral development banks need to initiate a series of demonstrations of implementation strategies. In the past, owing to the uncertainty of success, the banks have concentrated on studies of the potential for efficiency improvements, possible institutional frameworks for administering energy efficiency programmes, and the barriers to energy efficiency. Such studies should

c A unit of energy saved is generally equivalent to one produced, because it frees a unit of energy for use somewhere else.

continue, but actual implementation and demonstration programmes in developing countries are the only way to understand the problems *and* to develop solutions.

The experience in Thailand can serve in some regards as a model for the MDBs in establishing demonstration energy efficiency projects in the electric power sector. Initial reaction from The World Bank to the Thai DSM plan has been favourable, and it has also granted approval in principle for US$15 million in loan funds to supplement the Thai DSM effort.[d] These funds will be administered through the Global Environment Facility. An additional US$25 million soft loan has been offered to the Thai DSM effort, pending appraisal, by the Japanese Overseas Economic Cooperation Fund.

It is also encouraging that The World Bank has begun active pursuit of options for funding small-scale energy conservation and renewable energy projects. The Bank's FINESSE project studied the technical and financial potential for funding such projects in four Southeast Asian countries: Thailand, Malaysia, Indonesia, and the Philippines (World Bank 1991). As a follow-up to the initial FINESSE workshop, The World Bank has established an Asia Alternative Energy Unit. The Asian Development Bank has also shown interest in allocating loan funds for renewable and energy conservation projects. The latter institution is also implementing the UNDP study of least-cost greenhouse gas reduction strategies in Asian developing countries, including Thailand.

Conclusions

To avoid the sort of impacts estimated in the box opposite, all nations will have to find ways to reduce their contribution of greenhouse gases. The Climate Change Convention signed in Rio places the onus on everyone to limit carbon dioxide emissions, although the developed countries are expected to take the lead. While the implementation of this agreement is not yet clear, we have argued in this chapter that industrialized and developing nations can significantly contribute to efforts to halt global warming at relatively low cost - at least at the outset of such a programme.

Thailand's CO_2 emissions from fuel combustion will double over the next decade, from 24 to nearly 50 million tonnes annually. An aggressive demand side management effort in the power sector could reduce emissions by 2.5 million tonnes annually by the year 2001. The average cost of conserved carbon for these measures would be about *negative* US$190/tonne.

Thailand's nascent efforts in the area of electricity conservation can provide lessons for other developing countries. Clearly, in order to initiate effective DSM efforts in developing nations, industrialized nations will have to provide significant technical and financial assistance. Prime vehicles for

---

d US$15 million will be grant monies and US$15 million will be a soft loan.

Climate change impacts in Thailand

In Thailand, the warming under GISS 2 × CO_2 climate [a climate model scenario run assuming a doubling of CO_2 levels] is equivalent to a 3°C to 6°C increase in current mean annual temperature, a projection that is broadly in agreement with other GCMs [general circulation models]. There are, however, substantial differences between GCMs concerning changes in precipitation, which vary widely from normal but generally show a reduction under the GISS 2 × CO_2 scenario. Northern Thailand may be drier in most of the months except in July which is currently a dry period and this would appear to benefit cropping. However, August and September would experience only between 73 per cent and 89 per cent of present rainfall. Other GCMs, however, do not indicate such a reduction in rainfall and it is important to emphasize this uncertainty. Under the GISS 2 × CO_2 scenario winters are also drier but as very little rain is normally expected during that time of the year the adverse implications may be less.

Two particular aspects of the Thai economy were studied with respect to potential impacts from these projected changes in climate effects on rice production in Ayuthaya Province and effects of sea-level rise in Suratthani Province.

The CERES model was run for a 25 year set of daily climate variables (1964–1988). Model outputs for the current climate substantially exceeded observed values for transplanted rice and were lower than expected for yields of direct seeded rice. It was not possible, however, to conduct an adequate validation of the model and to re-tune it to observed data for Thailand. As a result, the consequences should be treated with caution.

The results indicate that under a change of climate projected for a doubling of CO_2 main crop rice cultivation in Ayuthaya Province would increase in the order of 8 per cent. These benefits would, however, be, in most cases, quite marginal because they are substantially less than the existing year-to-year variation. The modelled yields were also characterized by marginally greater yield variations. Off season rice, planted from mid-December to early February, exhibits a 5 per cent increase in average yield under the GISS 2 × CO_2 climate with concurrent increases in variation of 3–40 per cent. However, little value can be placed on these results because of lack of model validation. Indeed, the results are not consistent with those for Chiang Mai which were validated against observed data, and which indicate a decrease in rice yield of about 5 per cent under the GISS 2 × CO_2 scenario.

Thailand has approximately 2940 km of coastline, much of which contains important economic activities such as shrimp farming and rice farming. The study considered the potential impact of a 0.5 m and 1 m rise of sea levels in the Suratthani Province in southern Thailand. This region is characterized by a sand dune line which may mark an ancient shoreline and has a consistent elevation about 1 m above present sea level. It was therefore used as an indicative boundary to the area potentially affected by a 1 m sea level rise. The suggestion is that 7400 ha (37 per cent) of the study area would be affected by inundation under a 1 m sea level rise. About 4200 ha of productive agricultural land and large numbers of shrimp ponds would be lost.

Excerpt from M Parry, A Magalhaes, Nguyen H Ninh, *The Potential Socio-Economic Effects of Climate Change, A Summary of Three Regional Assessments*, UN Environment Programme, Nairobi, 1991, p 19.

providing this assistance are the multilateral development banks, which already shape the power infrastructure in developing nations. The banks must internalize the concepts of integrated resource planning and demand side management if they are to play a positive role in this regard. They will then be in a sound position to introduce loan recipients to these techniques and to assist them in developing national strategies for energy efficiency and carbon conservation.

References

Akbari, H and A Rosenfeld. *Conservation Supply Curves for Reducing CO_2 Emission*. Cost of CO_2 Emissions Reduction, Comments for a California Energy Commission Electricity Report Hearing (ER-90), Valuing Emission Reduction for Electricity Report 90, Sacramento, California, January 25, 1990

Atkinson, B, Atkinson, C, Koomey, J, Meier, A, Boghosian, S, and J McMahon. 'Supply Curve of Conserved Carbon: Emissions Reduction Potential Through Electricity Conservation in US Residential Buildings.' In conference proceedings from *Demand Side Management and the Global Environment*. Arlington, VA. April 1991

Bartels, C 'ECO Training: IIEC IRP/DSM Workshop III.' Summary of workshop coordinated by the International Institute for Energy Conservation. Bangkok. May 1991

Busch, J F. 'From Comfort to Kilowatts: An Integrated Assessment of Electricity Conservation in Thailand's Commercial Sector', Lawrence Berkely Laboratory Report LBL-29478, Berkeley, CA 1990

Cherniack, M and P du Pont. 'Demand Side Management for Thailand's Electric Power System: Five-Year Master Plan.' Prepared by the International Institute for Energy Conservation, for the Thai electric utilities. Bangkok. November 1991

du Pont, P and K Biyaem. 'A Walk on the Demand Side: Thailand Launches Its Energy Efficiency Initiatives.' Proceedings of ACEEE 1992 Summer Study on Energy Efficiency in Buildings. Asilomar, CA. September 1992

EGAT (Electricity Generating Authority of Thailand) (1990) 'General Information on EGAT Power Development Plan (PDP 90-03)' Systems Planning Department. Electricity Generating Authority of Thailand, October

EGAT (Electricity Generating Authority of Thailand). 'ECO's Input and Output Studied.' Data from IRP/DSM Workshop III coordinated by the International Institute for Energy Conservation. Bangkok. May 1991

EGAT (Electricity Generating Authority of Thailand). 'General Information on Power Development Plan (PDP 92-01(1))'. Prepared by the System Planning Department. Bangkok. April 1992

Faruqui, A and E Haites. 'Impact of Efficient Electricity Use and DSM Programs on United States Electricity Demand and the Environment.' In conference proceedings from *Demand Side Management and the Global Environment*. Arlington, VA. April 1991

IIEC (International Institute for Energy Conservation). 'Spreadsheet Exercises from the Thailand End-Use Analysis Workshop.' Bangkok. 1990

Monenco Consultants & Associates. 'Demand Side Management Working Paper No. 3.' Under contract to the National Energy Policy Office. Bangkok. July 1991

Nadel, Steve, 'Utility Conservation Programs' in *State of the Art of Energy Efficiency: Future Directions*. Edited by Edward Vine and Drury Crawley. American Council for an Energy-Efficient Economy. Washington, DC. 1991

NEA (National Energy Administration). 'Thailand Energy Situation 1990.' Ministry of Science, Technology, and Energy. Bangkok. 1991

NEPO (National Energy Policy Office). '1991 Load Forecast for the Thailand Electric System. Volume 1: Load Forecast Summary.' Prepared by the Load Forecast Working Group. Bangkok. September 1991

Parker, D. 'Residential Demand Side Management for Thailand.' Prepared for the International Institute for Energy Conservation. Bangkok. June 1991

Philips, M. 'The Least Cost Energy Path for Developing Countries: Energy Efficient Investments for the Multilateral Development Banks.' Prepared for the International Institute for Energy Conservation. Washington, DC. September 1991

SEI (Stockholm Environment Institute) 1992. (These numbers were supplied by the reviewer of my chapter.)

TDRI (Thailand Development Research Institute). 'Energy and the Environment: Choosing the Right Mix.' Prepared by T Chongpeerapien, S Sungsuwan, P Kritiporn, S Buranasajja, and Resource Management Associates for the 1990 TDRI Year-End Conference: 'Industrialising Thailand and Its Impact on the Environment.' Bangkok. 1990

TDRI (Thailand Development Research Institute). Updated projections of CO_2 emissions made by the Natural Resources and Environment Program. 1991

Tellus Institute. 'ECO: The Energy Conservation Model.' Model Documentation for Verion 91-1. Boston, MA. January 1991. TDRI (Thailand Development Research Institute). Updated data from Energy and Environment Project. Natural Resources and Environment Program. Bangkok. September 1991

TTCGE (Thai Technical Committee on the Global Environment). 'Thailand National Report to the United Nations Conference on Environment and Development.' Bangkok. November 1991

World Bank. 'FINESSE Workshop: Financing of Energy Services for Small-scale Energy Users.' Kuala Lumpur, Malaysia. October 1991

WRI (World Resources Institute). *World Resources, 1990–91*. Oxford University Press. New York. 1990

12

Carbon abatement in Central and Eastern Europe and the Commonwealth of Independent States

Stanislav F Kolar[a]

The countries of Central and Eastern Europe[b] consume more than 6 per cent of the world's primary energy and release more than 7 per cent of global carbon dioxide emissions. The republics of the former Soviet Union (FSU) consume 16 per cent of the world's energy, and release 15 per cent of the carbon. On a per capita basis, their consumption of energy is similar to that of the industrialized economies in Western Europe. Central planners in the former socialist bloc countries often measured success by the amount of energy produced and consumed by their economies. But they did not mention that their nations' per capita income was two to three times lower than in Western Europe. Simple arithmetic indicates that the economies of the former socialist bloc use two to three times as much energy to produce one unit of national income (see Figure 12.1).

The asymmetrical economic development in Central and Eastern Europe, initiated in 1928 in the FSU by the first five-year plan and in the late 1940s in Central Europe and the Balkans, is largely responsible for the region's high energy intensity. The economic theories of rapid industrialization that were implemented in these countries considered energy consumption as a means towards achieving strategic goals, such as producing a given amount of steel, cement, chemicals, and other products, many of them requiring vast inputs of energy. Fulfilling the production of these products had always been the first priority, and resources were allocated to ensure that such production could take place. To make the production cycle function as smoothly as possible, both energy production and energy prices were heavily subsidized.

a This chapter is based on country case studies by authors from Poland, Czechoslovakia, Hungary, Romania and the former Soviet Union. These case studies were initiated and carried out by William Chandler of the Pacific Northwest Laboratories, Battelle Memorial Institute, Washington, DC.

b Central and Eastern Europe is defined in this paper as Bulgaria, former Czechoslovakia, Hungary, Poland, and Romania. Where noted, the former East Germany is also included.

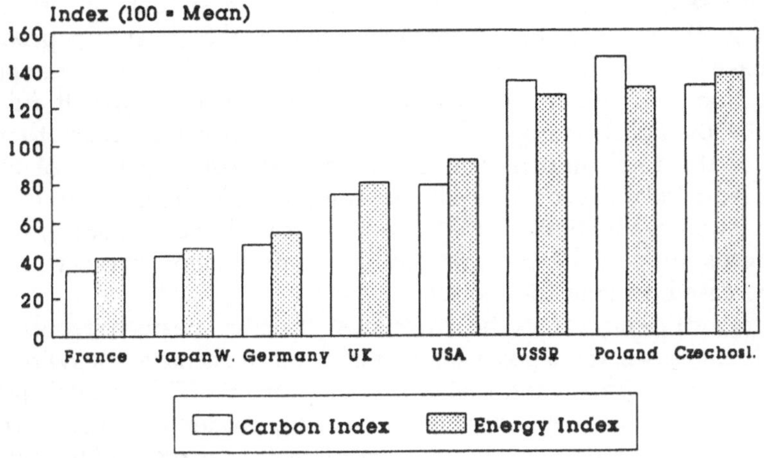

Source: Author

Figure 12.1 *Carbon and energy intensities compared, 1988 (indexes of carbon and energy/$GDP)*

Coupled with the lack of market incentives, this situation created obvious disincentives to use energy rationally.

Energy–environment nexus

The high energy intensity has had direct impact in the region on standards of living and environmental stability. It draws investment into the energy production sector in disproportionate amounts, and retards investments in other economic sectors. Only by increasing the energy efficiency in all activities can the peoples of Central and Eastern Europe hope to achieve living standards comparable to those enjoyed in Western Europe.

The immediate needs of the citizens of these countries, however, may seem to be far removed from global environmental concerns, such as greenhouse gas emissions. While the Western world contemplates setting national goals for the reduction of carbon dioxide emissions, the citizens of Central and Eastern Europe live in some of the most polluted areas of the world. Poland, Czechoslovakia, and Eastern Germany share a region with the highest sulphur dioxide depositions in the world. In the most affected communities in the former Czechoslovakia, bone growth in one-third of the children is retarded by 10 months or more. In Hungary, two-thirds of the drinking water is threatened by environmental hazards, and literally thousands of communities are exposed to unsafe drinking water.[1]

Many of the environmental threats in Eastern Europe are directly related to energy use. It is in Eastern Europe's self-interest to adopt national energy policies that increase efficiency as well as reduce carbon dioxide, the major greenhouse gas produced during the burning of fossil fuels such as coal, oil, and natural gas. Energy efficiency will free capital from energy production and make it available for other productive uses. In Poland, for example, coal production now requires double the annual investment that it did in the early 1970s. Polish energy-related investments consume more than one-third of the total industrial investments.[2] Almost a quarter of all steel produced in Poland is consumed by the energy sector, creating a vicious circle of industrial production for the sake of increased energy supply, which in turn feeds more of the inefficient industrial production that is required by the increased demand for energy.

Energy efficiency offers the single most effective means to reduce carbon dioxide emissions. It is a policy instrument that can achieve the twin goals of economic development and environmental protection. The capital and other resources that these countries invest in energy production could be reduced by substantial amounts if their energy-intensity was reduced to that of the economies of the European Community countries.

Scenarios for the future

Three scenarios were created for future energy supply and demand in Central and Eastern Europe by using the EPA energy end-use model.[3] A base case scenario (BCS) considers the future in the absence of special incentives to save energy, although it incorporates structural changes that may occur (and indeed, are already occurring) in these countries. An energy efficiency scenario (EES) was simulated to show the effects of cost-effective energy efficiency measures on energy supply and demand. And finally, an inter-fuel substitution scenario (IFS) was used to estimate the effects on carbon dioxide emissions of reducing reliance on coal as the predominant fuel, and increasing the use of natural gas and non-fossil energy sources. Data were drawn from a series of detailed energy studies for the FSU and Eastern Europe.[4]

Methodology

The EPA energy end-use model is a parametric model which estimates future energy demand as a function of economic growth, energy prices, price-, income-, and price cross-elasticities of demand, technical improvements in energy efficiency, and structural change in the economy. The user specifies initial base year demand for oil, natural gas, coal, and electricity for six major industrial categories by two-digit standard industrial classification (SIC) for the residential and commercial sectors, and for transportation. The user also provides initial base year activity levels for each sector. Using these data, the model calculates initial sectoral energy intensity coefficients and carbon dioxide emission levels.

The model projects future energy demand to the year 2030 in five-year increments, giving results for the major fuel types and electricity as well as future aggregate industrial energy intensity. It estimates energy demand on the basis of economic growth, structural change, price response, and technical energy efficiency improvements not attributable to price response. Carbon dioxide emissions are calculated for each sector and for each of the six industrial subsectors according to carbon coefficients for each fuel and the composite coefficient for electricity.

Structural change is characterized as the ratio of the growth of the five industrial (two-digit SIC) sectors to growth in overall GNP, plus an additional sector for general manufacturing. Structural change assumptions affect the energy intensity of the overall economy because shifting levels of energy-intensive activities (for example steel output as a share of GNP) change the energy required per unit of economic output. This rate should not be confused with energy-efficiency changes. Both GNP growth and sectoral growth rate ratios are exogenous assumption provided by the user.

The model projects future energy demand on the basis of three other important factors. First, it incorporates energy price response, which the user exogenously specifies by selecting rates of price change for oil, natural gas, coal, electricity, and the price elasticities of demand for each consuming sector. Second, the model modifies future energy demand estimates with a so-called technical factor, which is essentially a rate of change in energy intensity per unit of industrial activity – over and above the price response. This factor is justified empirically, and on the basis of case studies and is exogenous. Third, the model permits the user to specify a price cross-elasticity of demand for electricity which determines the rate of change in electricity demand as a function of the difference between fuel price and electricity price changes.

Marginal cost estimates for carbon dioxide emissions reduction were made using a simple spreadsheet model[5] that calculates levelized cost for each category of energy efficiency measures. These selected energy efficiency measures are based on a Czechoslovak report prepared by a government agency in Prague,[6] and applied to other countries of East Central Europe, except to Poland where separate estimates were developed by Polish experts.[7]

Modelling energy demand scenarios

The base case and energy efficiency scenarios were generated with the assumption that these economies will grow at an annual rate of 2.5 per cent. Population growth also varies from country to country, with Hungary having negative population growth and Poland the fastest growing. These inputs were specified for each country, though a common set of basic assumptions was used for selected, less country-specific variables. These include energy prices, and price and income elasticities of demand, as follows:

- *Energy prices:* real oil and natural gas prices grow by an average 2.5 per cent per year and coal and electricity prices by 1 per cent per year.
- *Price elasticities of energy demand:* the model uses -0.25 as the basic price elasticity of energy demand in all economic sectors. Most analyses of future global energy demand use price elasticities that range from -0.25 to -0.75. The more conservative estimate was used because the EPA model does not generate cross-elasticities among different energy carriers, except from primary fuels to electricity.
- *Income elasticities of demand:* most income elasticities of demand are specified exogenously based on case studies developed by country experts. However, income elasticities of demand for transportation reflect the belief that transportation demand is driven by certain common factors. These include:
  - truck travel - driven by average elasticities of service sector and heavy industry activities;
  - growth in number of heavy trucks - a function of GNP growth and average growth in heavy industry, chemicals production, and service sector activity;
  - bus travel - a function of population growth;
  - air travel - a function of GNP growth;
  - rail passenger travel - a function of population growth;
  - rail freight travel - a function of GNP growth.

Country results

I estimated the energy efficiency potential by determining first the cost-effective energy efficiency measures in several selected activities. Nine broad categories were identified where energy efficiency potential and costs were estimated (see Table 12.1). If implemented, these selected energy efficiency measures alone can reduce carbon dioxide emissions in the countries of Central and Eastern Europe by more than 70 million tonnes of carbon from current levels. This amount represents over 20 per cent of current carbon dioxide emissions in the region, and is measured against current economic activity levels.

This estimate of energy efficiency potential in selected industries and the residential and transportation sectors was used in the EPA energy end-use model to determine the emissions levels in Central and Eastern Europe in the year 2025. Allowing for an annual GNP growth rate of 2.5 per cent, and structural changes in these economies, a combined strategy of energy efficiency and decreasing share of fossil fuels in electricity generation can reduce total carbon dioxide emissions in Eastern Europe by 250 million tonnes from the projected level in 2025 (see Table 12.2 and Figure 12.2). An assumption was made that the share of coal in electricity production would decrease from the current 69 per cent to 50 per cent in 2005 and 30 per cent

Table 12.1 *Cost of CO_2 emission reduction in Eastern Europe, selected energy efficiency measures*

	Energy savings potential (PJ)	Capital cost ($ mill.)	Levelized cost ($ mill.)	Cost of conserved energy ($/GJ)	Total carbon emissions saved (mill. tons)	Cost of carbon emissions saved ($/ton)
Building insulation	850	2,781	305	0.35	18.5	-44.6
Boiler replacement	170	857	92	0.43	3.7	-41.5
Heating improvement	135	685	75	0.56	2.9	-36.1
Cogeneration	150	1,768	151	1.01	2.9	-11.1
Transmission and distribution losses improvements	500	6,071	588	1.15	10.9	-10.9
Existing industrial equipment improvement	700	8,928	980	1.40	13.5	7.7
Ferrous metals	363	5,250	576	1.59	7.0	16.8
New electrical motors in industry	400	11,071	950	2.38	7.7	54.8
Construction industry improvements[a]	150	4,286	368	2.45	2.9	58.5

a Mostly cement production

Source: Henel, Cabicar, and author.

Table 12.2 *Carbon dioxide emissions in Eastern Europe, energy efficiency scenario (MTC)*

	Buildings	*Industry*	*Transport*	*Total*
1985	108	210	38	356
2005	120	146	36	301
2025	159	117	47	310

Table 12.3 *Energy efficiency scenario for Eastern Europe electricity generation, 1987–2025 (%)*

	1987	*2005*	*2025*
Coal	69	50	31
Natural gas	7	10	20
Oil	2	2	2
Non-fossil	22	38	48

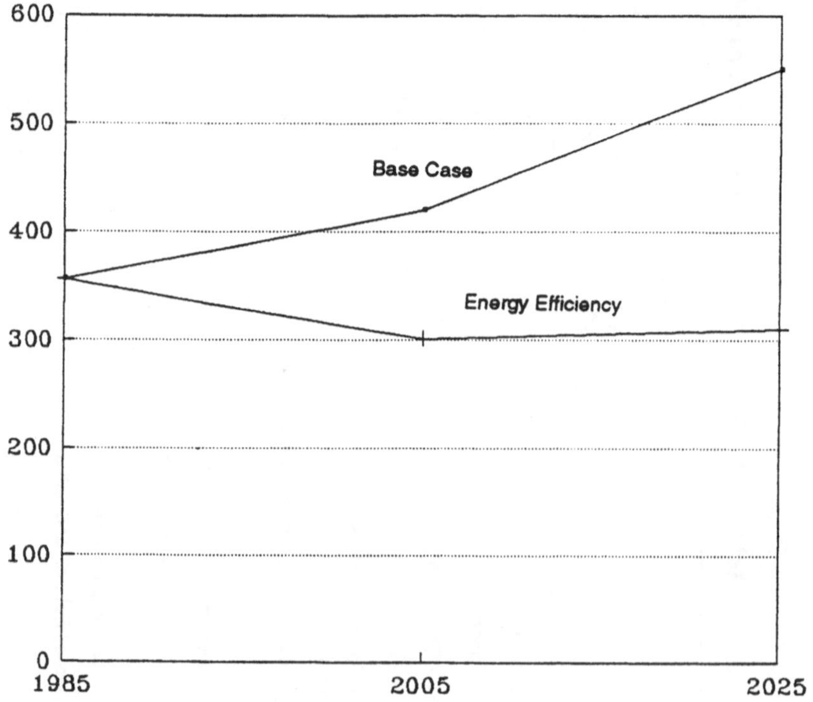

Source: Author

Figure 12.2 *Carbon dioxide emissions in Central and Eastern Europe, 1985–2025*

Table 12.4 *Selected Soviet energy efficiency measures, 1990–2005*

	Annual energy savings in 2005 (EJ)	*Total capital cost 1990–2005 (Bill. R)*	*Levelized cost (rubles)*	*Cost of conserved energy (rubles/GJ)*	*Total carbon emissions saved (mill. tons)*	*Cost of carbon emissions saved ($/ton)*
Shifting from harvesters to site threshing	0.3	0.2	0.19/0.22	7.32	5.83	-53.73
Switching small boilers to high-grade fuels	0.4	0.3	0.20/0.23	8.23	7.78	-52.94
Insulation of steam supply network	0.5	0.4	0.24/0.28	8.78	9.72	-52.47
Advanced technologies for industrial heating	0.2	0.2	0.22/0.27	10.98	3.89	-50.59
Insulation of cattle breeding buildings	0.2	0.4	0.42/0.42	21.96	3.89	-41.18
Automation of heating stations	0.2	0.4	0.52/0.59	21.96	3.89	-41.18
Efficient centralized boilers	0.6	1.2	0.47/0.55	21.06	11.67	-41.18
Change inefficient ovens to large boilers	0.3	0.7	0.53/0.54	25.62	5.83	-38.04
Regulated electric drive	1.4	3.7	0.7/10.82	29.02	27.22	-35.13
Control and measurement in energy use	0.5	1.7	1.11/1.41	37.33	9.72	-28.00
Low capacity multifuel boilers	0.7	3.3	1.09/1.27	51.76	13.61	-15.63
Reduction of electric transmission losses	0.2	1.3	1.85/2.29	71.37	3.89	1.17
Replacing wet cement clinker with dry method	0.2	1.4	1.93/1.86	76.86	3.89	5.88
Gas turbine and combined cycle plants	0.7	5.0	1.64/2.18	78.42	13.61	7.22
Efficient lighting	1.1	8.5	2.41/2.99	84.84	21.39	12.72
Improved brick production	0.1	0.9	1.80/1.80	98.82	1.94	24.70
Improved gas compressors in pipelines	0.3	4.7	3.38/3.95	172.01	5.83	87.44

Source: Alexei A Makarov and Igor Bashmakov, *Carbon Emissions Control Strategies: Case Studies in International Cooperation*, William U Chandler, Editor, World Wildlife Fund & The Conservation Foundation, Chapter 2: 'The Soviet Union' (Washington DC 1990).

in 2025. Conversely, the share of natural gas and non-fossil fuels increases proportionately (see Table 12.3).

Cost estimates for energy efficiency measures through 2005 in the former Soviet Union have been developed by two leading Russian energy analysts, A. Makarov and I. Bashmakov, (see Table 12.4). Almost 15 exajoules of energy, or approximately 25 per cent of current energy use, can be saved at net savings, that is below the cost of new energy supply. These energy savings translate into a reduction in carbon emissions of 250 million tonnes from current levels, that is 27 per cent of current carbon emissions in the former Soviet Union (see Figure 12.3). These measures span all sectors of economic activity, but, as in Eastern Europe, are concentrated in the industrial sector.

In spite of these reductions, however, the EPA energy end-use model projects that carbon emissions will increase absolutely in the FSU through 2025. Reducing reliance on coal, and increasing the use of natural gas and non-fossil energy sources, however, can further reduce carbon emissions by 400 million tonnes by 2025 (see Figure 12.3). Thus, fuel switching is necessary if the republics of the former Soviet Union are to reduce carbon emissions from current levels in 2025.

Economic restructuring

Economic restructuring is the most important aspect of devising policies for reducing energy use and carbon dioxide emissions in Central and Eastern Europe. The base case for each country assumes major changes in the

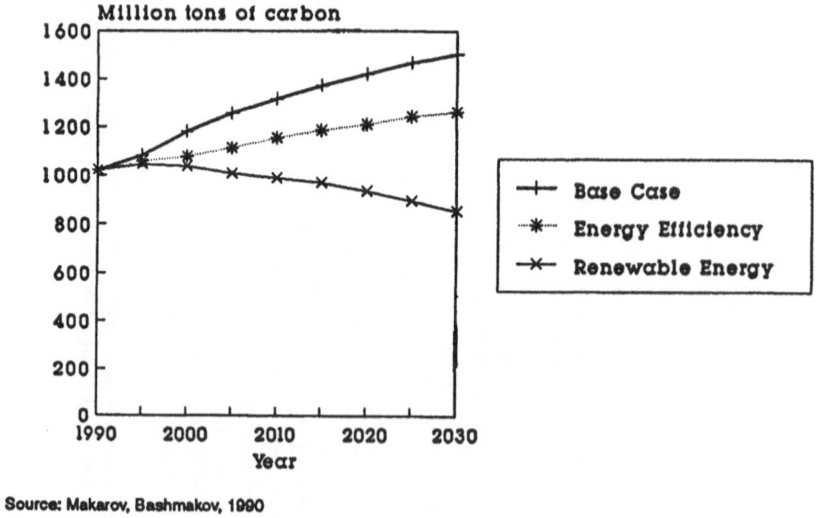

Figure 12.3 *Carbon emissions in the former Soviet Union, 1990–2030*

economies of these countries. This restructuring implies that as income levels approach West European levels, demand for energy amenities also increases to that in Western Europe. Demand growth for energy services, therefore, was modelled in part as a function of income growth. For example, the number of cars per person and living area per capita in the region was assumed to increase approximately to the current West European average when East European per capita income reaches that of Western Europe[8] (see Figures 12.4 and 12.5). The underlying assumption of economic restructuring is therefore the expectation that consumption as a share of gross national product will achieve current Western levels in Eastern Europe by the first quarter of the next century,[c] and that structural change will significantly decrease the share of the industrial sector in total energy demand (see Table 12.5).

The East European economies have invested asymmetrically in heavy industry at the expense of services and consumer goods. This imbalance means, for example, that their economies are very steel intensive compared to Western nations. The former Federal Minister of Economy of Czechoslovakia, and now the Czech Republic's Minister of Industry and Trade, recently remarked that every citizen in his country can have a tonne of steel under his/her bed, but they cannot eat it or drive it.[9] Indeed, Czechoslovakia produces one tonne of steel for every inhabitant each year, while the average in the OECD countries is less than half of that, and the European Community, a region with similar resource constraints as Czechoslovakia, produces only 40 per cent of the steel as Czechoslovakia on per capita basis.[10]

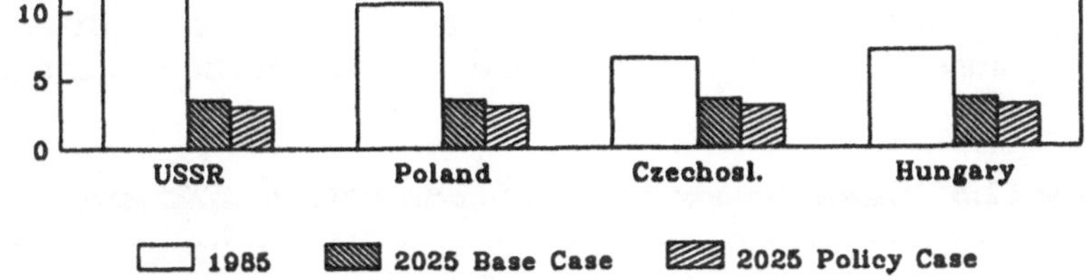

Source: Chandler, 1990

Figure 12.4 *Automobile ownership, country comparisons, 1985 and 2025*

c Consumption in Eastern Europe averages some 55 per cent of the East European GNP, whereas in the United States the corresponding figure is almost 70 per cent.

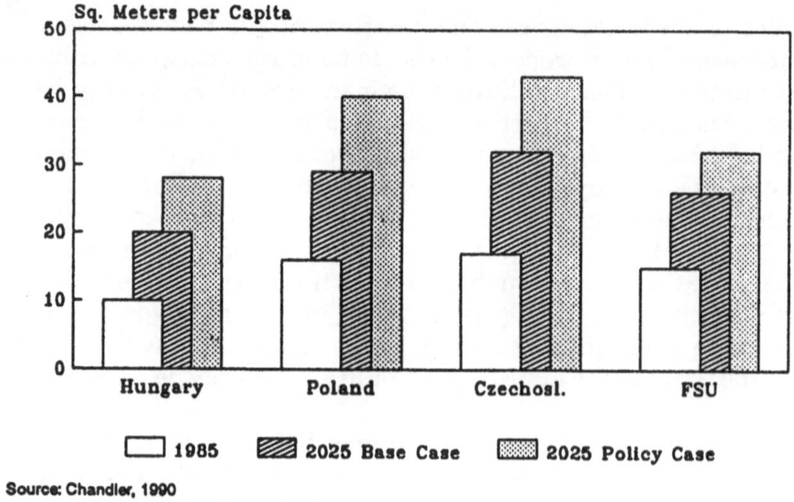

Figure 12.5 *Living area per capita in Eastern Europe, 1985 and 2025*

This concentration of heavy industry represents a structural imbalance in each of these economies deriving from the nature of central planning.

Just as demand for consumer goods was assumed to increase with growing incomes, demand for basic materials per unit of economic output was assumed to decline toward Western levels. The rates of decline assumed were also based on the expert judgements and recommendation of the case study participants, and vary from country to country. In the model, these changes were implemented through income elasticities of demand for each particular product or industry (see Table 12.6).

The industrial sector

On average, the industrial sector in Eastern Europe uses more than twice the energy to produce one dollar of output than do industries in the United States. Because of the large share of industrial production in the gross

Table 12.5 *Energy demand in Eastern Europe, 1985 and 2025 (exajoules)*

	Industry	*Buildings*	*Transport*	*Total*
1985 Demand	10.78	5.20	2.30	18.28
2025 Base case	13.66	12.46	4.99	31.11
2025 Energy efficiency	7.69	7.62	3.43	18.74

Includes Bulgaria, Czechoslovakia, former East Germany, Hungary, Poland, and Romania.

Table 12.6 *Structural change in Eastern Europe, 1985 and 2025*

	Eastern Europe	*European Community*	*United States*
GNP per capita ($ 1985)			
1985	5,482	10,507	16,494
2025	16,189	-	-
Industry share in GNP (%)			
1985	39.6	-	26.2
2025	30.0	-	-
Steel production (kg/$1000 GNP)			
1985	66.0	34	18.6
2025	28.2	-	-
Chemicals production[a] (kg/$1000 GNP)			
1985	14.5	4.4	3.9
2025	10.8	-	-
Living area (sq. metres/cap)			
1985	15	38	55
2025	37	-	-
Persons per automobile			
1985	7.0	2.8	1.7
2025	3.0	-	-

a Production of nitrogen- and potassium-based fertilizers only

national product, high industrial energy intensity is a dominant factor in the overall high energy intensity of those economies.

In the model, base-case industrial energy intensity (energy requirement per dollar of output) declines between 1990 and 2025 by 1 to 1.5 per cent in individual countries. This relatively high energy intensity reduction rate is due primarily to structural changes in each economy, and to a lesser extent to an assumed decrease (0.1 to 0.5 per cent per year) in the technical energy intensity of industrial production caused by capital turnover (see Table 12.7).

Industrial energy efficiency policy is assumed to increase the rate of energy intensity reduction beyond that of our Base Case – that is, beyond structural change – and will average 2.4 per cent per year for Eastern Europe as a whole (see Table 12.8).

The rate of energy intensity reduction will depend on the level of energy intensity in each country and on the relative costs of energy efficiency measures. The estimate is based on cost studies performed in Poland, Czechoslovakia, and Hungary.[11] To provide an approximation for Eastern Europe as a whole, those results were extrapolated to Eastern Germany, Romania, and Bulgaria. Separate analysis was developed for the Common-

Table 12.7 *Energy intensities, base case and efficiency case, 1985–2025 (% average annual change)*

	Residential	Commercial	Steel	Chemicals	Manufacturing	Transport
Czechoslovakia						
BCS	0.2	0.2	-0.3	-0.3	-0.5	-0.3
EES	-1.2	-1.2	-1.1	-0.5	-3.0	-1.3
Hungary						
BCS	0.0	0.0	0.0	-0.1	-0.5	-0.5
EES	-1.0	-1.0	-1.4	-1.9	-1.5	-1.5
Poland						
BCS	0.6	0.6	0.0	0.0	0.0	0.0
EES	-1.5	-1.5	-2.0	-2.5	-2.0	-1.0
Romania						
BCS	1.0	1.0	-0.1	-0.4	-0.3	0.0
EES	0.0	0.0	-2.6	-3.1	-3.9	-1.3

BCS = base case scenario
EES = energy efficiency scenario

wealth of Independent States (formerly the Soviet Union) by experts in Moscow.[d]

Industrial energy efficiency holds the most promise, simply because industry is the largest energy consuming sector in Eastern Europe. Despite major growth in the residential and commercial sectors – to match current West European energy use patterns – industrial energy use will continue to dominate the energy supply and demand picture in Eastern Europe and the former Soviet Union well into the twenty-first century. Compared to the base case, almost 6 exajoules of primary energy could be saved in the industrial sector in Eastern Europe by 2025 (see Table 12.5).

The reader is once again reminded that this potential is in addition to the energy savings embodied in the base case and therefore does not reflect the impact that economic reform will have on the energy intensity of national income. Economic reform in the region will result primarily in structural changes in each economy. Many experts in Czechoslovakia today, for example, call for reducing the production of steel from 15.5 to 7–8 million tonnes per year. Even more dramatic cuts have been recommended for non-ferrous metallurgy and chemicals production.[12] The structural change assumptions, embodied in both the base case and in the energy efficiency case, are based on the case studies completed in Poland, Hungary, and

d The countries of the former Soviet Union are included in this analysis, rather than the current members of the Commonwealth of Independent States (CIS). Lithuania, Latvia, Estonia, and Georgia are therefore included and are not members of CIS.

Czechoslovakia, and on consultations with experts in Romania and Bulgaria.[13]

The major effect of structural changes in Eastern Europe will be a reduced role for industry in producing national incomes, and conversely, in the increased role of services. This outcome will have profound consequences on standards of living in Eastern Europe and will best be manifested in the increased living area per capita, currently less than one-half of the West European average (see Figure 12.5).

Buildings

In the buildings sector, I assumed that as Eastern European incomes grow to match Western levels, so will living area per capita – and with it, energy demand in buildings. On the average, East European living area per capita will grow from 15 square metres today to over 37 square metres by 2025 (see Table 12.6).

Income growth will significantly increase the use of household amenities in Eastern Europe, including air conditioning. In all countries, except Hungary, base case energy intensities in the residential sector rise by between 0.2 per cent in the former Czechoslovakia to 1.0 per cent in Romania (see Table 12.7).

In the energy efficiency scenario, technical energy intensities of the residential sector vary considerably again among countries, from 0 per cent change in Romania to –1.5 per cent per year improvement in Poland. Because of the currently very low energy consumption in Romanian residences, no decline in energy intensity per square foot is expected in Romania, even if substantial energy efficiency measures are implemented. In Poland, coal of inferior quality provides the vast majority of the heat in residential dwellings, and replacing it with better quality coal or natural gas will account for a significant part of the high rate of energy intensity improvement. Carbon dioxide emissions will be reduced by an even greater rate than energy intensity, due to the lower carbon content of natural gas.

Energy consumption in buildings in Eastern Europe, unlike in industry, will increase even in the energy efficiency scenario, and will do so by nearly 50 per cent (see Table 12.5). Compared to the base case, however, energy efficiency measures in the residential and commercial sector can save nearly 5 exajoules of primary energy in 2025. These measures include, roughly in order of importance, building insulation, space heating efficiency improvements, and coal quality improvements.

Transportation

Energy service demand levels for transportation in Eastern Europe in 2025 are assumed to approach those in Western Europe today. Passenger and freight transport will increasingly employ private cars and large trucks, and

Table 12.8 *Eastern Europe final energy demand, 1985 and 2025 (exajoules)*

	1985	*Base case 2025*	*Efficiency case 2025*
Residential			
Coal	2.03	3.47	2.57
Oil	0.19	0.45	0.29
Gas	0.26	1.23	0.48
Electricity	0.40	1.23	0.65
Commercial			
Coal	0.37	1.00	0.64
Oil	0.10	0.23	0.15
Gas	0.15	0.36	0.22
Electricity	0.14	0.38	0.25
Iron & steel			
Coal	0.90	0.66	0.39
Oil	0.10	0.07	0.04
Gas	0.32	0.20	0.12
Electricity	0.19	0.15	0.10
Non-ferrous metals			
Coal	0.01	0.01	0.01
Oil	0.01	0.01	0.00
Gas	0.05	0.03	0.02
Electricity	0.07	0.06	0.04
Chemicals			
Coal	0.19	0.24	0.18
Oil	0.11	0.10	0.05
Gas	0.42	0.44	0.22
Electricity	0.31	0.53	0.29
Cement			
Coal	0.22	0.30	0.20
Oil	0.07	0.10	0.06
Gas	0.17	0.23	0.12
Electricity	0.06	0.08	0.05
Pulp & paper			
Coal	0.11	0.16	0.10
Oil	0.01	0.01	0.01
Gas	0.01	0.01	0.00
Electricity	0.04	0.05	0.03
Other industry			
Coal	1.79	2.25	1.48
Oil	0.95	1.18	0.71
Gas	1.24	2.01	0.93
Electricity	0.60	1.04	0.50

on average, the rate of car ownership per capita is expected to more than double (see Figure 12.4). Air travel in Eastern Europe is assumed to increase at a rate ten times the increase in miles travelled by buses, but railroads will remain important for both passenger and freight travel.

The base case assumes minimal improvement in automobile fuel economy. However, significant energy savings can be realized through policy measures such as standards for fuel economy. On the average, cars in Eastern Europe consume about 8.7 litres per 100 km. Increasing automobile fuel economy to 5 litres per 100 km – and implementing additional transportation energy savings measures, such as improving truck fuel economy, and converting the gasoline-powered trucks to diesel – can reduce transportation energy demand in 2025 by 1.5 exajoules compared to the base case (see Table 12.5). A net increase of 50 per cent over current levels would still be necessary even in the energy efficiency scenario, unless passenger transportation evolves completely differently to that in Western Europe. Policies to encourage or implement alternative transportation systems, however, have not been included in our energy efficiency scenario, despite the need of these countries to actively seek new transportation development patterns. Rather, I assumed that Central and Eastern Europe will develop a transportation system that is similar to the West European system.

Policy implications

The behavioural changes simulated in the base case scenario were designed assuming that the allocation of investment funds in the countries of Eastern Europe will change from the practices of forty years of central planning. The asymmetrical development of Eastern Europe has severely retarded the region's economic well-being, standards of living, and environmental stability. A direct consequence of the uneven economic infrastructure is high energy demand with a plethora of consequences. It is in the region's own interest to reduce its high energy intensity through a gradual change in investment allocation. This shift will occur naturally if those economies become open and market-oriented. The base case assumes that these economies will become fully integrated into the European market and that external economic developments – and basic dynamics of supply and demand in market economies – will influence how these countries produce their wealth.

Policy mechanisms for energy efficiency have been simulated in two separate ways, and were measured against the projected energy demand and carbon dioxide emission levels in the year 2025, that is against the base case. First, I assumed that investments – particularly in the buildings and transportation sectors – and energy efficiency standards were put in place to reduce the intensity of energy services to levels of efficiency deemed cost-effective. For example, automobile fuel usage should not exceed 5 litres per

100 km. Such analysis can readily be implemented in the EPA model for all economic sectors and subsectors.

Second, I assumed that economic reforms will cause energy prices to escalate from their current subsidized levels to equilibrium levels – that is, to world prices. Electricity and heat prices are still subsidized to most consumers. The EPA model calculates energy demand with an annual increase in real energy prices of between 1.0 and 2.5 per cent for different fuels and electricity. This method assumes that prices matter to enterprises, as well as to residential and commercial building consumers, and that energy users respond to prices as in the West.

In a sensitivity test, energy price elasticities were increased to -0.6. In this case, the necessary price increase required to simulate the entire energy conservation potential for Eastern Europe would have to be only half as steep. Interestingly, the results from the price reform case resembled those from the regulatory case through the next two decades. The price elasticity of demand was assumed to be only -0.3, exceedingly low for long-term elasticities by Western standards. The reader should note here that either the regulatory or the price reform cases may be simulated independently or conditionally in the EPA energy end-use model. Simulating one or the other case, or both, can serve to ascertain the necessary measures needed to achieve a level of energy efficiency in the national economy deemed cost-effective. In the real world, both energy prices and regulatory policies will have to be used as tools for reducing energy intensity.

Considering the results of the model runs for the price reform case and the regulatory case, it appears that the East European countries have a lot of flexibility in using price reform and regulatory policy to achieve their energy efficiency potential. The fact that energy prices in the region will eventually equal world market prices is virtually certain so long as these countries move toward the free market. What is yet to be ascertained is their resolve to make careful use of market regulation to balance their national interests and free market capitalism.

Conclusion

Central and Eastern Europe's self-interest requires that it spends less resources on energy production and consumption, and more on generating marketable products and services – at reduced costs, including reduced energy costs. Results from the case studies prepared by experts from these countries clearly indicate that it is often cheaper to save energy – and improve productivity at the same time – than to produce more energy. Doing so will also reduce the risk of climatic change by reducing the emission of greenhouse gases, particularly carbon dioxide.

The technological and economic changes that will be necessary in Central and Eastern Europe in order to achieve these goals will be enormous. To illustrate, the nation of Germany is subsidizing the region of the former East Germany - a region of only 16 million inhabitants - with about $90 billion per year. Still, many experts believe that it will take one generation before the former communist East Germany is culturally, economically, politically, and morally integrated into Germany as a whole. What resources will be required in the rest of Central and Eastern Europe, if that region wants to become a first class citizen of Europe? Clearly, no foreign government can play the leading role in such an undertaking. These countries must themselves take charge and lead the way. By the same token, Central and Eastern Europe, at once the most polluting and the most polluted part of the world, cannot expect that allowances be made always to accommodate their exceptional difficulties. These nations cannot disregard standards, environmental or otherwise, that are expected of those in Western Europe.

Reducing energy intensity and carbon dioxide emissions will require concerted effort on the part of these nations. They must implement economic and political reforms; assure non-inflationary growth; and create new forms of ownership of private property.

The end of the Cold War certainly presents a unique opportunity for Central and Eastern Europe and for the CIS to learn how best to create conditions for healthy market economies. Such conditions are necessary for their transition to economic and environmental well-being. Western aid agencies such as the European Bank for Reconstruction and Development, the World Bank, the International Monetary Fund, and bilateral aid organizations can only stimulate the process of reconstruction and restructuring. Their efforts, however important, cannot change the ways of a half billion people. Technology and know-how will be transferred in a lasting way only if Western commercial interests are rooted in the economies of these nations.

The effects of reducing energy intensity in the region will have an immediate impact on the macroeconomic level. Production and consumption patterns of goods and services will change dramatically to reflect the prevailing patterns in Western Europe. This shift will be most immediately apparent in Hungary, the former Czechoslovakia, and Poland, countries with energy and resource constraints similar to those of Western Europe. The role of the industrial sector will gradually decrease in the overall GNP balance, which will require large adjustments in the workforce as services become more important than industry.

In Romania, Bulgaria, and the FSU, these changes will most likely take longer than in Central Europe. The lack of market-economy tradition and the slower pace of economic and political reforms have already put these countries behind Hungary, the former Czechoslovakia, and Poland in the race to become fully-fledged partners in the European economic and political union, and thus delay their transformation into modern industrial nations. If any meaningful acceleration of pace is to take place in the economic transformation of these nations, the European Community must extend market access and economic cooperation to these nations. All of these changes are preconditions for an effective energy efficiency strategy that can reduce carbon emissions in these transitional economies.

References

1 T Fleischer and J Vargha, eds, 'The Most Important Tasks of Environmental Protection in Hungary,' ISTER, East European Environment Research, Budapest, 1989

2 S Sitnicki, *et al.*, Chapter 3: 'Poland,' in *Carbon Emissions Control Strategies: Case Studies in International Cooperation*, William U Chandler, ed.

3 This model is the 'EPA Energy End-Use Model' developed by Irving Mintzer, *Projecting Future Energy Demand in Industrialised Countries: An End-Use Oriented Approach*, World Resources Institute, October 1988, and modified by W U Chandler of Battelle, Pacific Northwest Laboratories for the US Environmental Protection Agency with the assistance of Stanislav Kolar, PNL, and the advice of Jean-Charles Hourcade and Richard Baron, CIRED, Paris, France

4 The studies on which this chapter is based are: William U Chandler, Stanislav Kolar, Adrian Gheorghe, and Stanislaw Sitnicki, 'Climate Change and Energy Policy in Eastern Europe: Two Scenarios for the Future,' *Energy*, Vol. 16, No. 11/12, pp 1423–1435, Pergamon Press, 1991; S Sitnicki, *et al.*, Chapter 3: 'Poland,' in *Carbon Emissions Control Strategies: Case Studies in International Cooperation*, William U Chandler, ed., World Wildlife Fund and The Conservation Foundation, Washington, DC, 1990; Alexei A Makarov and Igor Bashmakov, *Carbon Emissions Control Strategies: Case Studies in International Cooperation*, Chapter 2: 'The Soviet Union,' in William U Chandler, ed., World Wildlife Fund and The Conservation Foundation, Washington, DC, 1990; Tamás Jászay, Chapter 4: 'Hungary,' in *Carbon Emissions Control Strategies: Case Studies in International Cooperation*, William U Chandler, ed., World Wildlife Fund and The Conservation Foundation, Washington, DC, 1990; Marie Košťálová, Jiří Suk and Stanislav Kolar, *Reducing Greenhouse Gas Emissions in Czechoslovakia*, Pacific Northwest Laboratory, Richland, Washington, December 1991

5 This model was developed by the author

6 M Henel and B Cabicar, *Rentabilita statnich prostredku, vlozenych na podporu vyssiho zhodnocovani paliv a energie narodnim hospodarstvi – ekonomicke zduvodneni a oblasti pusobnosti*, VUPEK (Research Institute of the Fuels and Energy Complex), Prague, September 1991

7 S Sitnicki, *et al., Carbon Emissions, op. cit.* (endnote 4)

8 For data, see, variously, Economic Commission on Europe, *An Energy Efficient Future*, New York: United Nations Economic Commission on Europe, 1983; D Shonak, *et al.*, *Transportation Energy Data Book*, Oak Ridge: Oak Ridge National Laboratory, 1989

9 Opening remarks, US Electric Power Technologies Conference, Prague, Czechoslovakia, July 7, 1992

10 For data see, Central Intelligence Agency, *Handbook of Economic Statistics, 1989*, Directorate of Intelligence, Washington, DC, September 1989

11 See references in note 4 above

12 Communications with Jiří Suk, Forecasting Institute, Czechoslovak Academy of Sciences, Prague, Czechoslovakia, April 1990

13 See references in note 4 above

13

Greenhouse gas emission abatement in Australia

Hugh Saddler

Greenhouse gas emissions have been at the forefront of public policy debate in Australia for over three years. In October 1990 the federal government adopted an 'interim planning target' – to reduce emissions of greenhouse gases to levels 20 per cent below 1988 emissions by 2005. The reduction of 20 per cent relative to 1988 levels was first proposed at an international conference held in Toronto, Canada in 1988; it is referred to as the Toronto target in this chapter.

Gases controlled by the Montreal Protocol on Substances that Deplete the Ozone Layer, that is, chlorofluorocarbons and related compounds, were not included, but the government had previously announced that the use of CFCs would be eliminated by 1995. The undertaking with respect to other greenhouse gases was subject to the important qualification that it would not proceed with measures which have net adverse economic impacts nationally or on Australia's trade competitiveness.

The eight Australian state and territory governments have important policy powers relating in particular to the electricity and gas industries and to the control of pollution. Their concurrence and participation is virtually a prerequisite for the realization of the federal target. The eight governments have agreed to participate with the federal government in the development of a national greenhouse response strategy. A draft strategy document was released for public comment in June 1992.

Although it is normally viewed as a developed country, Australia's economy is heavily dependent on exports of raw and partly processed commodities. Among the most important Australian exports are coal, liquefied natural gas (LNG), and alumina and aluminium metal smelted with coal-fired electricity. Consideration of greenhouse response strategies has therefore been strongly influenced by concerns about the possible effects of any abatement measures on Australia's international competitiveness as a supplier of fossil fuel intensive commodities.

A steady stream of reports and studies from government, business and non-governmental organizations has provided the material for an enthusiastic public policy debate. Many of the reports have sought to estimate the costs to the Australian economy of reducing carbon dioxide emissions by changes in energy supply and use, focusing in particular on the costs of meeting the Toronto target. This narrow emphasis on a single goal reflects partly the concerns of special interest groups who fear the impact of achieving the target on their activities. Most studies have sought to estimate the macro-economic effects of a carbon tax on fossil fuel use that would suppress demand for fossil fuels in 2005 to the Toronto target level. The narrow focus of so many of these studies has had two unfortunate effects. First, it has encouraged an 'all or none' view of the desirability of implementing emission reduction policies. And second, it has led to neglect of policy instruments other than a carbon tax.

Nevertheless, the studies have greatly improved our understanding of the workings of the Australian energy system and its interaction with the wider economy. Very much less is known about non-energy related sources of greenhouse gas emissions, which are the principal sources of the other important anthropogenic greenhouse gases such as methane and nitrous oxide.

In this chapter, I examine the cost and scope of emission abatement measures available in Australia. I also review estimates of the effect on the Australian economy of achieving various levels of abatement.

Abatement of energy sector emissions

Estimated carbon dioxide emissions from the use of energy in Australia in 1989, and 'business as usual' projections for 2005 are shown in Table 13.1. The energy use figures shown are the quantities of energy consumed at the

Table 13.1 *Estimated energy-related CO_2 emissions in Australia*

Economic sector	*Energy (PJ)*		*Carbon dioxide (MT)*	
	1989	*2005*	*1989*	*2005*
Residential	302	456	44.8	61.7
Commercial	143	248	30.2	47.9
Manufacturing	959	1343	105.9	145.4
Transport	997	1247	77.0	96.0
Agriculture, mining etc.	228	311	23.0	32.3
TOTALS	2629	3605	281	383

Data are for financial years running respectively from July 1988 to June 1989 and from July 2004 to June 2005. Carbon is elemental carbon as carbon dioxide, that is, 0.273 of the mass of carbon dioxide.

Source: Australian Commission for the Future (1991).

point of end use, whereas the carbon dioxide emission figures embody the allocation of emissions arising from energy transformation processes (such as electricity generation and oil refining) required to deliver energy to the respective end uses. However, emissions associated with the transformation of energy for export (such as those associated with liquefaction of natural gas) are included in the agriculture and mining sector.

These projections imply high rates of growth in energy consumption. The main assumptions which underlie this projection include: continuing high rates of net immigration and hence population growth; continuing structural shifts in the economy towards energy-intensive materials processing activities, particularly primary aluminium and paper; and rather modest increases in energy use efficiency.

Ecologically sustainable working groups

During 1991 the federal government coordinated a national process under the rubric of 'Ecologically Sustainable Development' (ESD). The ESD process involved collaboration between all levels of government, business and national environmental groups. It aimed to provide policy advice on how to implement ecologically sustainable development in Australia. As part of this process, the government commissioned a series of studies of the technological potential to improve energy use efficiency in selected energy services in the residential, commercial and manufacturing sectors (Ecologically Sustainable Development Working Groups 1991a). These are the only set of detailed 'bottom up' or 'engineering' studies of the potential for emission abatement by demand side measures so far undertaken for Australia as a whole.

The studies examined the present stocks of energy-using equipment relevant to the respective processes, assessed the energy savings available from the use of selected technologies for each service, and estimated the associated costs and benefits. The energy services assessed were as follows:

Residential:	Hot water
	Refrigerators and freezers
	Washing machines and dish washers
Commercial:	Heating, ventilation and air conditioning (HUAC)
	Lighting
	Other services (hot water, cooking, office and other electrical equipment)
Manufacturing:	Metal smelting
	Electrolytic processing
	High temperature firing
	Metal melting and other high temperature metal processing
	Electric motors and drives

The ESD studies estimated that in 1988 these energy services accounted for emissions of about 118 million tonnes (MT) of carbon dioxide, that is, about 42 per cent of the total from the Australian energy system. They also projected future energy services. They assumed 'frozen efficiency,' that is, no change in 1988 technology. This assumption, plus various others as to economic and demographic trends, determined that emissions from these energy services in 2005 would total about 192 MT.

ESD Methodology

These studies assessed a wide variety of technical options for each of these services or groups of services, including process changes and fuel substitution as relevant efficiency improvements. They evaluated the economic benefits and costs of the various options exclusive of any economic benefit of emission abatement itself. Thus benefits included reduced energy consumption, and in some cases associated savings in maintenance costs, increased productivity of capital equipment, etc. They measured the benefits of energy savings in terms of a schedule of projected resource costs for electricity, gas, petroleum products and coal. Costs were taken to be the additional capital costs of the new equipment. All studies performed the analyses in terms of investment programmes over the period 1992 to 2005 affecting the present stock of equipment in the Australian economy. They assumed that benefits would continue to accrue beyond 2005, until the end of the economic life of the equipment concerned. They further assumed that any cost savings achieved by the use of more efficient equipment would be realized in the form of reduced energy consumption, and hence carbon dioxide emissions, rather than increased output. Finally, they discounted the streams of costs (mainly capital costs) and benefits (mainly energy savings) back to 1990 at a real discount rate of 8 per cent.

The studies were confined to measures which either show a net benefit or a relatively modest net cost. Technical options involving, for example, prematurely scrapping and rebuilding major industrial plant were not considered. Indeed, for many of the technical options, the ESD found that costs are relatively small (and hence net benefits are positive) because the technical improvements are realized only as new plant is built and new equipment purchased, either as replacement for existing obsolete plant equipment or in the course of expanding total output. Thus, there is an inverse relationship between average emissions per unit of output and total emissions in such activities as aluminium smelting. If the industry grows rapidly (several new smelters), then total emissions will grow strongly, but emissions per tonne of aluminium will fall. Conversely, with no growth in the industry, there would be no change in total emissions, but also no improvement in emissions per tonne of aluminium.

The studies were concerned with the technical potential for emission reductions, not with the practicalities of achieving that potential. Thus, in most cases very high rates of penetration of optimally efficient technologies in-purchases of new plant and equipment were assumed, even in cases (such as electric motors) where most new purchases are of low efficiency motors which are not the economically optimal choice at a discount rate of 8 per cent. No administrative or incentive costs for programmes which might be needed to stimulate the economically optimal choice of equipment were included in the cost-benefit assessment. For this reason, the estimates of net benefits/cost of emission abatement could be interpreted as understating the costs. On the other hand, in a number of the studies - notably those dealing with residential sector energy services - the assumed resource cost for electricity was substantially less than the true cost at the customer's meter, thereby understating the benefits of energy savings.

ESD results

A supply curve of the abatement measures associated with each of the eleven energy services studies is shown in Figure 13.1. It should be noted that for each service, the potential abatement shown as a single block is in turn the sum of a number of individual measures, each having its own net abatement cost which may be higher or lower, sometimes by a substantial margin, than the average of all measures applicable to the particular service. Table 13.2 summarizes the abatement potential and costs of each measure reflected in Figure 13.1.

The estimated total emission reduction potential from all of the energy services studied was 61 MT of carbon dioxide, assuming no change in the present mix of supply technologies. On this basis, emissions associated with

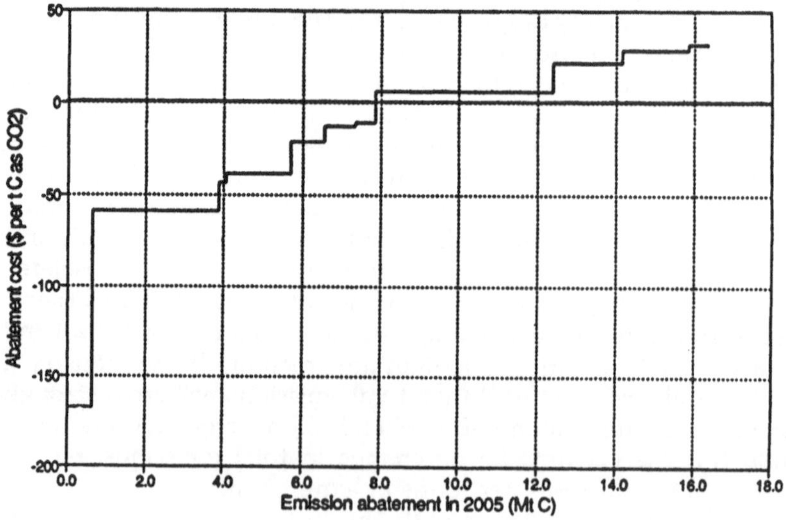

Figure 13.1 *Cost of CO_2 emission abatement in Australia, selected energy efficiency measures*

Table 13.2 *Emission abatement potential of selected energy efficiency measures in Australia*

	Emissions (MTC as CO_2)				
	Base 1988	*Frozen efficiency 2005*	*Saving 2005*	*Cumulative saving*	*Cost ($/TC)*
Commercial miscellaneous	1.2	2.2	0.6	0.6	-168
Commercial HVAC	4.6	8.7	3.2	3.9	-59
Industrial metal processing	0.6	0.8	0.2	4.1	-44
Electric motors and drives	5.5	8.8	1.7	5.7	-38
Industrial high temperature	2.3	3.4	0.8	6.5	-21
Smelting	4.4	5.3	0.8	7.3	-12
Industrial electrolysis	4.8	8.7	0.5	7.9	-11
Residential hot water	3.8	6.0	4.5	12.4	7
Commercial lighting	1.9	3.7	1.8	14.2	22
Residential refrigeration	2.0	3.2	1.7	15.9	28
Residential major appliances	1.0	1.6	0.5	16.4	32
TOTAL	32.2	52.4	16.4		
Percent of frozen efficiency 2005:	62%		31%		

All dollar values here, and throughout this chapter are Australian dollars unless otherwise stated.

the services studied in 2005 would be 13 MT higher than 1988 emissions of 118 MT, that is, an increase of 11 per cent. This level of emission abatement is equal to about 32 per cent of the projected frozen efficiency emissions from the services studied.

Unfortunately, the ESD studies did not provide any reliable, consistent estimate of per centage abatement relative to a 'business as usual' projection. Moreover, the potential of the measures studied may not be representative of the energy services not included in the studies, the most important of which are:

- all uses of energy in agriculture, fishing etc.;
- all uses of energy in mining;
- all uses of energy in transport;
- low temperature process heat in the manufacturing sector;
- residential space heating and cooling;
- residential cooking;
- electronic and other miscellaneous (mainly electrical) residential appliances.

The activity levels associated with some of these services, notably agriculture and transport, will grow rather more slowly than the activities included in the studies.

ESD transport sector study

The ESD Working Groups also studied the potential for carbon dioxide emission abatement in the transport sector (Ecologically Sustainable Development Working Groups 1991b). They considered a wide variety of technical and behavioural changes to transport in Australia. The existing transport system is dominated by road and air which account respectively for 85 per cent and 7 per cent of total carbon dioxide emissions. Any technical improvements in the fuel efficiency of road and air transport in Australia will be almost entirely dependent on imported technology. Consequently, the potential for efficiency improvements in these modes will be largely determined by international developments in motor vehicle and aircraft technology. The study made two different assumptions about the rate at which fuel saving technologies would be incorporated into mass produced vehicles and aircraft on a global basis, and incorporated these into 'low' and 'high' emission abatement scenarios. For rail transport, the main options for fuel consumption improvements derive from upgrading the permanent rail system on heavily trafficked routes and improved train signalling and control systems, based largely on indigenous technology. The study also incorporated assumptions about changes in urban passenger transport systems, which currently account for about 45 per cent of total Australian transport energy use. The measures considered included changes in urban form (consolidation of cities to increase residential densities) and greater use of mass transit, bicycles and walking. 'Low' and 'high' levels of abatement were again considered.

Estimated carbon dioxide emission abatement under these two scenarios were respectively 12 per cent for the 'low' case and 27 per cent for the 'high' case below 'business as usual' projections for 2005. However, since the 'business as usual' projection is for an increase in emissions of 40 per cent, these figures still represent higher emissions than in 1988 by respectively 25 per cent and 2 per cent.

The transport study did not analyse the economics of the various measures considered and of the two scenarios as a whole. As noted earlier, the economics of technical measures for efficiency improvements will depend heavily on international technologies and will likely mirror the economics of such measures in Japan, North America and Europe, appropriately adjusted for the local price of fuel and spatial densities. The projected changes in urban transport systems would require very substantial redirections of investments in all types of urban infrastructure, that is, not just transport infrastructure. Most Australian cities have spare capacity in at least some components of their urban infrastructure. Measures to consolidate urban form would result in substantial capital savings from the reduced requirement for new infrastructure at the urban periphery. There would be a need for significant additional investments in public transport infrastructure, but these would be offset, at least in part, by reduced investment in urban arterial roads. Thus the overall outcome could be a net saving in capital investment (McGlynn, Newman and Kenworthy 1991).

Other sectors

The remaining categories of energy service have not been scrutinized systematically on a national basis as the services already described. Information about residential space heating and cooling and cooking is particularly deficient. Some studies have drawn general conclusions about emission reduction potential in the very large low-temperature industrial process heat category. A detailed engineering study of a representative sample of industrial plants in the food processing, paper and other industries concluded that cost-effective savings of the order of 10 to 20 per cent might be available on average (Warren Centre 1991). An abatement of 15 per cent would be equivalent to about 8 MT of carbon dioxide in this end use in 2005. Opportunities for greater use of cogeneration are also associated with the use of low temperature process heat. The ESD Working Groups estimated also that additional cost-effective opportunities for gas-fired cogeneration could yield an emission abatement of about 4 MT carbon dioxide (Ecologically Sustainable Development Working Groups 1992).

Aggregate carbon conservation potential

Because the various studies referred to have not been performed on a single consistent set of baseline energy consumption data, and because definitions of 'business as usual' are not consistent, it is difficult to sum the results to give a single figure for potential emissions abatement. However, subject to some assumptions, it can be estimated that the technical potential of these measures is for a carbon dioxide emission reduction in 2005 of between 71 and 86 MT, depending on which transport scenario is chosen, relative to the projected 'business as usual emission' level of 383 MT shown in Table 13.1.

To this total, one could add a few additional million tonnes from savings in the unexamined residential sectors (space heating and cooling, cooking, appliances). However, even making allowance for these savings, the general form of the result is clear. Although very substantial reductions in carbon dioxide are technically available through the adoption of cost-effective or close-to-cost-effective measures to increase the efficiency of energy use, the reductions are somewhat less than the 102 MT required to stabilize emissions in 2005 at 1988 levels.

A recent critique of these studies has claimed that they overstate the potential and understate the cost of achieving such levels of emission reduction by demand side measures (ACIL Australia 1992). The critique points to imprecision and confusion in the definition of the 'business as usual' case against which savings are measured and argues that the estimates of potential savings exaggerate what is achievable.

In essence, this criticism simply restates arguments about the existence, nature and causes of the 'efficiency gap' (Grubb 1990). As such, the criticism misunderstands the purpose of the studies, which are concerned with the technical potential for savings, not the savings achievable under the prevailing economic and policy environment. The savings identified are those which have been assessed as being cost-effective at a discount rate of 8 per cent, which is generally accepted as an appropriate rate for determining social costs and benefits, but is considerably lower than the discount rate commonly used for private and business decisions about purchases of energy-using equipment. To achieve the technical potential for energy efficiency improvements will require quite large and rapid changes in purchase decisions by energy users and will in turn require the implementation of a variety of new policies and programmes by governments, energy utilities and other parties. No estimate of the cost of incentives and other measures, likely to form part of such programmes, is included in the analysis.

Additional policy measures

It will be apparent from the figures cited, that – even assuming the full achievement of technical potential for efficiency improvement – stabilizing carbon dioxide emissions at 1988 levels by 2005 will require changes in other factors which affect the level of emissions from the energy system. Such changes could include:

1 a much lower level of immigration, and hence of population growth, affecting the absolute (as opposed to the per capita) rate of growth in economic output;
2 a great reduction in output from highly energy intensive export-oriented industries, notably aluminium production;
3 a reduction in carbon dioxide emissions associated with the energy supply system.

The first two of these options are not considered to be politically desirable or acceptable by the majority of Australians. In any case, unilateral action by Australia would have very little effect on global carbon dioxide emissions, since the growth in overall economic activity and aluminium production, and associated growth in carbon dioxide emissions, would simply occur elsewhere.

The third option, of reducing emissions associated with the energy supply system, is obviously the preferred method for achieving further reductions in carbon dioxide emissions. Australia has very large reserves of both black coal and brown coal (lignite), which are favourably located in relation to the major centres of population and economic activity. Consequently the electricity supply industry is heavily dependent on coal, which in 1988 accounted for 76 per cent of total electricity generated (Ecologically Sustainable Development Working Groups 1991c). Between 1978 and 1988 emissions of carbon dioxide per MJ of electricity generated decreased from about 310 tonnes to 270 tonnes, an improvement of about 13 per cent. This decrease was achieved largely by replacing older generating plant by large new 500 MW and 660 MW generating units. While there is probably some scope for modest further improvements in thermal efficiency of conventional coal fired generation, more substantial changes would require significant changes in technology.

A systematic review of available generation options, focusing on cost and carbon dioxide emissions, concluded that conventional black and brown coal fired steam turbine generation is the lowest cost, but most carbon dioxide intensive, option (Ecologically Sustainable Development Working Groups 1991c). Combined cycle gas turbine technology would be slightly more expensive and is a realistic alternative, making use of large uncommitted gas resources located off the south east and the north west shores of the country. Australia's natural resource endowments also make it well placed to utilize nuclear, wind, solar thermal and photovoltaic generation technologies, but these would all be significantly more expensive than coal and gas fuelled technologies in most parts of Australia.

The review also calculated abatement costs for these technologies, relative to conventional black coal fired steam turbine technology. Considerable uncertainty surrounds all the cost estimates because none of the alternative technologies are deployed on a commercial scale in Australia. The cost of gas is also uncertain.

For combined cycle gas turbines, the best estimate is a few tens of dollars per tonne of carbon abated. For wind and nuclear, the costs are between $100 and $300 per tonne. Advanced coal combustion technologies are among the least cost-effective options, with abatement costs of up to $500 per tonne. The cost is so high because of the relatively modest emission abatement available by use of these technologies. The economics of abatement are much more favourable relative to conventional brown coal fired steam generation, because this technology is intrinsically both more costly and more carbon dioxide intensive than black coal fired steam generation. It should be noted that these comparative assessments of generation technologies are based on costs at the power station busbar only. They do not take account of interconnected system characteristics which will affect the proportion of total demand which can be supplied by particular technologies, and the economics of doing so.

Two recent studies, using different modelling methodologies and somewhat different data estimates and assumptions, have concluded that the Toronto target for carbon dioxide abatement could be met. This goal would be achieved by: extensively substituting gas turbine combined cycle plants for coal fired steam generation technology; and by using some renewables such as wind and/or solar thermal energy technologies (Jones 1992;

Australian Commission for the Future 1991). Both studies model the effects on carbon dioxide emissions of progressive introduction of more efficient energy using technologies and the replacement of coal by gas and renewables in electricity generation. Of course, as the electricity sector becomes progressively less carbon dioxide intensive, the emission abatement achieved by using electricity more efficiently is reduced. Because the two studies differ somewhat in their estimates of extent of demand side efficiency gains, they differ also in their estimates of the extent and cost of adjustments required on the supply side, particularly in electricity generation.

Given the relative costs of the respective technologies, there would be a net cost to the economy (excluding the benefit attributable to greenhouse gas emission abatement itself) of adopting this strategy. A large proportion of the current Australian coal fired generating capacity has been commissioned since 1975, and would normally be expected to have a life of 25 to 30 years. Meeting the target for emission abatement by 2005 would require prematurely scrapping a number of power stations, with an additional economic penalty. The penalty would not apply if the same, or even a more stringent abatement reduction, were to be met by a later date, say about 2015. This approach, however, incurs the additional ecological impact of releasing carbon dioxide emissions during the ten years from 2005 to 2015.

Economic impact of abatement strategies

One of the studies referred to above included estimates of the macro-economic consequences of the proposed emission reduction measures (Australian Commission for the Future 1991). The ACF study is the one in Australia to date which integrates the results of 'bottom up', technology-specific energy modelling with a 'top down' macro-economic model. Two levels of abatement were analysed: the demand side measures only, falling somewhat short of emission stabilization; and the demand and electricity supply measures combined, achieving the Toronto target, as described above. The major impact on the economy in both cases comes from the surge in investment expenditure needed to pay for the new, more efficient energy-using and energy-supply equipment. This strategy requires diverting economic resources from consumption to investment, with the result that consumption expenditures fall.

Since the ACF study found that most of the demand side measures were cost effective relative to expected energy prices, it is not surprising that aggressive deployment of these measures alone yields net macro-economic benefits over the long term, as measured by a slightly higher rate of GDP growth than in the base case. Achieving the Toronto target requires that Australia diverts a higher proportion of resources to energy-related investments and that the average unit cost of energy is higher than in the base case. Thus, GDP growth is lower than in the base case by between a half and one per cent, depending on assumptions about how the economy adjusts to change. This outcome is obviously a relatively small change, much less than the changes associated with the normal ups and downs of economic activity. One reason for the change being so small is the study's assumption that Australia takes a unilateral decision to limit greenhouse gas emissions. Domestically produced fossil fuel resources, particularly coal and petroleum, are therefore available to be exported if not required for domestic consumption.

Other economic studies have sought to model the effects of measures to reduce greenhouse gas emissions on the Australian economy or particularly vulnerable sectors of it, most of which have been summarized by the Industry Commission (1991). These studies share the following common features:

1 a 'top down' approach to modelling the Australian energy economy;
2 the use of a carbon tax as the policy instrument by which emission abatement will be achieved;
3 a focus on the cost of achieving only the Toronto target, with no analysis of the costs of lower levels of abatement;
4 no allowance for improvements in technology which could increase the efficiency of energy supply and use without increasing costs.

This approach assumes that all markets for energy services are perfectly competitive, that is, energy is used throughout the economy with optimal technical efficiency and no costless opportunities for efficiency improvement are available. It follows that an increase in energy prices by means of a carbon tax is needed to induce any change in technical efficiency. Of course, this assumption varies sharply from the findings of the sectoral energy use studies described above.

Disagreement over this issue is by no means confined to Australia, but is a persistent theme around the world in debates over energy and greenhouse emissions (see for example Manne and Richels 1990; Williams 1990; Nordhaus 1991). Some have characterized this as a disagreement in perspective between economists and engineers. It could perhaps be more accurately described as a disagreement between those who sit behind desks and theorize about the economy; and those who go out to visit factories and building sites and talk to people who are making decisions about energy-using equipment.

The most detailed study of this kind was undertaken by the Industry Commission (IC) itself (1991). The IC estimated that a carbon tax of A$80 (1988 prices) per tonne of carbon (equivalent to about US$60 per tonne) would be required to reduce emissions to the Toronto target. A tax at this rate was found to reduce GDP by 2 per cent. The effect on the output of the energy industries was considerably greater – the output of the coal industry fell by 26 per cent, that of the electricity industry by 11 per cent, and of the gas industry by 19 per cent. This last result diverges strikingly from the results of the bottom up, technology oriented studies, which project an increase in gas industry output as gas is substituted for coal and electricity. It would appear that the substitution elasticities (between factors of production and between fuels) used in the macro-economic model used for this study do not accurately represent the realistic technical possibilities available in the Australian energy system. The IC recognizes the inadequacies of the modelling approach it has used, commenting that the work has been undertaken for 'illustrative purposes'.

Not surprisingly, given the different assumptions, this estimate of the GDP cost of abatement is somewhat higher than that from the previously cited ACF study. The Industry Commission study assumed that the revenue raised by the carbon tax would offset direct taxes (income tax and company tax). Somewhat surprisingly, it did not model the effect of offsetting the carbon tax by reducing another consumption tax or related tax, such as payroll tax, which is currently the focus of some policy debate in Australian politics. A recent study of the US economy concluded that a moderately severe gasoline tax (a form of one-sector carbon tax) would depress GDP and consumption if the revenue were used to reduce either the budget deficit or direct taxes, but would have virtually no effect on economic activity if used to reduce payroll taxes (Brinner et al, 1991).

Non-energy emission abatement

The other important anthropogenic greenhouse gases emitted in Australia are methane and nitrous oxide. Compared with carbon dioxide, little is known about either sources of, or possible abatement measures for these gases, particularly nitrous oxide. The most important source of methane, thought to account for about two-thirds of total emissions, is domestic livestock, principally cattle, sheep and pigs. Landfill (municipal garbage) is also an important source. Small quantities of methane are released as a result of coal mining and from the natural gas distribution system. Agricultural activities, in this case soil denitrification associated with both the application of nitrogenous fertilizers and the use of legumes in improved pastures, is thought to be the main source of nitrous oxide (Ecologically Sustainable Development Working Groups 1992). Measures which it is thought could contribute to reducing these emissions include: the use of rumen modifiers (anti-bloat capsules) with intensively reared cattle; a modest decrease in stocking rates on some pasture types used for extensively reared cattle; aerobic, rather than predominantly anaerobic treatment of piggery waste; and the optimization of application rates of nitrogen fertilizers. The ESD did not estimate the scope, let alone the cost of these measures if applied on a large scale.

Abatement of atmospheric carbon dioxide levels by increased tree growth in Australia has been analysed superficially. Over the last few years growing concern about deforestation and soil erosion stimulated a variety of government and privately supported programmes to reverse the trends of two centuries of European colonization. These programmes could perhaps stop and perhaps reverse the continuing emission of carbon dioxide previously sequestered in biomass in trees and in the soil. As such they can legitimately be seen as an important part of national activities to curb greenhouse gas emissions, although that is not the reason they were initiated. From a greenhouse perspective, therefore, they are costless measures. However, much more far-reaching tree plantation programmes would be required to make a significant contribution to offsetting carbon dioxide from energy related activities. No estimates are available yet of the possible cost and scale of such programmes in Australia.

Australia's international role

As previously noted, Australia is a large producer and exporter of fossil fuels and fossil fuel intensive commodities, notably primary metals, which depend on the use of coal and coal-fired electricity. For example, less than 10 per cent of Australian aluminium smelting capacity relies on hydro-electricity. The imposition of a carbon tax would have a drastic effect on the cost of energy used in the production of these commodities, and hence on the cost structures of producers.

Australia is an efficient, low cost producer of these commodities. It is expected that continued and growing output of these commodities will be very important for the country's economic future. Hence, policy makers are concerned that a carbon tax imposed only in Australia would damage – possibly severely – the competitive position of Australian producers in world markets. That is why the Australian government qualified its adoption of the Toronto target as an 'interim' planning target with reference to net adverse economic impacts and trade competitiveness. The potential effects of a multilateral carbon tax are less clear, however. For example, such a tax would probably suppress international demand for thermal coal, but, as a result of fuel substitution, might increase demand for LNG. For primary metal exports, the effect of a multilateral carbon tax depends on both the demand for metal and the effect of the tax on competing producers.

To examine this question, the Industry Commission (IC) developed a model which focused on Australia's place in the global economy. Again, a carbon tax was the chosen policy instrument to achieve emission reductions. The IC concluded that a global carbon tax of the size estimated to be required to achieve a 40 per cent reduction in emissions by 2005 (estimated to be somewhat less than the abatement corresponding to the Toronto target) would reduce Australian GDP by between 1 and 3 per cent. The range depends on assumptions about substitution elasticities. The effect on GDP is thus roughly the same as that generated by applying a similar sized carbon tax to Australia only. However, the modelling results suggest that many other countries might suffer somewhat smaller GDP reductions than Australia.

The effects of such a carbon tax on individual industry sectors in Australia was highlighted in a study released in early 1992 (London Economics 1992). This study concluded that both the steel and aluminium industries would be driven into irreversible unprofitability and forced to shut down by a carbon tax, whether imposed unilaterally by Australia, by OECD countries or globally. A number of the assumptions used in this study appear dubious, notably those relating to international demand, and alternative, competing international suppliers of aluminium. However, the broad thrust of the conclusion is not unexpected, given the choice of a carbon tax as the policy instrument for achieving emission abatement.

Carbon taxes, externalities and other policy instruments

Aggregate economic models typically assume a uniform price elasticity of demand for energy across broad groups of energy users, if not the whole economy, and also assume a uniform elasticity across all sizes of price changes. The case of the aluminium industry demonstrates the invalidity of these assumptions. Aluminium smelters typically embody state-of-the-art technology at the time they were built. Apart from trivial adjustments, efficiency improvements can only be made by building a new smelter. Thus a smelter's price elasticity of demand for electricity is virtually zero up to a certain size price increase, while beyond that size, the elasticity becomes infinite. That is, the smelter shuts down because it is uneconomic to operate.

It is important to distinguish between a carbon tax which is imposed as a proxy for environmental costs not yet internalized; and a carbon tax imposed as an instrument to achieve a pre-determined level of emission abatement to respond to climate change. The two are identical only in a world of perfect markets. It is difficult to argue in principle against the full internalization of environmental costs. But the cost of climate change is potentially so pervasive that the costs of the greenhouse effect (and the economic benefits of avoiding it) simply cannot be expressed in monetary units. These fundamental issues cannot be pursued here, where the point is simply that different criteria may be used to assess policy instruments and options for achieving greenhouse gas abatement.

Given the nature and extent of the 'efficiency gap' in the market for energy services, it is quite likely that a level of carbon tax sufficient to put half of Australia's aluminium smelters out of business would still not be sufficient to induce some small-scale industrial and commercial businesses to make efficiency improvements having a payback of only a few months. Such an outcome would be neither fair to the aluminium smelters nor efficient for the Australian economy. Carbon taxes have many advantages (Pearce 1991), particularly when careful consideration is given to their place in overall fiscal policy But in the real, imperfect world, the carbon tax should be combined with other, more precisely aimed policy instruments to achieve emission abatement of greenhouse gases.

References

ACIL Australia, 1992. 'An assessment of the achievability of an Australian commitment to stabilise energy-related CO_2 emissions.' in *Two studies pertinent to Australia's decision on the terms of participation in a global convention on climate change.* ACIL Australia, Canberra

Australian Commission for the Future, 1991. *Energy futures.* Melbourne

Brinner, Roger E, Shelby, M G, Yanchar, J M and Cristofaro, A, 1991. 'Optimizing tax strategies to reduce greenhouse gases without curtailing growth'. *The Energy Journal* 12 (4), 1–14

Ecologically Sustainable Development Working Groups, 1991a. *Final report – energy use.* Canberra

Ecologically Sustainable Development Working Groups, 1991b. *Final report – transport.* Canberra

Ecologically Sustainable Development Working Groups, 1991c. *Final report – energy production.* Canberra

Ecologically Sustainable Development Working Groups, 1992. *Greenhouse report.* Canberra

Grubb, M, 1990. *Energy policies and the greenhouse effect. Volume one: Policy appraisal.* Dartmouth Publishing Company, Aldershot

Industry Commission, 1991. *Costs and benefits of reducing greenhouse gas emissions* (2 vols). Canberra

Jones, B P, 1992. 'The UN Convention on Climate Change: effects on Australia's energy sector'. *Agriculture and Resources Quarterly* 4 (2), 186–95

London Economics, 1992. *The impact of global warming control policies on Australian industry.* London

McGlynn, G, Newman, P and Kenworthy, J, 1991. *Transport energy scenarios for Australian cities.* Institute for Science and Technology Policy, Murdoch University, Perth

Manne, Alan S and Richels, R G, 1990. 'CO_2 emission limits: an economic analysis for the USA'. *The Energy Journal* 11 (2), 51–74

Nordhaus, William D, 1991. 'The cost of slowing climate change – a survey'. *The Energy Journal* 12 (1), 37–66

Pearce, D, 1991. 'The role of carbon taxes in adjusting to global warming'. *The Economic Journal* 101, 938–48

Warren Centre for Advanced Engineering, University of Sydney, 1991. *Energy management in the process industries.* Sydney

Williams, Robert H, 1990. 'Low cost strategies for coping with CO_2 emissions limits'. *The Energy Journal* 11 (4), 35–60

Part IV

Conclusion

14

Constructing a global greenhouse regime

Peter Hayes

The construction of a resilient, global greenhouse gas regime requires that state elites share a consensus as to common values and norms of behaviour; submit their respective states to observe the rules and procedures of the regime; and participate in the institutional arrangements established under the regime.

The signing of the Climate Change Convention is *prima facie* evidence that most national leaders agree that the global climate system must be conserved and restored, and that it must be used rationally in ways that are compatible with ecological imperatives. The rest of the interlocking elements of the regime have been left largely unspecified. Some analysts believe that states so hedged their commitments under the Convention that it merely maintains 'polluter sovereignty' and is so vague as to be almost meaningless.[1]

I am less pessimistic, however. I am convinced that a fully fledged regime can be constructed that will achieve the goals of the Climate Change Convention. Achieving this goal will first and foremost mean activating the norms implicit in the Convention as to mutual reciprocity between, and differential responsibilities of, the rich and poor nations. Specifically, the articles on financial assistance and technology transfer (see Chapter 1) must be elaborated in protocols and implemented.

Donors may place conditions on financial transfers to the South to abate greenhouse gases. In this chapter, I outline the contrary positions held by potential parties to a Climate Change Convention on conditionality and additionality of resources for greenhouse projects in developing countries. The biggest likely demand on resources provided by the North to the South will be to fund technology transfer.

The desirability and even the meaning of technology transfer are contentious, however. I outline two important positions that have emerged in debates on this topic before turning to three aspects of multi-pronged technology transfer strategy for greenhouse gas reduction. These are: cost reduction; technical assistance and training; and information dissemination and transnational research collaboration.

For these commitments to be implemented, parties will require confidence that other beneficiaries under the Convention are not cheating. Procedures must be established to monitor and to verify compliance with convention commitments along with mechanisms to resolve disputes and to enforce compliance.

Transferral of financial resources and technologies on a scale envisaged in this study would transform North–South political–economic relations. I conclude by asking whether 'geoecological' issues such as climate change may strengthen the South's bargaining position in its geopolitical and geoeconomic relationships with the North. My answer to this question is only a qualified 'possibly'. Nonetheless a global greenhouse regime may succeed for three reasons:

1 the likelihood of continued technological innovation and associated reduction in the cost of emissions abatement;
2 the contribution of science on comprehension of the greenhouse effect;
3 the influence of social movements on governmental policy.

Conditionality and additionality

The language of the Convention implies – although it is nowhere specifically stated – that the commitment of parties to the Convention should encompass the costs of actually reducing greenhouse gas emissions in developing countries. (As developing countries carefully avoided committing themselves to abatement at this stage, they did not obtain a precise matching commitment from the developed world to fund their abatement costs but only vague statements of intention; see Chapter 1.)

A variety of conditions on providing or accepting financial assistance to abate greenhouse emissions or to ameliorate the impacts of climate change have been advanced. These include the possibility that donors will tie the aid to greenhouse abatement activities and/or require that recipients reform their energy prices and institutional structures; and on the part of possible recipients, that acceptance of such aid in no way infringes on the exercise of national sovereignty in determining the best use of development assistance and that its provision in no way reduces existing flows of official development assistance.

Some analysts have argued that transfers to the South ought not to be for any purpose. Michael Grubb, for example, suggests that the transfer should only be convertible into provision of technical assistance and equipment needed to abate greenhouse gas emissions, and not cash.[2] Conversely, the Group of 77 have argued that funds provided for incremental abatement costs 'will be to a great extent of a compensatory nature.'[3]

There are strong arguments that compensatory payments should be linked closely with greenhouse gas reduction activities. Most important, the resource transfer calculated in Chapters 5 and 6 was based on an obligation-to-pay index applied to marginal carbon abatement cost curves. This estimate did not define marginal benefits of avoiding greenhouse gas induced climate change. Thus, it does *not* represent an estimate of compensation that the North might owe the South for having pre-empted atmospheric space or for climate change induced damages.

In practical terms, it is also impossible to calculating monetary values for compensation. There are other advantages to providing compensation in the same dimension as the damages imposed by climate change. An in-kind approach makes explicit the nature of the international transaction. It thereby avoids any connotations of a buy-off of recipient elites.[4]

From a pragmatic perspective, technology transfers financed by funds placed at the disposal of developing countries would increase economies of scale in supplier countries, thereby reducing the cost of supplying the aid in the first place.[5] Linking compensation to specific uses may also help to circumvent elite corruption in recipient countries. It could also increase political support in donor countries.

However, linking resource usage reaches its limits when dealing with the most vulnerable states where urgent local development priorities require that resources be applied across a variety of economic and welfare projects, not limited to greenhouse projects. The most advanced expression of this non-linkage is the proposal for an insurance fund for the most-affected island and coastal states as outlined in Chapter 7. Some mix of compensatory and linked-compensatory resource transfer therefore seems inevitable.

Other conditions that might be placed on the transfers studied above include sectoral policies aimed at reforming energy prices or institutional arrangements in a recipient country. In China, for example, coal prices bear little relationship to supply and demand. In the Soviet Union, subsidized natural gas prices foster a high rate of leakage of methane out of transmission pipes. Conditions related to project self-financing by the recipient state, to environmental performance, and to expanding the role of the private sector (including permitting foreign investment in abatement activities) might also be included by donors.[6] If financial and resource efficiencies are to improve in developing country energy utilities, many deep-seated causes of poor institutional performance must be rectified, including: overstaffing; inappropriate skill mixes; shortages of middle level and technical staff; low wages; rigid and politicized hiring and firing practices; political interference, graft and corruption in procurement and billing activities; and inadequate training facilities.[7]

As international energy expert Russell deLucia puts it, 'The primary problems are associated with institutional matters and market structure,' not technology or know-how.[8] In short, the energy sector in developing countries is often so irrational and inefficient that donors will be very reluctant to provide substantial technical or financial assistance unless prices are revised upward to reflect cost, and energy utilities are privatized.[9]

Developing countries assert strongly that they will not accept greenhouse-abatement financial assistance unless such transfers add to rather than substitute for current official development assistance (ODA) flows. The ODA recipients are concerned that existing ODA will be diverted from local development priorities to global environmental concerns, to redress problems created by the donor countries in the first place. They also hold that it is impossible to distinguish between development projects that benefit only their own country, and those that damage or restore only the global environmental protection.[10]

'Additionality', however, is a deceptively simple word. ODA flows fluctuate from year to year in most donor budgets; some donor countries have announced a gradual increase in ODA up to about 1 per cent of annual GNP; others are allowing their ODA to slide to even lower levels. There is no simple (or even complicated) way to ascertain what would have happened to ODA if donors do not fund greenhouse projects in recipient countries. Furthermore, many of these projects are justified in traditional developmental terms with or without consideration of global environmental concerns of the donors.[11] The issue is largely symbolic, therefore, with the United States declaring in 1991 that its contribution to one important mechanism for greenhouse funding, the Global Environment Facility, comes from existing ODA flows on the one hand; and Norway creating a new budget line separate from existing ODA to fund the Facility on the other.[12]

Defusing this issue requires two things to happen. First, overall ODA levels (minus identifiable greenhouse-related aid) should not fall but rather should remain constant or increase. Second, greenhouse-related projects in China, India and elsewhere (especially those funded by the GEF) should deliver enough developmental benefits to allay fears that greenhouse-related projects are an environmental diversion that benefit only the donor states.

Countries like China have also insisted that they will only participate in a Climate Change Convention that does not impinge on their national sovereignty.[13] This stipulation presumably includes retaining control over setting of priorities for the use of external assistance in domestic development projects.

Finally, donor countries may insist on using bilateral rather than multilateral aid mechanisms to transfer greenhouse abatement assistance. In part, this preference would arise from the sheer scale of the transfers discussed above that would exceed the total current United Nations budget and fears that a bloated, inefficient international bureaucracy could not hope to meet the challenge in a timely fashion. Again in part, it would follow from the pursuit of narrower national interests in tying the aid to their own suppliers of greenhouse abatement goods and services as in traditional bilateral aid relationships. In so far as greenhouse projects are funded multilaterally, donor countries prefer the World Bank's Global Environment Facility rather than setting up a new mechanism for climate change (as suggested by China). Developing countries have objected strongly to the non-representative and World Bank dominated decision-making system at the Facility. Given the strong statement in the Convention on reforming the GEF (see Chapter 1), it seems inevitable that either the South must be allowed to participate in the decision-making at GEF on an equal basis, or bilateral funding will dominate the global climate change arena.

Technology transfer

There is little dispute that technology transfer from the technologically powerful to the technologically deficient countries will require major expenditure by the North. Unless this transfer is achieved, there is little chance that the South can abate to ecologically acceptable levels as defined in Chapter 5. Technology transfer refers to human- and paper-embodied knowledge (such as operating procedures and manuals), known as technique, as well as knowledge embodied in physical equipment and plant.

Although the chorus of consensus is deafening on this score, the terms of the transfer have been a major sticking point in negotiations over climate change. Issues such as intellectual property, the role of transnational corporations, and the investment climate in recipient countries have all been hotly disputed.[14] Before treating these issues, however, this section outlines two qualifications as to the desirability of large-scale technology transfer to reduce greenhouse gas emissions.

Do nothing

Some influential environmentalists have argued that current patterns of political and economic power between and within states virtually preclude any significant progress toward greenhouse gas reductions via resource and technology transfer.[15] Patrick McCully, for example, argues that a massive influx of new aid would exacerbate the plight of the impoverished majority rather than achieve greenhouse gas reductions.[16] The history of aid, he contends, is one of corruption, failed projects, waste, neocolonial control, and increased debt and dependency. He expects that a climate fund, whether a new entity – as called for by China – or administered by the World Bank's Global Environment Facility – as called for by the OECD countries and so designated in the Convention as an interim measure – will behave no differently to existing aid agencies. He cites Ian Smillie of Intermediate Technology in London to the effect that energy aid in the 1970s left a legacy of 'windmills that didn't turn, solar water heaters that wouldn't heat, and biogas experiments that were full of hot air before they started.'[17]

Greenhouse gas abatement projects, in McCully's view, are just one more of a long list of development fads that employ mostly first world, hit-and-run consultants who increase the South's technological dependency on the

North. Sinking billions of dollars into greenhouse aid would create a perpetual technological dependency machine.

I will not take issue here with McCully's critique of aid nor whether social relations which block technology adoption in many developing societies must be realigned before much can be achieved by way of aid-supported development.[a,18] Rather, I will analyse what could happen if nothing is done, as McCully seems to suggest.

In Figure 14.1, I show two IPCC emission curves, and the permitted (post-reduction) emission trajectories developed from the efficiency scenario for the world and the South (see Chapter 5).[b] The South's projected emissions would exceed the IPCC case E global permitted total in about 2100 (at 1.6 gigatonnes[c] of carbon) if the status quo in 2030 is simply extrapolated.[d] It *already* exceeds the global permitted emissions in IPCC case F (at 1.9 gigatonnes of carbon).

Left to itself, therefore, the South eventually exceeds the global permitted total that is defined in relation to putative acceptable rates of ecological damage associated with climate change induced by emission of greenhouse gases (see Chapter 5). In short, the rest of the world can't afford to leave the South to its own devices. 'Do nothing' is not a viable option, however great the obstacles to change in the South or in North–South relations.

Do more but differently

Martin Bell of the Science Policy Research Unit at Sussex University has levelled a more penetrating criticism against the notion of massive technology transfer to reduce greenhouse gases in the South. He notes that energy efficiency (and related carbon abatement) is obtained from pervasive, non-energy-saving technical change throughout an economy.[e] 'It is therefore impossible,' he avers, 'to identify any distinct category of "CO_2 emission-

---

a The fact that Russia and Japan earlier this century, and the 'newly industrialized countries' in recent decades, obtained and absorbed major quantities of foreign technology and generated considerable amounts of indigenous technology demonstrates that technological dynamism is not precluded by the global status quo.

b IPCC case E is the global emission curve that is projected to increase atmospheric carbon dioxide to 50 per cent greater than the pre-industrial level by the year 2000. IPCC case F is the global emission curve that would keep this concentration at today's level by the year 2000.

c A gigatonne is a billion (10^9) tonnes.

d Neither post-2025 increases in per capita carbon intensity nor the South's population growth after 2025 were taken into account in this extrapolation. If these factors had been included, then the South's emissions would grow rather than decline slightly and would exceed global permitted emissions even earlier than is shown in Figure 14.1.

e This energy-reducing technical change is referred to as autonomous energy efficiency increase in Chapter 5 and comes about due to product and process redesign; the impact of sensor technology and control systems on efficiency; and many other developments that reduce the energy-intensity of production.

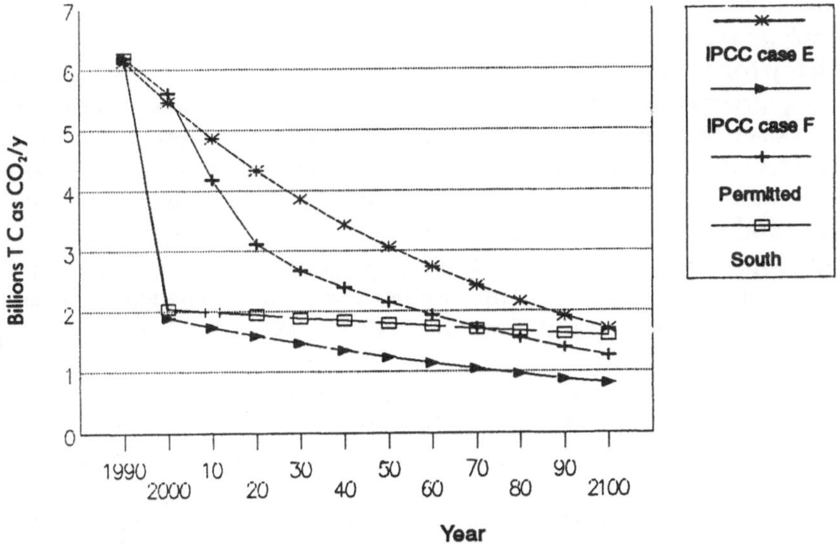

Figure 14.1 *Projected South emissions vs IPCC and study world totals*

reducing technology" which might be the focus for new initiatives concerned with international technology transfer.'

Obviously a technology transfer fund could focus on major 'energy saving' technologies, but that approach neglects many technologies and techniques with as much or more carbon-abatement potential. That is, the transfer of 'greenhouse' technologies is not likely to be blocked only by political and economic barriers. Transferring 'greenhouse' technology alone would achieve far less abatement than is desirable and achievable on economic *and* ecological grounds.

Bell argues that a more broad-ranging approach is needed that accounts for a whole spectrum of technology used by developing countries. Transfer should encompass not merely know-how (the construction and operation of transferred plant and equipment) and know-why (the research, development, design, demonstration and deployment of technology). It should also include the techniques of incremental learning involved in fine-tuning existing plant and equipment, and in managing the organizational changes that foster such learning.[19] The very notion of technology transfer therefore needs to be recast and the content of the phrase expanded and deepened.

This emphasis on the organizational aspects of technology transfer places the onus for realizing the potential benefits primarily on the recipient countries. Nit Chantramonklasri, for example, found that Thai firms differed greatly in their internal technological capabilities, and that technologically innovative firms were both market competitive and more energy efficient.[20]

The incremental learning and managerial capabilities that occur after technology transfer are as or more important to increasing organizational capability than transferral of skills during a discrete project. This on-going learning process, however, requires managerial effort and allocation of scarce skilled staff and time. As Sanjaya Lall states, every new application of a given technology requires adaptive engineering work.[21] In contexts where policies motivate such organizational learning, Bell argues that donors should fund transnational corporations to transfer to developing countries the managerial and engineering techniques required to learn incrementally and continuously. The resultant human assets are expensive, invisible, mobile, enduring, scarce and therefore extremely valuable in most developing countries.[22]

Organizational innovation, however, also requires a favourable macroeconomic structure. Thus, Frances Stewart argues that there is nothing inevitable about the undesirable dependency fostered by current modes of technology transfer.[f,23] On the contrary – faulty macroeconomic policies and resultant malign decision-making incentives that face organizations in many developing countries explain the South's inability to become technologically self-reliant rather than the problems associated with transfer *per se*.[24] Pervasive factors that are determined at a macroeconomic level often thwart micro-level technological changes. Such obstacles include distorted energy prices; regulated fuel supplies; capital scarcities; uncompetitive markets; government procurement policies; stagnant scientific and technological infrastructure; protectionism; low investment in education; and lack of information programmes to overcome market failures.[25]

These considerations imply that transferring only the latest energy efficiency technologies in core energy transformation systems used in industry or the energy sector will result in much less abatement than is possible. It is also pointless to transfer state-of-the-art plant if it is operated as poorly as are many productive enterprises in the developing world. Rather, human and organizational resources must be developed first to improve capacity factors and product quality of existing plant in all sectors of the economy. These same human capabilities can ensure that transferred technology is adapted to operate at high efficiency rather than reverting to past practices.[26] Importantly, much of the technology for controlling greenhouse gas emissions is already in the public domain and often already accessible to developing countries.[g] Investing in human resources that enhance technological and managerial capacity is particularly attractive because the required training is often relatively cheap, entails little or no

f Stewart refers to the high cost of technology imports; loss of decision-making control; inappropriate technology; and underdeveloped scientific and technological infrastructure in developing countries.

g China and Brazil, for example, both manufacture efficient compact fluorescent light bulbs.

capital expenditure, and often yields economic and resource-saving benefits almost immediately.[27]

This emphasis on building endogenous, self-reliant technological capabilities does not condemn the South always to lag behind the technological frontier of the latest hardware. Instead, it responds to two imperatives that will otherwise overwhelm the ability of any plausible transfer of narrowly defined technology to contribute meaningfully to greenhouse gas reductions in the South.

First, the vast demographic transition of the South's population growth combined with immense urbanization implies a primary reliance on informal, self-help development. Only an enormous proliferation of local technological research and development institutions can generate and deliver sufficient adapted, appropriate technology to end users. Without this local capability, environmentally benign technological alternatives will often not fulfil local needs and will not attract local users.[28] Foreign technology can only supplement and never substitute for local technological *capabilities* that support the modernization process.

Second, developing countries confront the likelihood of a 'green' technological revolution in North in bioengineering, waste control, recycling, and product and process engineering early in the twenty-first century. This reformation will be as epochal as were steam motive power, electrical and then electro-mechanical technology, and electronics in their time.[29] The green 'techno-economic' paradigm is driven by the need to preserve, conserve and restore ecosystems at local, national, regional and global levels. Consequent technological innovation in the North may devalue many of the traditional commodity exports and current manufacturing strengths of the developing countries (most notably, of the fossil fuel exporters).

Leaders in developing countries must be alert therefore to the opportunities that arise in a greenhouse world to obtain the best terms for technology transfer. To this end, they must nurture a highly receptive local environment to gain the most benefit from this transfer. Vendors of carbon abatement services, for example, could link their services to offsetting transfers of techniques and technology by buyers of abatement.

Multi-pronged approach

A judicious and careful blend of technology transfer from abroad combined with policies aimed at stimulating the virtuous circle of local technological development and competitiveness are essential to avoid the vicious circle of technological dependency and stagnation. A multi-faceted greenhouse technology transfer strategy will address at least three priorities, namely, reducing the cost of transferred technology; technical assistance and training; and information dissemination and technological collaboration.

Cost reduction

Two major determinants of the cost of transferred technology to developing countries are the rate of local and foreign innovation on the one hand, and whether the importing state is informed so as to enter bargaining on an equal footing with technology suppliers.

Many of the technological needs of developing countries – especially in the rural and urban-informal sectors – are poorly served by foreign technology suppliers. Indigenous centres of scientific and technological research are critical to expand the supply and reduce the cost of generating new technology and adapting imported technology that fulfils local needs. Demonstration programmes are badly needed that address the technical as well as market and non-market barriers to successful technological development in developing countries. Donor support for indigenous research centres and demonstration programmes should be expanded greatly to redress the imbalance in current research foci on technological needs that emanate from the industrial North rather than the modernizing South.

Transnational corporations are very important agents of technology transfer to developing countries.[30] Aid/technology recipients will need to re-examine their traditional technology import policies to stimulate the corporate conduit of technology flows, including reforms of pricing controls, taxes, income repatriation policies, more liberal licensing arrangements, and less stringent ownership limits in joint ventures.[31] Developing countries badly need to increase their flexibility to deal with transnational corporations if they are to 'stay in the loop' of the international technology alliances.[32] Such strategic corporate alliances to develop new technology will likely predominate in the first wave of the greenhouse-driven technological revolution.

The system of creating and protecting intellectual property rights is closely related to the cost of technology generation and imports in developing countries. It is also a vexed issue in the GATT and UNCTAD[h] fora that is unresolved in relation to possible technology transfer protocols in the Climate Change Convention. The jury is still out as to the net costs and benefits of strengthening intellectual property rights in developing countries. In all probability, there will be big winners (some of the technologically developed and technology exporting developing countries might gain substantially) and big losers (countries with absent or weak

h GATT is the General Agreement on Tariffs and Trade; UNCTAD is the UN Conference on Trade and Development.

domestic scientific and technological infrastructure could pay more for technology imports and reap little in return).[i,33]

A consensus on this issue may prove to be a precondition for implementing the Climate Change Convention.[34] Developing countries have demanded that environmentally sound technology be transferred to them on a concessional and preferential basis and that patents be transferred on a non-commercial basis.[35] But developing countries need not wait until this global standoff ends before obtaining more technology from transnational corporations involved in greenhouse projects. South Korea, for example, had extensive licensing arrangements at the same time concurrently with a loose intellectual property regime. Its electric utility also used 'turn-key' plant contracts to unpackage skills and to train its own engineers initially in know-how skills and later in know-why skills.[36]

Technical assistance and training

Along the lines of the latter 'trick of the trade', Martin Bell has proposed that donors direct new resources to offset the costs of developing human and organizational capabilities to generate and manage technological change.[37] Transnational companies already participate in such transfers provided they recover their costs. Driven by international competition, there appear to be few proprietary barriers to companies transferring such skills, even in 'state-of-the-art' technologies. To support this skills transfer, donors would need to accept longer time horizons and invest in long running training programmes rather than with traditional, discrete aid projects.

Information dissemination and transnational collaboration

Donors should also support information dissemination programmes that serve energy efficiency programmes of governments and non-governmental agencies. Relatedly, increased scientific work on climate change monitoring and analysis should be supported in the South. Independent scientific and research communities should be strengthened in developing countries if their leaders are to negotiate in an informed fashion on an equal footing with their counterparts from wealthier societies.[38] Such

---

i Carlos Primo Braga lists the costs of stronger intellectual property rights (such as patents, copyright, trademarks and secrets) as including: administrative and enforcement costs; increased royalty payments; displacement of 'pirates'; opportunity cost of over-investment in research and development; and anti-competitive effects. Benefits include: increased domestic research and development; disclosure of new knowledge; global technological dynamism and increased attention to research priorities of developing countries; increased technology transfer; increased direct foreign investment; a variety of other benefits such as better training, spillover effects; avoidance of trade retaliation; and access to possible *quid pro quos* provided by the technologically powerful countries in return for observance of GATT rules governing trade in intellectual property.

information can rectify the bargaining deficiencies of importing governments as well as facilitate collaboration between autonomous loci of research and development activity within the South. South–South networks of energy utilities and non-governmental networks of scientists and technologists concerned with energy efficiency should be fostered to hasten the pace of technological development and diffusion.

The transport sector exemplifies the need for an expanded role for government as well as South–South collaboration. Travel, car ownership, and freight are growing faster than income in all developing countries. Greenhouse emissions from transport systems are determined by population growth, travel and freight per person, and greenhouse emissions per passenger- and tonne-kilometre. These latter items are largely determined by economic choices which are constrained in the short run by existing settlement patterns, activities, and transportation infrastructure. The ability to design efficient cities and transport infrastructure plus the youthful vintage of vehicular stocks make it possible to combine big increases in transport services with advanced technologies for greenhouse friendly transport systems.

In poor countries, walking and animal powered carts are the main modes of transport for most people. As incomes rise, bicycles, motorcycles, light three-wheeled and various forms of utility and van-based public transport systems emerge. Due to the small number of privately owned light vehicles in developing countries, the combined total of the carbon emissions from fossil fuel used in transport in South and East Asia (excluding Japan), China, Africa, Latin America and the Middle East amounts to about 18 per cent of the world's transport sector carbon emissions.[j] By 2025, however, one projection shows that their transport emissions will have increased from their current 0.3 gigatonne per year (about 20 per cent of the world total) to about 0.5–0.8 gigatonne of carbon as CO_2, or between 30 and 40 per cent of the world's transport emissions.[k,39] Reducing this emission by 25 per cent by 2025 would save between 10 and 20 per cent of the South's projected permitted emission in that year in this study (see Figure 14.1).

Even in the wealthy countries, mere improvements in new vehicle efficiency will not by themselves significantly reduce overall carbon dioxide emissions from the transport sector if growth in overall use continues on current trends. Changes in modal balance, urban density, and regulation and market policy instruments will all have to be used to curb the transport sector's greenhouse contribution.[40]

---

j I ignore here the greenhouse impact of CFCs used in foam products for new vehicles and for air conditioning systems.

k The difference between the projection's low and the high transport 2025 emission rates in the South – about one third of a gigatonne – arises from the different assumptions about incomes and energy prices that promote or inhibit efficiency gains combined with low/high population growth rates.

In developing countries, the bulk of the passenger and freight transport is on off-road and rural tracks on traditional transport systems. Policy instruments and technologies transferred from the wealthy to the poorest countries may be of some use in cities (as in Singapore and Hong Kong) but have little bearing on the central transport problems. There is an urgent need for these countries to collaborate in research and development of these traditional transport systems. Wealthy societies have little recent experience and existing technological capability relevant to these issues. A bullock cart, for example, has evolved over thousands of years to operate in rough terrain. Local technicians used immediately available materials to make and maintain the carts. Adding pneumatic tyres or creating hard roads without redesigning the whole cart can greatly reduce its resilience and lifetime. Improving a bullock cart and upgrading rural roads is far more complicated than designing a high technology motor vehicle from advanced materials to run in a predictable highway system.

Three priorities for collaboration and information dissemination in the transport sector are:

1 increased technical and financial assistance for producing intermediate means of transport, especially for human and animal-powered freight;[l]
2 establishing local organizational capabilities to construct, maintain, and rehabilitate roads;
3 creating low-cost, rural-urban transport links.[41]

So far, I have reviewed the critical issues arising from the Convention that pertain to the realization of mutual reciprocity which is at the heart of the implicit North–South contract in the treaty. In the next section, we move from norms of behaviour to procedures relating to implementation and regulation of behaviour of parties to the Convention.

Implementation procedures

All institutions based on international cooperation face free riding by signatories who obtain the benefits of an international agreement while avoiding the costs by non-compliance.[42] The likelihood that signatories to a Climate Change Convention might try to avoid meeting their commitments poses the question of monitoring, verification and enforcement of compliance.[43]

l A typical woman in rural Ghana spends about 990 hours per year to deliver about 46 tonne-kilometres of freight (to meet household crop marketing, grinding, harvesting, fuel, and water needs); and a typical man about 340 hours per year to deliver about 12 tonne-kilometres of freight.

Monitoring and verification

Analysts concur that a multi-gas agreement will be much harder to monitor and likely impossible to verify. Taxes, tradeable permits, and abatement services all require monitoring to ensure that the terms and conditions of the scheme to fulfil commitments under a Climate Change Convention are being met. Monitoring, however, must be scientifically credible. Monitoring of non-CO_2 greenhouse gases such as methane from paddy fields or from diverse, mobile point sources such as cattle is not feasible due to the uncertainty as to emission rates.[44] While rice production is relatively well known, methane emissions vary greatly with soil type, nutrients, light, and temperature. Estimates for rice paddy in Spain and Italy vary by more than 100 per cent.[45]

Also, sinks for and terrestrial reservoirs of greenhouse gases must also be monitored, both to ensure that sinks endowed as property rights are maintained, and to verify any claims made as to additional carbon fixation.[46] Yet rates of re- and de-forestation are highly contentious, and satellite-based remote sensing cannot yet provide adequate monitoring and verification of biotic carbon sinks. The deforestation rate in Brazil, for example, is highly controversial (estimates range by a factor of five).[47] Similar arguments apply to other greenhouse gases such as nitrous oxide.

For all these reasons, therefore, I conclude that it is only meaningful to cost monitoring and verification of carbon dioxide released from fossil fuels at the outset of the implementation of the Convention.

In Chapter 5, however, I assumed that property rights are created in proportion to national carbon sinks that are the basis of determining permissible emissions in future years. It is possible to monitor and verify the status of the forest stocks and thus carbon *reservoirs* in those forests although it is not feasible to track the carbon *flows* to and from them. A monitoring system is feasible that would use remote sensing and *in situ*, ground-based validation to determine the fulfilment of commitments made to maintain or to expand these reservoirs. I assume, therefore, that biotic stocks of carbon will be included eventually under a protocol for monitoring and verification, although not the carbon emissions from these sources.

Verification is the international control of compliance with agreed measures and behaviour by means of tools and procedures agreed upon in an instrument of international law – for example, a protocol on compliance to a Climate Change Convention. Verification can be defined as having different densities depending upon the level of distrust between parties to the agreement and the technical difficulty of obtaining information with an adequate level of confidence on the other.

I assume that the verification procedures that are adopted in a verification protocol will be multilateral rather than bilateral in implementation (although they may rely heavily on national/unilateral monitoring and verification capabilities such as satellite systems). Assuming that all parties will be accorded equal treatment in the protocol, it is reasonable to suppose that all parties also will be subject to monitoring and verification by an implementing organization established under the Convention.

Nature of emission sources

Anthropogenic sources of CO_2 vary greatly with respect to characteristics that affect greatly their suitability for monitoring and verification. Some are stationary, emit copiously and continuously, and are suitable for direct, quantified monitoring. Power stations and large factories exemplify this type of emitter. Other sources are stationary and numerous but only emit intermittently very small quantities of gas. Fireplaces and open fires are typical. There are also very many mobile point sources that are sporadic emitters such as vehicles and livestock. Finally, there are very diffuse sources such as non-commercial fuels based on animal wastes.

Only the first category is suitable to direct monitoring and verification. There are, for example, well-developed techniques for determining gaseous emissions such as ultrasonic instruments in the off gas stack which measures the effluent density and velocity and thereby volume to within 3–5 per cent accuracy. The other sources all exhibit characteristics that would make information collection enormously onerous due to their number or the lack of observational methods.

Determining emissions

Good statistics are available for energy production and consumption balances for most countries. However, to convert these energy data to emissions on an international basis, energy balances must be made more complete and accurate, carbon content and conversion into emissions must be made more precise, and statistics must be collected according to consistent and compatible ground rules. The parties to the Convention must agree on the types of data, required disaggregation and detail, and common reporting rules for national reports.[48] Fortunately, the Intergovernmental Panel on Climate Change has already produced a set of guidelines of this nature, which are being refined and updated.[49] In particular, rules are needed to determine whether emissions are based on energy production or consumption. The latter is particularly problematic because of the difficulty of ascertaining conversion losses.

The implementing organization must be able to verify the accuracy of data such as ash content of coal or oxidation rates of conversion in power plants, etc. It cannot hope to collect the requisite data independently but only to analyse the data supplied by parties to the convention by cross-checking the reported fuel cycles and conversion to emissions with 'spot' checks including on-site visits to 'check the books' of very large, stationary emitters. In most nations, however, the latter checks would only cover 1–5 per cent of a given country's emissions. It is crucial to an effective greenhouse regime that verification be conducted routinely by subjecting national reports to independent, critical scrutiny and assessment, treated as an expert technical rather than political process.[50]

A verification system that combines data analysis with spot checks would likely enable the implementing organization to detect an emission infringement of the Convention by a party that deviates 10 per cent or more from the party's commitments. Similarly, a verification system that uses remote satellite and air-based sensing with local inspections should be able to detect departures from declarations to maintain or to expand biotic carbon pools such as forest reserves, to within a five per cent deviation from commitment. (The verification protocol would have to define the ground resolution at which it requires monitoring, the calibration and interpretational rules to be followed, and the density of selective observation needed for confidence to exist that parties are complying with their commitments.)

Verification cost

The cost of the verification system will consist of the direct costs of the implementing organization engaged in checking the annual national reports of compliance and field inspections, plus the indirect costs of obtaining independent sources of information needed to cross-check national claims about emissions or the status of carbon stocks or sinks.

It is reasonable to assume that the implementing organization will not have to meet the capital or direct operating costs of remote sensing satellites. Rather, these costs will be covered in the budgets of the space agencies in Europe, Japan, and the United States which already pay for the huge cost of earth-observing satellites.

By way of comparison, the International Atomic Energy Agency's (IAEA) safeguards department currently consists of 450 persons including 190 field inspectors. The implementing organization for verifying a Climate Change Convention would probably require about twice as many staff given the much larger number of facilities to be visited and much broader international scope of the verification system compared with that applied in the nuclear field.

In 1987 the IAEA safeguards applied to about 230 tonnes of plutonium, 30,000 tonnes of enriched uranium, and 50,000 tonnes of depleted uranium, thorium or uranium. In 1983, the IAEA safeguards agreements applied to a total of 881 installations such as power reactors and other fuel cycle facilities.[51] A carbon monitoring system will apply to billions of tonnes of fuels, and millions of hectares of forest at hundreds of thousands of sites – a much bigger task.

The safeguards surveillance and materials balance inspectorate system for the sensitive nuclear materials run by the IAEA costs about US$30 million per year. The cost of a system that verifies compliance with a Climate Change Convention might therefore approach $100 million per year. It is doubtful that more than three times the IAEA's budget would be provided to the implementing organization at a time when the IAEA already finds it difficult to obtain funds for such a politically sensitive field of concern to great powers.

The source of the funds for the implementing organization and its activities would either be charges that follow the UN scale of payments or a special formula similar to that developed by the IAEA in 1971 (and later revised) that levies states on a per capita income basis (with a ceiling) and a cap on contributions by poor states.[52]

Verification or confidence building?

The previous sections have argued that the greenhouse arena is characterized by complexity due to multiple gases (unless limited to carbon dioxide); an effectively infinite number of point and mobile pollution sources; mostly national information on energy use which is subject to distortion, withholding, and differing reliability, varying analytical methods, and underlying assumptions; reliance on extrapolation from existing energy statistics rather than new monitoring of greenhouse gas emissions; and a long lead time before an effective monitoring system and verification could be created.

In Chapter 5, I analysed three mechanisms to achieve agreed reductions and to fund the South's 'excess' incremental abatement costs: carbon taxes, tradeable permits, and trade in abatement services. Each of these mechanisms poses different demands on a verification system. A carbon tax system, for example, requires that a baseline emission be set and updated each year to confirm that states are reducing emissions to agreed targets. A tradeable permit system within an overall global emissions target demands that trading be monitored continuously in addition to establishing national emissions relative to an agreed baseline. Trade in abatement services requires that claimed reductions by one country actually have been achieved in another country. Monitoring compliance of such claims could be politically difficult for an international monitoring system.

In all three cases, achieving a high degree of certainty seems to require an extensive monitoring system and bureaucracy. Yet most states do not (yet) perceive the stakes in the greenhouse regime to demand monitoring and verification like that imposed on flows of special nuclear materials.[m] Only a small international bureaucracy based on national reports and data cross-checking seems politically feasible at this time.[53]

The history of international arms control and environmental agreements offers six important lessons for a greenhouse verification system. First, environmental costs and benefits do not accrue as fast as the costs and benefits of abandoning arms control agreements and the stakes are not perceived as central to the immediate security of the state nor (usually) to regime survival in that state. States may therefore be less demanding of a

m The exceptions are those states the existence of which is threatened by sea level rise or other climatic effects on aridity etc.

verification system for environmental agreements than in other domains. Moreover, when states coordinate because of self-interest, there is little reason to defect or cheat and little or no verification or enforcement is needed. If the costs of carbon abatement are as low as suggested in the studies reported in this book – at least for the first 20 per cent reduction and therefore the first decade or two of an agreement – then verification measures are needed mostly to build confidence in the regime rather than to raise the question of non-compliance and enforcement.

Second, it is inevitable and proper that enforcement responsibility will be lodged primarily at the same level as implementation responsibility, that is, within nation states. The bulk of the monitoring and verification should be conducted at this level, rather than internationally.

Third, most military control regimes were created in confrontational contexts under conditions of secrecy and with little or no participation. These characteristics led to many problems of implementation for arms control agreements. This experience implies that a greenhouse regime should strive for maximal transparency and openness, including a strong role for non-governmental organizations in monitoring compliance.[54]

Fourth, some states have skillfully used verification issues in the past to block international agreements (most notoriously, the United States with regard to the Complete Test Ban Treaty).[55] If the analysis in previous sections is correct, then this problem should not arise in the greenhouse gas arena.[n]

Fifth, there are important precedents for monitoring and verifying international atmospheric agreements, at the regional level in Europe, and globally in ozone depletion convention. This experience should provide some good signposts for the greenhouse regime, especially for regional (in Europe) and subregional greenhouse gas agreements (in Northeast Asia) that could be developed to supplement a global greenhouse regime.[56]

Sixth, the history of international monitoring and inspection of nuclear power provides some useful lessons. The IAEA's history suggests that an international secretariat should be created to audit national reports and ensure that they are bona fide, consistent and follow internationally recognized procedures. An independent technical committee could be appointed to define the reporting requirements of states to the parties to a Climate Change Convention. The same technical committee could also explore with states qualitative anomalies (such as refusal to allow an on-site inspection of emission rates or a claimed efficiency improvement) and quantitative discrepancies (such as inconsistencies between national reports and international statistics) that might arise from time to time.[57]

---

n Exceptions may arise in the case of pariah and paranoid states worried that the public (or even confidential) release of a detailed greenhouse gas inventory will provide their enemies with a surrogate input–output balance profile of their economy. North Korea springs to mind in this regard.

Disputes and enforcement

Article 14 of the Convention states that disputes between parties should be settled by negotiation or by any peaceful means that they care to select, including arbitration by the International Court of Justice, and/or in accordance with procedures yet to be adopted by the parties to the treaty.[58] Thus, the Convention provides little guidance as to what methods of dispute resolution should be incorporated into a protocol. It casts no light at all on the appropriate means of enforcing compliance with treaty commitments.

At a meeting in The Hague in 1989, twenty-four national leaders called for a 'new institutional authority' to set and implement environmental standards. Currently, however, only national institutions can implement standards authoritatively. Moreover, there is no compulsory dispute settlement jurisdiction relating to multilateral environmental regimes. Invariably, agreement by disputants is required before it is submitted to third party adjudication. The major stumbling block had been the socialist bloc rejection, and more recently, US rejection of compulsory third party arbitration.

In relation to disputes involving the failure of developing countries to comply with emission reduction targets, it would likely be highly counterproductive to try to enforce agreed targets by economic sanctions. These would worsen the very technological and economic difficulties that cripple many developing countries' ability to comply.[59] It would also impose substantial costs on states that meet their commitments and thereby reduce the benefits to new signatories considering joining the Convention. Moreover, using the trade system to enforce climate change policies would likely prove to be unmanageable because every product and service that is traded internationally results in greenhouse gas emissions.[60]

Consequently, alternatives to supranational regulation have emerged. States recognize each other's licensing rather than ceding licensing powers to an international authority. Such reciprocal recognition schemes operate in many areas including phytosanitary certificates for exports, shipping oil pollution prevention certificates, marine waste disposal permits, hazardous materials trade, and trade in endangered species. States also harmonize standards and standard-setting laws and procedures, often by adopting models from overseas. (For example, the environmental impact assessment, green labelling and pollution taxes).[61]

Many states have also committed themselves, outside treaties, to providing early warning and notification, for example, with respect to banned chemicals, exchange of standards, or adjustments to trade-controlled items (such as endangered species). Regimes also create transnational networks of lower level 'operational' national officials who short-circuit hierarchical communications across boundaries, or who communicate directly with international organizations that may then reintroduce environmental information at much higher political levels in the same nation state. Such structures can defuse and even prevent disputes arising in the first place.

Non-adversarial techniques have also developed to resolve international disputes from escalating to interstate conflicts. Local legal challenges have been mounted across borders thereby achieving settlement without involving the states themselves. (In Europe and North America, this technique requires that legal systems grant status to foreign parties in local judicial or administrative procedures). The filing of complaints and the launching of infringement hearings are two other techniques that have been used (in the Montreal Protocol and in the European Economic Community, respectively).

Reporting requirements are also an important means of imposing national 'discipline' on treaty parties, especially when combined with international expert auditing and public debate in committees or annual conferences. Such procedures are already well developed in the occupational health and safety agreements administered by the International Labour Organisation. Similar procedures are used by the International Monetary Fund, and multilateral funding agencies also conduct national and sectoral audits as preconditions for or requirements of development loan agreements. Environmental auditing, however, has a weak tradition although the Montreal Protocol requires substantial reporting to permit monitoring of compliance and administration of various aspects of the agreement.

As noted earlier, it seems inevitable that the national reporting requirements of a Climate Change Convention will be the heart of a monitoring and verification system, and the key to effective enforcement through self-regulation. National reporting will the core confidence-building measure that will build widespread commitment to the regime. It can be supplemented by international auditing of the kind referred to above, but such reports (as occurs with the International Energy Agency reports) buttress only the normative power of domestic proponents of fulfilling treaty commitments and have no direct legal connotations.[62]

These measures may be supplemented by other measures that increase the incentives of signatories to comply. Large emitters, for example, can offer to match the abatement of new signatories. They can also threaten to punish offenders by reducing their own abatement by as much or more than that of the defector, thereby reducing the benefits of free riding by increasing the costs of climate change. States can also lock in their own commitments so as to reduce the uncertainty facing potential signatories as to whether they will reap the benefits of reduced climate change.[63]

Regional building blocks

A global greenhouse regime will take a decade or more to construct. Regional efforts will likely be the building blocks on which a global regime will be built.

In this section, I examine the potential of regional approaches for funding the incremental costs of developing countries. I follow this with a review of the hopeful trend toward sub-regional cooperation on climate change scientific research.

Regional greenhouse trade initiatives

Given the limited past experience with transfer schemes based on carbon taxes, tradeable permits, or the sale of abatement services, it is unlikely that they will successfully commence quickly. It is more likely that these schemes will be tested first in national and regional experiments.[64] In this section, I examine the potential for such a scheme in the Asian Pacific region.

In many situations, regional cooperation may be more cost-effective than national initiatives in abating greenhouse emissions due to expansion of available resources of technology and information, economies of scale achieved through trade, and reduction in information and administrative costs.[65] There is no doubt that immense scope exists in the Asia Pacific region for improving energy efficiency and reducing greenhouse emissions.[66] One Asian Development Bank survey (based on 160 energy audits in Thailand, the Philippines, and South Korea) found that energy savings were economically justified in most industry groups.

In Thailand, energy savings of 12–13 per cent were found to be justified except for chemicals and non-metallic products where the potential was much higher. In the Philippines, energy savings of 18 per cent in industry were identified plus another 16 per cent potential by substituting natural gas and biomass for oil. And in Korea, energy savings were found to be 5.5 per cent with big potential in kilns and furnaces and additional potential for cogeneration and district heating.[67]

In Table 14.1, I show the cumulative projected emissions, required reductions, and incremental abatement cost (calculated at the high marginal cost schedule) for four major industrial and three major developing countries in Asia Pacific in 1995, 1995–2004/5, and 1995–2024/5 as calculated in Chapter 5. The industrial countries have about forty per cent more projected cumulative emissions over the scenario's thirty year period, but are responsible for about three times as much required reduction as the developing countries. Because the industrial countries quickly move to the higher echelons of marginal abatement cost, they also spend much more on abatement than the developing countries over the same period. These figures indicate that markets in tradeable permits and abatement services could emerge quickly in the Asia Pacific region as the industrial and developing countries appear to have complementary capabilities and needs that could reduce the cost of abatement in the North while transferring substantial resources to the South.

In Indonesia, for example, the institutional capability to sell abatement services sought by overseas firms or utilities already exists. KONEBA, a

Table 14.1 *Asia-Pacific regional carbon abatement services trade potential*

	1995	*1995–2004*	*1995–2025*
A CO_2ff projected emissions, efficiency adjusted scenario (MTC)			
Industrial Asia/Pacific			
USA	1447	14814	47060
Canada	154	1581	5022
Japan	344	3503	10983
Australia	70	710	2176
Subtotal	2015	20607	65241
Developing Asia Pacific			
China	748	7877	27084
India	294	3163	11652
Indonesia	60	644	2304
Subtotal	1102	11683	41040
Ratio Dev./Ind.	0.55	0.57	0.63
B Required reduction, efficiency adjusted scenario (MTC)			
Industrial Asia/Pacific			
USA	92	4278	26684
Canada	10	456	2847
Japan	21	1002	6206
Australia	4	200	1222
Subtotal	127	5936	36960
Developing Asia Pacific			
China	19	1014	8003
India	8	447	3655
Indonesia	2	88	705
Subtotal	29	1548	12363
Ratio Dev./Ind.	0.23	0.26	0.33
C Incremental cost, efficiency adjusted scenario (Million $)			
Industrial Asia/Pacific			
USA	4576	396324	18773438
Canada	488	42293	52040
Japan	1075	92382	29345
Australia	215	18260	26684
Subtotal	6354	549259	18881506
Developing Asia Pacific			
China	956	52366	748062
India	415	24282	350828
Indonesia	82	4663	66961
Subtotal	1453	81310	1165851
Ratio Dev./Ind.	0.23	0.15	0.06

Millions of current dollars, not present valued
Part C uses incremental cost calculated with the Nordhaus marginal cost curve

quasi-private firm established by firms in the fertilizer sector backed by a World Bank loan, is marketing shared energy savings contracts with firms in the large industrial and commercial sector. KONEBA is also seeking energy efficiency contracts in Sri Lanka, the Philippines, and Malaysia.[68]

In China, key industries such as iron and steel, aluminum, brick-making, glass and ammonia are very energy inefficient, as are buildings and cooking stoves. One analyst estimates that a 30 per cent reduction in projected energy use can be achieved by 2025.[69] Major increases in energy efficiency will require replacement rather than retrofit of existing and obsolete equipment. Yet China lacks foreign exchange, has unreliable supplies of more efficient fuels such as diesel, operates at small scales of production due to poor transport infrastructure and decentralized economic activity, and accords a low priority to maintenance and repair of equipment that are crucial to energy efficient operations.[70] China has an especially irrational energy price structure that, as the World Bank puts it, appears to be used 'primarily to generate and distribute revenues rather than to influence supply and demand.'[71] China also requires externally funded projects to generate foreign exchange to repay the loans. Projects that increase energy efficiency do so indirectly if they reduce oil use allowing increased oil exports, but crediting the energy efficiency project with the foreign exchange earning will require China to adjust its internal procedures.[72]

These practical obstacles imply that emission abatement programmes in China will have to cast a very wide net in order to address the real constraints on improving energy efficiency and shifting from coal to natural gas. Fortunately, China is one developing country that gives a high priority to energy efficiency. The Energy Conservation Company of the State Energy Investment Corporation spends about $300 million per year to provide project matching funds and technical advice through 156 energy management centres with 5,000 employees at the state and regional levels.[73]

Yet China has limited ability to substitute natural gas for coal. Nuclear power and hydroelectricity are both site-constrained. China already invests about 10 per cent of total public investment in the energy sector into efficiency. The Asian Development Bank is preparing a major loan project to follow up its technical assistance project on energy efficiency in China (funded by Japan and matched by the UN Development Programme).

Projects could be undertaken through existing bilateral and multilateral channels to reorient traditional lending practices. China, for example, purchased cast-off Japanese factories to make highly inefficient appliances such as refrigerators in China – equipment that will require many extra power plants that will cost China much more than it would have to buy factories to make efficient appliances. A compact fluorescent lamp factory provides energy at about one tenth the cost of a new power plant – yet donors find it much easier to package and fund gigawatt size power plants than they do energy efficiency projects.[74]

Greenhouse initiatives at a regional level must be sensitive above all to local conditions. The islands of Asia Pacific, for example, have very different energy economies and needs to those of the big Asian states. The islands are greatly dependent on oil for commercial energy although some also use substantial quantities of fuelwood. One cost-effective and culturally appropriate measure in many of the island microstates will be to engage in coastal reforestation and agro-forestry projects based on indigenous plants of high cultural utility to Pacific islanders.[75]

Regional efforts are also needed to enhance the flow of technology transfer. Regional trade liberalization, for example, may stimulate competition that increases the transfer of energy efficient technology. In Thailand, for example, local refrigerator manufacturers have been locked into using inefficient and obsolete compressors made by a Thai–Australian joint venture that licensed technology from a US firm by high duties and bans on imported compressors. When another firm finally received permission to make a decade-old Japanese compressor, the Thai–Australian joint venture introduced a more recent US compressor. Increased competition thereby improved compressor efficiency in Thailand by 15–20 per cent.[76]

Public and private initiatives to explore the potential for firms and utilities to undertake abatement projects in developing countries of Asia and the Pacific are needed urgently to demonstrate the viability of such activities. A state level utility in Australia, for example, could approach a counterpart in Asia, and then propose to part-fund an abatement project which would be credited to Australia's own national abatement in the Climate Change Convention. The project partners could seek third party private financing, and approach the World Bank/UNDP Global Environment Facility for support. Innovative financing mechanisms such as the Australian utility buying discounted private debt owned by a country like the Philippines could be used to finance the domestic costs of the abatement project.

The Asian Development Bank would act as honest broker, identifying the potential projects, preparing the loan documents, arranging for private co-financing, and monitoring and evaluating performance. The maxim 'first in, first served' will play powerfully in emerging markets for tradeable permits and abatement services.

Potential for sub-regional collaboration

Regional approaches may be attractive and productive for reasons other than the minimizing of cost. Because greenhouse gases are universal on the one hand, and because adjoining countries share common features and interests on the other, a regional programme on climate change may both facilitate the implementation of the Convention and foster regional cooperation. In Northeast Asia, for example, a regional environmental consultative forum is likely to be established in 1993. Initial steps to create this forum became possible in mid-1990 when the Cold War began to thaw rapidly. It will likely encompass climate change issues as well as regional oceans management, resource management, environmental technology research and development, standards, and so on.

Of course, the six nations of Northeast Asia already cooperate on some environmental issues, but mostly on a bilateral basis. The South Korean Forestry Research Institute, for example, is establishing formal cooperative arrangements with its Chinese counterpart in the Chinese Academy of Sciences. The North Korean Academy of Sciences has long cooperated with Chinese institutes. And the Mongolian Academy of Sciences has conducted collaborative research with the Chinese Academy on grasslands ecology and desertification. None of these efforts cover the whole sub-region, however.

The UN Development Programme is the only existing sub-regional cooperative framework that includes all the regional states. It has sub-regional programmes on transboundary acid rain, clean coal technology, new and renewable sources of energy, and the development of the Tumen River area, all of which are salient to the greenhouse gas issue. But these activities are still at an early stage of development and offer little that contributes directly to regional activity on the greenhouse gas issue.

A sub-regional climate change programme could address a variety of greenhouse related issues. Acid rain fallout, for example, affects carbon emissions/sinks. Acid rain from coal burning is transported from Manchuria and North Korea onto Japan and South Korea; and from Russia onto Mongolia.[77] Methane emissions from the production and use of coal is another possible research priority in Mongolia, and North and South Korea. China has substantial experience in methane monitoring from coal mining and from rice paddy fields, and could provide expertise and share surplus equipment with other countries in the region.[78]

Perhaps the most inclusive, urgent, and long term of all the tasks relates to the need to conduct regional climate modelling in Northeast Asia. All the developing countries in the region lack the basic capability to generate scenarios of climate change impact on which to base adaptation studies and national response strategies which parties to the Convention are obliged to produce. At this time, global circulation models offer poor resolution for Northeast Asia, let alone for a single country. Nations in the region are reduced to relying on qualitative scenarios of climate change with consequent uncertainty as to the range of possible costs and benefits of climate change to which they must respond.

A regional effort to develop a regional climate model is therefore an urgent priority. Such a project would nest a regional model within existing global circulation models rather than attempting the extremely costly exercise of building a new global model for use within the region from scratch. The choice of regional climate modelling parameters, the data requirements to validate the model, the interpretation of the model's results, etc. are all items that require a regional rather than a national approach.

The United Nations Development Programme could convene a regional scientific collaboration along these lines that would:

1 create a regional scientific advisory board for this activity;
2 identify a regional network of collaborating institutional and scientific participants
3 convene a planning session to identify the appropriate scientific approach to the various tasks;
4 develop a detailed budget for the work for both short and long term work.

Such joint work could lay the groundwork for a regional climate centre, as has been proposed by the Academy of Sciences in Beijing. A regional climate change centre in Latin America, and a regional scientific network on climate change scientific cooperation in Southeast Asia – both of which are in advanced planning stages – provide two models from other regions that could be drawn upon in Northeast Asia.[o,79]

Thus far, I have analysed the arrangements for building confidence in the greenhouse regime and the major difficulties pertaining to resource and technology transfer for greenhouse-related activities. In the next section, I explore the extent to which the North–South conflict may affect the ability of parties to the Convention to negotiate meaningful protocols. The North–South stand-off has been institutionalized since 1964 when the UN Conference on Trade and Development (UNCTAD) was established.[80] Is it really possible to erect a global greenhouse regime in spite of North–South antagonism as to the nature of aid, the terms of trade, and technology transfer? To what extent does a global environmental issue such as climate change portend the emergence of a new organizing principle of interstate relations on a par with geopolitical and geoeconomic concerns? Is geoecology on the international agenda?

North–'South' conflicts

The archetype of southern politics – UNCTAD – was built on polarized political blocs organized around adversarial and conflict-prone economic relations between the poor and wealthy states. In the 1980s, however, the unity of the South's Group of 77 (G77) had been severely stressed already by the rise of the newly industrialized countries in Asia which heralded the end of the 'third world'.[81] The end of the second Cold War combined with the

---

o The Southeast Asian network will have its secretariat in Singapore and will meet first in Manila in March 1992. A regional climate change programme for North Africa will likely begin soon with European Community Support. Australia proposes to establish an Asia-Pacific Center for Climate Change Research. The Chinese Academy of Science is also convening a Global Change Research Network of Temperate East Asia to meet in Beijing in May 1993.

steady decline of the 'South' as a unified entity may make it easier to de-link old debates from new agendas such as climate change.

Thus, even the concepts of 'South' and 'North–South' conflict may be obsolete and irrelevant in the climate change arena. Indeed, since late 1991, negotiations over climate change have resulted in a fractured and pragmatic set of political axes. Completely new alignments of cooperation and conflict emerged that are still fluid but no longer mirror the old North–South cleavage. Thus, in 1991 Anne Kristin Sydnes identified at least five groups from the 'South' in the climate change negotiations. These were:

1 radical, like-minded activist states (eg Bangladesh and Maldives) which view climate change as a major threat to their national existence;
2 potential, like-minded activist states (eg Mexico) which see climate change as a good way to extract additional concessional aid;
3 energy-consuming, hardliner states (eg Brazil, China, and India) which support more research but object to costly commitments and actions;
4 energy-exporting, hardliner states (OPEC states led by Saudi Arabia and supported by Australia) which object to potential market shrinkage and trade impacts;
5 unpredictable 'transition' states (eg Taiwan, South Korea, and some Eastern European states) which are already ambivalent as to the North–South cleavage given their position in the world economy and international hierarchy of states.[82]

By December 1991, the G77's unity had virtually collapsed at the greenhouse negotiations in Geneva.[83] A breakaway Group of 24 (G24) proposed that developing countries consider acting on greenhouse issues while awaiting action by the OECD.[p,84] Two other southern strains also emerged in addition to the centrist G24, namely, a group of energy exporters which backed the United States in stalling agreement; and the AOSIS island states which joined the European Community, Australia, Canada, and New Zealand in calling for a strong convention. The AOSIS states believed that the dilatory and ideological stance of the hardline, big poor states in G24 jeopardized their chance of obtaining any resources from the OECD. A fourth group of still uncommitted countries emerged, including Argentina and Mexico. These divisions continued up to the signing of the Climate Change Convention in June 1992 and are reflected in its text. It is difficult to believe that the state elites of the G77 can reconstruct their solidarity while negotiating protocols to the Convention now that they have discovered that their interests diverge fundamentally in relation to climate change.

p The G24 included: Algeria, Bangladesh, Brazil, Cameroon, Chile, China, Colombia, Cuba, Gambia, Ghana, Guinea, India, Indonesia, Kenya, Malaysia, Namibia, Nigeria, Pakistan, Philippines, Sao Tome et Principe, Tanzania, Thailand, Uganda, and Zimbabwe.

Geoecological power?

The fragmentation of the South places developing countries in a weak bargaining position on the central issues of financing, technology transfer, and compensation payments that are still to be addressed in the Convention. The greenhouse issue exemplifies a general dilemma that developing countries face in global environmental politics. Global and regional environmental predicaments present them with new demands on scarce resources for regime and national survival as well as new bargaining opportunities with the OECD states. It remains to be seen exactly how southern elites will respond to these pitfalls and opportunities in the greenhouse arena.

On the negative or threatening side of their security, environmental problems could shift the priorities of wealthy trade and aid partners away from political and social stability in the South to global dilemmas of less concern to the southern elites. They also confront new and unruly domestic social movements often aligned transnationally with powerful counterparts in OECD states.

Vulnerable states could launch ideological campaigns against environmental issues in an attempt to wrest the political initiative away from the OECD states in the international arena. Polarization around issues such as climate change between the big, poor states and the big rich states block rather than foster international cooperation.[85] Conflicts at a global and regional level on environmental issues could spill over into geoeconomic and geopolitical dimensions of interstate relations salient to climate change, thereby gridlocking ongoing negotiations.

Big, poor states may also use environmental issues to extract concessions from the OECD states in long-standing geoeconomic and geopolitical arenas. The greenhouse issue is unique in that the South influences a global asset that is greatly valued by the North: Earth's climate. Negotiations to date have been stalled by ideologies transposed from prior North–South conflicts into the greenhouse arena. But the elites of big, poor states have also tried to play a climate destruction card in a slow motion game of global climate change poker.

This strategy may fail, however. As Dallas Burtraw and Michael Toman have explained, most negotiations have two phases. The first phase is the bargaining over terms and content of agreement, which was partly completed at Rio. The second phase now underway is concerned with ratifying and implementing the agreement. 'Any proposed outcome that cannot be credibly implemented in the second phase of the game,' they note, 'cannot be credible in the first phase of the game.'[86]

Due to weak administrative and market institutions, states such as India, China, or Brazil may be unable to abate in accordance with global commitments to reduce greenhouse gas emissions – with or without transfers of resources from the wealthy nations. In addition, poor states are more vulnerable to the economic and social impacts of climate change than wealthy states. For these reasons, it is likely that the leaders of the OECD will be unmoved by implicit threats from the developing countries to destroy the earth's climate system unless the rules of international commerce are reformed.

Ironically, demands by big, poor states for massive resource transfers undermines the credibility of such threats for two reasons. First, it suggests to donors that the problems are so large that aid recipients may not be able to deliver the abatement, even if they receive additional support on a large scale. Second, to the extent that large-scale resource transfer achieved abatement and stimulated development, it would increase the dependency of southern elites on this source of external support. Recipients who defected from the regime would be sawing off one of the branches supporting them. It is not surprising therefore that the elites of the big, poor states are not persuaded that cooperating in the climate change arena helps rather than harms their prospects of staying in power.[q]

A new power game

The potential economic impacts of mitigation and adaptation strategies have elevated environmental concerns from low to high politics, on a par with traditional economic and military preoccupations of the great powers. In contrast to the nuclear arms race, for example, no single state or group of states so predominates in emissions or abatement capability that it can impose an international regime on everyone else. The potential candidate – the United States – abdicated from its potential hegemonic role in this regard at Rio by refusing to commit itself to reduction targets.[87] Moreover, unlike the geopolitical and geoeconomic domains of interstate relations, there are as yet no widely accepted ideologies that frame geoecological issues such as climate change.

Consequently, ecological alignments in international relations remain fluid and unpredictable. No single state or group of states can lead or coerce other states to join a greenhouse regime or to build an oppositional grouping. It remains an open question whether governments will construct a meaningful greenhouse gas regime. Faced with this agnostic prognosis, there are three reasons to be optimistic about the medium- and long-term future of a greenhouse regime: technological innovation, the contribution of scientists to elite and popular understanding of climate change issues, and social movements.

Technological innovation

Rising energy efficiency is closely associated with technological dynamism,

q This analysis ignores the impacts of climate change on regional interstate relations. Some states may welcome the possibility that climate change induced damages could weaken local adversaries.

in turn an attribute of competitive firms and economies. Domestic and international competition drives technological innovation that will reduce the cost of greenhouse gas reductions, even at high levels of abatement. Governments can impede or encourage this phenomenon, but they can't stop it in the long run.

Nonetheless, an international climate change agreement that sets the ground rules for investors and states would enhance this phenomenon. Relatedly, bilateral and regional initiatives to demonstrate the technological feasibility and economic attractiveness of greenhouse gas reduction will be an important immediate step toward a greenhouse regime. It is crucial to identify the costs of abatement above the 20–30 per cent level of reduction for which data is available today. Only when the true cost of preserving the world's climate is known will political leaders be able to respond meaningfully.

Scientific research

Scientists will continue to develop a common stock of scientific knowledge on greenhouse issues out of which political elites can forge consensus on policy issues over time. Environmental regimes reflect not only interests, including the influence of domestic stakeholders in international affairs, and legal authority flowing from or ceded to international institutions, but also different world views. The smaller the common understanding and commitment to shared values, the weaker the regime. Thus, consensual knowledge is critical to overcoming the divisions of interest, authority, and belief systems that militate against international agreement.

Political scientist Peter Haas calls this influence 'epistemic' because scientists have been able to shape the images held by politicians and diplomats as to what is at stake in negotiations. He notes that scientists have already provided an ecological basis to international agreements in the Mediterranean Action Plan and the Vienna Convention to stop ozone depletion.[88] There is little doubt that the Intergovernmental Panel on Climate Change has and will continue to play exactly this role in goading governments to grapple with the problem of global climate change.

Social movements

The inability of governments to concur in immediate stringent greenhouse gas reductions under the Convention may stimulate even greater efforts by social movements to address the issue of climate change on a global basis. In the developed countries, these organizations prefigure emerging social trends, are paradoxical in that they represent contradictions in the social order, and often transform the status quo.[89] In many developing countries, non-governmental organizations are among the few wellsprings of social activity that are not dominated by government or administrative structures.[90] Non-governmental organizations are able to pioneer creative and innovative solutions to many problems that stymie governments.

Although local citizen groups are the bedrock of social movements aimed at increasing accountability and participation in decision making of governments, they have invented new ways of communicating across cultural and institutional barriers, both within and between countries. Citizen groups provide a unique interface at the intermediate level of society to link national governmental policies and programmes with local realities via a host of social, economic and political organizations at the provincial and district level, including federations of cooperatives, trade unions and businesses, institutes, and churches.[91] In many developing countries, this level of civil society is weak and must be strengthened to complement efforts to decentralize national public bureaucracies onto local, autonomous governmental institutions.[92]

Another hallmark of environmental politics is the role of strong national, regional, and transnational social movements concerned with environment, development and social justice.[93] The Climate Action Network exemplifies this trend. Established first in Western Europe, North America and Australia, the Network now includes vibrant and self-reliant regional networks in Southeast Asia, South Asia, Latin America, and Eastern Europe. The Network marshalls scientific information to present policymakers with strong recommendations and backs up these positions with strong political pressure on national governments and in the course of climate change negotiations.[94]

Many citizen groups are starting carbon abatement projects without waiting for governments to reach international agreement. They are the key to reaching the millions of decision makers and billions of people who must change their daily routines if greenhouse gases are to be reduced to ecologically acceptable levels. They can inspire, complement and (when necessary) circumvent governments to initiate shifts in popular and elite world views. Non-governmental networks increasingly cross national boundaries to generate common positions on issues that divide their respective governments, including the old North–South divide. They can also monitor the implementation of the agreement and trumpet loudly when governments fail to meet their commitments. The production of independent inventories of greenhouse gas emissions is an important contribution in this regard which has already had an impact on international negotiations on climate change.[95]

An important component of an international greenhouse strategy in the short- to medium-run is to increase the participation of non-governmental organizations in private and public international financing of energy and environmental investment projects. The participation of non-governmental organizations in the project cycle of the World Bank's Global Environment Facility (GEF) is an important first step in this direction. In Mexico, the EMIR (*Eficientacion Mexicana de Iluminacion Residencial*) project began in 1991 when the

International Institute for Energy Conservation working with US scientists and the *Comision Federal de Electricidad* (the Mexican electricity utility) proposed to the GEF that it lend $10 million to Mexico to improve the efficiency of residential lighting by promoting CF (compact fluorescent) lamps. Over its life, each 16 watt CF lamp that costs $10 to install will save about $33 of electricity and $9 of incandescent lamps, and will reduce greenhouse gas emissions by about one fifth of a tonne of carbon.[96]

Citizen groups are also pushing for a direct role in the funding decisions of multilateral banks. At the NGO Global Forum that was held at the same time as the governmental 1992 Earth Summit, they committed themselves to urge the governments of developed countries to provide adequate, new and additional funds on concessional terms to developing countries, and to ensure that they participate in expenditure decisions and implementation of funded projects to ensure that these resources are well spent.[97]

The creation of an NGO consultative committee by the GEF is an important step in this direction, and one that other organizations, especially the regional developments banks, should emulate. Indeed, by December 1992 citizen groups had produced already a positive, even visionary reform agenda for the GEF.[98]

Conclusion

Market-driven technological innovation, increased popular participation in decision-making, non-governmental mobilization for sustainable development, and the role of the scientific community in policy formation may impel governments to overcome all the barriers to agreement. The first steps toward creating a global greenhouse regime are likely to be small rather than large, bilateral rather than multilateral, and regional rather than global. The transfer of resources is likely to be pragmatic, linked closely to abatement activity, and largely additional to existing aid flows.

Initially, therefore, a greenhouse regime must be flexible enough to demonstrate what is possible rather than to strive for final policy commitments that are simply ignored. The low (or possibly negative) net cost of abatement for the next one or two decades will grant the world a breathing space in which to explore the frontiers of social and technological possibility.

As Ralph Buultjens has written, no other issue has the ability to bring together so many people and nations as does climate change.[99] The negotiations to create a global greenhouse regime are a rare opportunity to form a global coalition of interests that transcends national boundaries and historical antagonisms. It is perhaps the first time in history that the poor in the developing countries have a powerful ally among influential citizen groups and even some governments in the developed world. Thus, Greenpeace International's greenhouse gas scenario uses a development scenario that favours the poor countries rather than simply projecting the unequal global status quo into the future, as did the Intergovernmental Panel on Climate Change.[100]

It is conceivable, therefore, that humanity will not march over the precipice of climate change, but will stop, look down, and will head instead toward a sustainable future.

Notes and references

1 See J Berreen and A Meyer, 'A Package Marked "Return to Sender," Some Problems with the Climate Convention,' *Network '92*, Centre for Our Common Future, Geneva, no. 4, June–July, 1992, p 7

2 Michael Grubb, 'The Greenhouse Effect: Negotiating Targets,' *International Affairs*, volume 66, no. 1, 1990, pp 82–83

3 Ghana on behalf of the Group of 77, 'Transfer of Environmentally Sound Technology,' informal paper to Intergovernmental Negotiating Committee for a Framework Convention on Climate Change, 4th session, December 1991, p 5

4 *Ibid*, p 23

5 *Ibid*, p 21

6 R deLucia, *Sustainable development and rural poverty alleviation: evolving perspectives on needed new thinking and approaches*, report to International Federation for Agricultural Development, World Rural Poverty Study, deLucia and Associates, Cambridge, Massachusetts, September 1990, p 2–7

7 G Schramm, 'Issues and Problems in the Power Sectors of Developing Countries,' in UN Department of Technical Cooperation, *Report on the Stockholm Initiative on Energy, Environment and Sustainable Development (SEED): Strategies for Implementing Power Sector Efficiency*, Stockholm, November 1991, p 58

8 R deLucia and M Lesser, *Natural Gas and New Power Generation/Cogeneration Technologies: Implications and Opportunities for Some Developing Countries*, paper to the International Association for Energy Economists' 11th Annual International Conference, Caracas, deLucia and Associates, Cambridge, Massachusetts, June 1989, p 7

9 UN Economic and Social Commission for Asia and the Pacific, *Energy Policy Implications of the Climatic Effects of Fossil Fuel Use in the Asia-Pacific Region*, ESCAP Symposium, Paper NR/SCE/1, September 1990, Tokyo, December 12, 1990, p 122

10 G Porter, 'Reaching a Consensus on Financial Resources,' *Network '92*, Centre for Our Common Future, Geneva, no. 4, June–July, 1992, p 1

11 O Kjorven and A Kristin Sydnes, *Funding for the Global Environment: The Issue of Additionality*, Report 4, Fridtjof Nansen Institute, Lysaker, Norway, 1992, provide an excellent analysis of this issue

12 O Kjorven, *Facing the Challenge of Change: The World Bank and the Global Environment Facility*, Report 3, Fridtjof Nansen Institute, Lysaker, Norway, 1992, pp 63–66; A Kristin Sydnes, *Developing Countries in Global Climate Negotiations*, Report 4, Fridtjof Nansen Institute, Lysaker, Norway, 1991, pp 9–11

13 As stated in People's Republic of China, 'China's Principled Position on Global

Environment Issues,' chapter 5, *National Report of the People's Republic of China on Environment and Development*, August 1991, pp 53–54
14 See UN Centre on Transnational Corporations, *Climate Change and Transnational Corporations: Analysis and Trends*, ST/CTC/111, United Nations, New York, 1991
15 P Adams and L Solomon, *In the Name of Progress, The Underside of Foreign Aid*, Earthscan, London, 1991; and P Adams, *Odious Debts, Loose Lending, Corruption and the Third World's Environmental Legacy*, Earthscan, London, 1991
16 P McCully, 'The Case Against Climate Aid,' *The Ecologist*, volume 21, no. 6, November–December 1991, pp 244–257
17 *Ibid*, p 248
18 On the latter issue, see P Hayes, 'Social Structure and Rural Energy Technology,' in Nautilus Inc, *ed*, *Southern Perspectives on the Rural Energy Crisis*, Conference of NGOs and the Environment Liaison Centre, Nairobi, 1981, pp 37–48
19 M Bell, *Continuing Industrialisation, Climate Change and International Technology Transfer*, Science Policy Research Unit, Sussex University, December 1990, pp 75–80
20 N Chantramonklasri, 'The Development of Managerial and Technological Capability in the Developing Countries,' in M Chatterji, *ed*, *Technology Transfer in the Developing Countries*, MacMillan, London, 1990, pp 38, 44
21 S Lall, 'Transnationals and the Third World: Changing Perceptions,' in S Lall, *Multinationals, Technology and Exports*, St Martin's Press, New York, 1985, p 72
22 M Bell, *Continuing Industrialisation*, *op cit*, endnote 19, p 84
23 F Stewart, 'Technological Dependence,' in F Stewart, *Technology and Underdevelopment*, Westview Press, Boulder, Colorado, 1977, p 123
24 F Stewart, *Macro Policies for Appropriate Technology in Developing Countries*, Westview Press, Boulder, Colorado, 1987
25 G Anandalingham, 'Energy Conservation in the Industrial Sector of Developing Countries,' *Energy Policy*, August 1985, p 338
26 R Kaplinsky, 'Technology Transfer, Adaptation and Generation: A Framework of Evaluation,' in M Chatterji, *ed*, *Technology Transfer*, *op cit*, endnote 20, pp 22–23
27 UNCTAD Secretariat, *Technology Policy in the Energy Sector: Issues, Scope and Options for Developing Countries*, United Nations Conference on Trade and Development, Geneva, June 1989, p 101
28 A Barnett, 'The Diffusion of Energy Technology in the Rural Areas of Developing Countries: A Synthesis of Recent Experience,' *World Development*, volume 18, no. 4, pp 539–553; and other essays on specific country or technology experiences in the same issue
29 C Freeman, *Technology Policy and Economic Performance*, Pinter Publishers, London, 1987, pp 64–79
30 J Granger, *Technology and International Relations*, W H Freeman, New York, 1979, p 62
31 M Grubb, 'Technology Transfer and the Global Environment: Motives and Mechanisms' (mimeo), Royal Institute of International Affairs, London, September 12, 1991
32 A Mody, *Staying in the Loop, International Alliances for Sharing Technology*, World Bank Discussion Paper 61, World Bank, Washington DC, 1989, p 2

33 C Primo Braga, 'The Developing Country Case For and Against Intellectual Property Protection,' in W Siebek, *ed*, *Strengthening Protection of Intellectual Property in Developing Countries*, World Bank Discussion Paper 112, Washington DC, 1990, pp 68–87

34 L Lunde, *The North/South Dimension in Global Greenhouse Politics, Conflicts, Dilemmas, Solutions*, Report 9, Fridtjof Nansen Institute, Lysaker, Norway, 1990, p 20

35 P Wexler, *International Negotiations on Climate Change*, Center for Global Change, University of Maryland, February 1992, p 18

36 Lee Jin-loo and J N Sharan, *Technological Impact of the Public Procurement Policy: The Experience of the Power Plant Sector in the Republic of Korea*, Geneva, July 1985, p 14

37 M Bell, *Continuing Industrialisation, op cit*, endnote 19, p viii

38 L Lunde, *Science or Politics in the Global Greenhouse, A Study of the Development Towards Scientific Consensus on Climate Change*, Report 8, Fridtjof Nansen Institute, Lysaker, Norway, 1991, p 127

39 E Parson, *The Transport Sector and Global Warming*, Paper E-90-11, JFK School of Government, Global Environmental Policy Project, Harvard University, Cambridge 1990, pp vii–viii

40 J Mackenzie and M Walsh, *Driving Forces: Motor Vehicle Trends and their Implications for Global Warming, Energy Strategies, and Transportation Planning*; World Resources Institute, Washington DC, December 1990, p 38

41 UNCTAD, *Policies and Mechanisms for Achieving Sustainable Development*, UNCTAD TD/B/1304, Geneva, August 15, 1991, p 25 and Figure 2.1

42 S Barrett, 'Free Rider Deterrence in a Global Warming Treaty', First Draft, London Business School report to Environment Directorate OECD, Paris, May 1991

43 For a general treatment of these issues, see C Russell *et al*, *Enforcing Pollution Control Laws*, Resources for the Future, Washington DC, 1986

44 W Fischer *et al*, *A Convention on Greenhouse Gases: Towards the Design of a Verification System*, Forschungszentrum Julich GmbH, Julich, Germany, October 1990. The following section draws heavily on this study

45 D Victor, 'Limits of market-based strategies for slowing global warming: The case of tradeable permits,' *Policy Sciences*, volume 24, 1991, p 210

46 *Ibid*, p 207

47 Using satellite data, A Setzer and M Pereira, 'Amazonia Biomass Burnings in 1987 and an Estimate of their Tropospheric Emissions,' *Ambio*, volume 20, no. 1, February 1991, pp 19–22, estimate Brazil's forestry related CO_2 fossil fuel emissions at 1.7 gigatonnes in 1987; World Resourches Institute, *World Resources, 1991*, Oxford University Press, New York, 1990, p 346, estimated the emissions at 1.2 gigatonnes that year; A Agarwal and S Narain, *Global Warming in an Unequal World: A Case of Environmental Colonialism*; Centre for Science and Environment, New Delhi, 1991, p 4, argue that the figure should be based on a decadal average, not 1987, and should be reduced to between 0.38 gigatonnes. Thus, estimates for an important emitter range differ by a factor of five

48 J Lanchberry *et al*, *Verification and the Framework Convention on Climate Change*, Verification Technology Information Centre, London, May 1992, p 25

49 Organisation for Economic Cooperation and Development, *Estimation of Greenhouse Gas Emissions and Sinks*, final report to the IPCC from OECD Experts

Meeting, February 1991, Revised August 1991
50 J Lanchberry *et al, Verification and the Framework Convention, op cit,* endnote 48
51 A Krass, *Verification, How Much is Enough?,* Taylor and Francis, London, 1985, p 94
52 B Schiff, *International Nuclear Technology Transfer, Dilemmas of Dissemination and Control,* Rowman and Allenheld, Totowa, New Jersey, p 113
53 See J Wettestad, 'Verification of International Greenhouse Agreements: A Mismatch between Technical and Political Feasibility,' *International Challenges,* volume 11, no. 1, 1991, pp 41–47
54 J Ausubel and D Victor, 'Verification of International Environmental Agreements' (mimeo), November 13, 1991, forthcoming in *Annual Review of Energy and Environment*
55 P Lewis, 'Experiences in Verification - What Can Be Learned for a Greenhouse Gas Convention,' in J Primio and G Stein, *eds, A Regime to Control Greenhouse Gases,* Forschungszentrum Julich GmbH, Julich, Germany, 1992, p 54
56 M Efinger and H Breitmeier, 'Verifying a Convention on Greenhouse Gases: A Game-Theoretic Approach,' in J Primio and G Stein, *eds, A Regime to Control, ibid,* p 66
57 D Feldman, 'Some Lessons of the IAEA's Nuclear Non-Proliferation Regime for Confidence-Building Under a Greenhouse Gas Convention,' in J Primio and G Stein, *eds, A Regime to Control, op cit,* endnote 55, pp 79–84
58 Intergovernmental Negotiating Committee for a Framework Convention on Climate Change, *Climate Change Convention,* UN Document A/AC.237/18 (Part II)/Add.1, May 15, 1992, as revised at the June 1992 UN Conference on Environment and Development
59 M Grubb, 'The Greenhouse Effect: Negotiating Targets'; *International Affairs,* volume 66, no. 1, 1990, p 76
60 K von Moltke, 'International Trade, Technology Transfer and Climate Change,' in I Mintzer, *ed, Confronting Climate Change, Risks, Implications and Responses,* Cambridge University Press, 1992, p 302
61 P Sand, *Lessons Learned in Global Environmental Governance,* World Resources Institute, Washington DC, June 1990, pp 25–27
62 T Simmons, 'The IEA Energy Data System,,' in J Primio and G Stein, *ed, A Regime to Control, op cit,* endnote 55, 119–129; and discussions of IEA in annexes to *Pledge and Review Processes: Possible Components of a Climate Convention,* Royal Institute of International Affairs, Energy and Environment Program report, August 2, 1991, London
63 S Barrett, 'Free Rider Deterrence,' *op cit,* endnote 42, pp 9–14
64 A Markandya, 'Global Warming, The Economics of Tradeable Permits,' in D Pearce, *ed, Blueprint 2, Greening the World Economy,* Earthscan, London, pp 59–61
65 Asian Development Bank, *Environmental Considerations in Energy Development,* ADB, Manila, May 1991, p 96
66 *Ibid,* p 100; see also J Topping, A Qureshi and S Sherer, *Implications of Climate Change for the Asian Pacific Region,* Climate Institute, paper for the Asian-Pacific Seminar on Climate Change, Nagoya, Japan, January 1991
67 V V Desai, K Nyman, *Industrial Energy Conservation: Notes on Three Country Studies,* Energy Planning Unit, Asian Development Bank, Manila, circa 1986, pp 2–4
68 M Philips, *Energy Conservation Activities in Asia,* International Institute for

Energy Conservation, Washington DC, September 1990, p 12–13
69 M Levine *et al.*, *Energy Efficiency, Developing Nations and Eastern Europe*, A Report to the US Working Group on Global Energy Efficiency, April 1991, p 20
70 G Doyle, 'Future Coal Use in the Asia Pacific Region,' in T Siddiqi and D Streets, *Responding to the Threat of Global Warming, Options for the Pacific and Asia*, Argonne National Laboratory and Environment and Policy Institute, East West Center, Workshop Proceedings, ANL/EAIS/TM-17, June 21, 1989, Honolulu, p 3–21
71 World Bank, *China: Socialist Economic Development*, Annex E, The Energy Sector, Report 3391-CHA, June 1981, p 11
72 M Philips, *Energy conservation activities, op cit*, endnote 68, p 3
73 *Ibid*
74 International Institute for Energy Conservation, 'Factory Lost Opportunities Project,' International Institute for Energy Conservation, Asia Regional Office, Bangkok, 1990; see also A Gadgil and G Jannuzzi, *Conservation Potential of Compact Fluorescent Lamps in India and Brazil*, LBL-27210 Rev., Lawrence Berkeley Laboratory, University of California, Berkeley California, September 1990
75 R Thaman, 'Coastal Reforestation and Agro Forestry as Immediate Ameliorative Measures to Address Global Warming and to Promote Sustainable Habitation of Low Lying and Coastal Areas,' in T Siddiqi and D Streets, *Responding to the Threat of Global Warming, Options for the Pacific and Asia*, Argonne National Laboratory and Environment and Policy Institute, East West Center, Workshop Proceedings, ANL/EAIS/TM-17, June 21, 1989, Honolulu, pp 4-37–45
76 S Myers *et al*, *Energy Efficiency and Household Electric Appliances in Developing and Newly Industrialized Countries*, LBL-29678 UC-350, Lawrence Berkeley Laboratory, University of California, Berkeley California, December 1990, p 44
77 Asian Development Bank, *Environmental Considerations in Energy Development*, ADB, Manila, May 1991, p 77
78 B Leach, *China and Global Change, Opportunities for Collaboration*, US National Academy Press, Washington DC, 1992
79 See 'Development of an Inter-American Institute for Global Change Research,' workshop report, San Juan, Puerto Rico, July 15, 1992; and 'Regional Co-Operative Activities to Support Global Change Research in ASEAN Countries as a Component of START,' October 20, 1992, annex B to 'Minutes of Second Meeting of the Southeast Asian Regional Committee for START (SARCS),' Ministry of Science, Technology and the Environment, Kuala Lumpur, Malaysia, October 9, 1992
80 S Krasner, *Structural Conflict, The Third World Against Global Liberalism*, University of California Press, Berkeley, California, 1985; M Williams, *Third World Cooperation, The Group of 77 in UNCTAD*, St Martin's Press, New York, 1991
81 N Harris, *The End of the Third World, Newly Industrializing Countries and the Decline of an Ideology*, Penguin Books, London, 1986
82 A Kristin Sydnes, *Developing Countries, op cit*, endnote 12, pp 7–8
83 T Hyder, 'Climate Negotiations: The North/South Perspective,' in I Mintzer, *ed*, *Confronting Climate Change, op cit*, endnote 60, p 330
84 *Ibid*, p 8
85 L Lunde, *The North/South Dimension, op cit*, endnote 34, p 2

86 *Ibid*, p 16
87 *ECO* 'Last Change for Climate Treaty, US Intransigence Still Hinders Negotiations' (New York), April 24, 1992, p 1
88 P Haas, *Saving the Mediterranean, The Politics of International Environmental Cooperation*, Columbia University Press, New York, 1990, p xxii; and P Haas, 'Ecological Epistemic Communities and the Protection of Stratospheric Ozone' (mimeo), Political Science Department, University of Massachusetts in Amherst, January 1991, forthcoming in *International Organization* special issue on Knowledge, Power and International Policy Coordination
89 A Melucci, *Nomads of the Present, Social Movements and Individual Needs in Contemporary Society*, Hutchinson Radius, London, 1989; C Jennett and R Stewart, *eds*, *Politics of the Future, The Role of Social Movements*, MacMillan, Melbourne, 1989
90 P Ekins, *A New World Order, Grassroots Movements for Global Change*, Routledge, London, 1992
91 J Holmbert, *Making Development Sustainable, Redefining Institutions, Policy, and Economics*, Island Press, Washington DC, 1992, pp 56–7
92 See International Centre for Integrated Mountain Development, *Economic Policies for Sustainable Development in Nepal*, report to Asian Development Bank, Khatmandu, May 1990, p 373
93 J Clark, *Democratising Development, The Role of Voluntary Organisations*, Earthscan, London, 1991; and G Leach and R Mearns, *Beyond the Fuelwood Crisis, People, Land and Trees in Africa*, Earthscan, London, 1988, pp 100–122
94 N Dubash and M Oppenheimer, 'Modifying the Mandate of Existing Institutions: NGOs,' in I Mintzer, *ed*, *Confronting Climate Change*, *op cit*, endnote 60, p 275; and Climate Change Network, 'A Force for Change' (mimeo), December 1992
95 See 'Atmosphere and Climate,' chapter 24 of World Resources Institute, *World Resources, 1992–1993*, Oxford University Press, New York, 1992, pp 345–355; and S Subak *et al*, *National Greenhouse Gas Accounts: Current Anthropogenic Sources and Sinks*, Stockholm Environment Institute, Boston, 1992
96 Center for Building Science, 'Mexico Large-Scale Compact Fluorescent Lamp Project' (mimeo), Lawrence Berkeley Laboratory, undated
97 'Alternative Non-Governmental Agreement on Climate Change,' in *Alternative Treaty-Making Process* from the International Non-Government Organization Forum, Rio de Janeiro, June 1–14, 1992, p D-3
98 'NGOs: Conventions Must Drive GEF,' *ECO*, December 1992, Geneva, p 7
99 R Buultjens, 'Years of Waste, Call for Action,' *Earth Summit Times*, volume 3, no. 2, August 27, 1992, p 14
100 Stockholm Environment Institute, *Towards Global Energy Security: The Next Energy Transition, An Energy Scenario for a Fossil Fuel Free Energy Future*, draft report to Greenpeace International, Boston, May 1992; J Parikh, 'IPCC strategies unfair to the South,' *Nature*, volume 260, December 10, 1992, p 507

Appendix: The Climate Change Convention

Introduction

After negotiations which spanned 15 months, the UN Framework Convention on Climate Change was finalized in May 1992. It was opened for signature at the UN Conference on Environment and Development - the Earth Summit - in Rio de Janeiro, Brazil, on 4 June 1992.

As of mid-October 1992, 158 countries had signed the Convention, including the European Community. In order for the Convention to become law, it must be ratified by national legislatures of 50 countries, a process that may take two years.

The aim of this agreement is to stabilize atmospheric concentrations of greenhouse gases at levels that will prevent human activities from interfering dangerously with the global climate system. In signing the Convention, Governments agree to reduce emissions of the warming greenhouse gases to "earlier" levels by the end of the decade. States are required to report periodically on their level of emissions and efforts to slow climate change. The target of reducing carbon dioxide emissions to 1990 levels by the end of the decade - advocated by the European Community, Japan and most other countries but opposed by the United States - is stated as a goal to be met voluntarily.

To enable developing countries to meet their obligations under the Convention, developed countries agree to provide "new and additional" financial assistance. Such assistance is, for the time being, to be channelled through the Global Environment Facility, a fund administered jointly by the World Bank, the UN Development Programme and the UN Environment Programme (UNEP).

Background

The groundwork for the Framework Convention began in 1988 when the United Nations General Assembly adopted resolution 43/53 recognizing climate change as a common concern of humanity. That year, UNEP and the UN World Meteorological Organization (WMO) established the Intergovernmental Panel on Climate Change (IPCC) to investigate the potential severity and impact of global climate change and to suggest possible policy responses. The IPCC's First Assessment Report was published in August 1990 and discussed at the second World Climate Conference later that year.

The report noted, among other things, that the 1989 session of the UN General

Assembly had agreed that existing legal instruments and institutions dealing with climate change were insufficient and that a framework convention on climate change was needed. As a "framework", the Convention would outline a set of general principles and obligations in various areas. Subsequent negotiations would produce specific targets and quantitative reductions that would be added as protocols to the framework convention.

In December 1990, the General Assembly set up the Intergovernmental Negotiating Committee for a Framework Convention on Climate Change (INC), to be supported by UNEP and WMO. Negotiations began in February 1991 and ran parallel to the work of the Committee preparing for the Earth Summit in the hope that a convention would be ready for signing by Governments in Brazil.

Climate Change Convention

The Parties to this Convention

Acknowledging that change in the Earth's climate and its adverse effects are a common concern of humankind,

Concerned that human activities have been substantially increasing the atmospheric concentrations of greenhouse gases, that these increases enhance the natural greenhouse effect, and that this will result on average in an additional warming of the Earth's surface and atmosphere and may adversely affect natural ecosystems and humankind,

Noting that the largest share of historical and current global emissions of greenhouse gases has originated in developed countries, that per capita emissions in developing countries are still relatively low and that the share of global emissions originating in developing countries will grow to meet their social and development needs,

Aware of the role and importance in terrestrial and marine ecosystems of sinks and reservoirs of greenhouse gases,

Noting that there are many uncertainties in predictions of climate change, particularly with regard to the timing, magnitude and regional patterns thereof,

Acknowledging that the global nature of climate change calls for the widest possible cooperation by all countries and their participation in an effective and appropriate international response, in accordance with their common but differentiated responsibilities and respective capabilities and their social and economic conditions,

Recalling the pertinent provisions of the Declaration of the United Nations Conference on the Human Environment, adopted at Stockholm on 16 June 1972,

Recalling also that States have, in accordance with the Charter of the United Nations and the principles of international law, the sovereign right to exploit their own resources pursuant to their own environmental and developmental policies, and the responsibility to ensure that activities within their jurisdiction or control do not cause damage to the environment of other States or of areas beyond the limits of national jurisdiction,

Reaffirming the principle of sovereignty of States in international cooperation to address climate change,

Recognizing that States should enact effective environmental legislation, that environmental standards, management objectives and priorities should reflect the environmental and developmental context to which they apply, and that standards applied by some countries may be inappropriate and of unwarranted economic and social cost to other countries, in particular developing countries,

Recalling the provisions of General Assembly resolution 44/228 of 22 December 1989 on the United Nations Conference on Environment and Development, and resolutions 43/53 of 6 December 1988, 44/207 of 22 December 1989, 45/212 of 21 December 1990 and 46/169 of 19 December 1991 on protection of global climate for present and future generations of mankind,

Recalling also the provisions of General Assembly resolution 44/206 of 22 December 1989 on the possible adverse effects of sea level rise on islands and coastal areas, particularly low-lying coastal areas and the pertinent provisions of General Assembly resolution 44/172 of 19 December 1989 on the implementation of the Plan of Action to Combat Desertification,

Recalling further the Vienna Convention for the Protection of the Ozone Layer, 1985, and the Montreal Protocol on Substances that Deplete the Ozone Layer, 1987, as adjusted and amended on 29 June 1990,

Noting the Ministerial Declaration of the Second World Climate Conference adopted on 7 November 1990,

Conscious of the valuable analytical work being conducted by many States on climate change and of the important contributions of the World Meteorological Organization, the United Nations Environment Programme and other organs, organizations and bodies of the United Nations system, as well as other international and intergovernmental bodies, to the exchange of results of scientific research and the coordination of research,

Recognizing that steps required to understand and address climate change will be environmentally, socially and economically most effective if they are based on relevant scientific, technical and economic considerations and continually re-evaluated in the light of new findings in these areas,

Recognizing that various actions to address climate change can be justified economically in their own right and can also help in solving other environmental problems,

Recognizing also the need for developed countries to take immediate action in a flexible manner on the basis of clear priorities, as a first step towards comprehensive response strategies at the global, national and, where agreed, regional levels that take into account all greenhouse gases, with due consideration of their relative contributions to the enhancement of the greenhouse effect,

Recognizing further that low-lying and other small island countries, countries with low-lying coastal, arid and semi-arid areas or areas liable to floods, drought and desertification, and developing countries with fragile mountainous ecosystems are particularly vulnerable to the adverse effects of climate change,

Recognizing the special difficulties of those countries, especially developing countries, whose economies are particularly dependent on fossil fuel production, use and exportation, as a consequence of action taken on limiting greenhouse gas emissions,

Affirming that responses to climate change should be coordinated with social and economic development in an integrated manner with a view to avoiding adverse impacts on the latter, taking into full account the legitimate priority needs of developing countries for the achievement of sustained economic growth and the eradication of poverty,

Recognizing that all countries, especially developing countries, need access to resources required to achieve sustainable social and economic development and that, in order for developing countries to progress towards that goal, their energy consumption will need to grow taking into account the possibilities for achieving greater energy efficiency and for controlling greenhouse gas emissions in general, including through the application of new technologies on terms which make such an application economically and socially beneficial,

Determined to protect the climate system for present and future generations,

Have agreed as follows:

Article 1. Definitions[a]

For the purposes of this Convention:

1 "Adverse effects of climate change" means changes in the physical environment or biota resulting from climate change which have significant deleterious effects on the composition, resilience or productivity of natural and managed ecosystems or on the operation of socio-economic systems or on human health and welfare.

2 "Climate change" means a change of climate which is attributed directly or indirectly to human activity that alters the composition of the global atmosphere and which is in addition to natural climate variability observed over comparable time periods.

3 "Climate system" means the totality of the atmosphere, hydrosphere, biosphere and geosphere and their interactions.

4 "Emissions" means the release of greenhouse gases and/or their precursors into the atmosphere over a specified area and period of time.

5 "Greenhouse gases" means those gaseous constituents of the atmosphere, both natural and anthropogenic, that absorb and re-emit infrared radiation.

6 "Regional economic integration organization" means an organization constituted by sovereign States of a given region which has competence in respect of matters governed by this Convention or its protocols and has been duly authorized, in accordance with its internal procedures, to sign, ratify, accept, approve or accede to the instruments concerned.

7 "Reservoir" means a component or components of the climate system where a greenhouse gas or a precursor of a greenhouse gas is stored.

8 "Sink" means any process, activity or mechanism which removes a greenhouse gas, an aerosol or a precursor of a greenhouse gas from the atmosphere.

9 "Source" means any process or activity which releases a greenhouse gas, an aerosol or a precursor of a greenhouse gas into the atmosphere.

Article 2. Objective

The ultimate objective of this Convention and any related legal instruments that the Conference of the Parties may adopt is to achieve, in accordance with the relevant provisions of the Convention, stabilization of greenhouse gas concentrations in the atmosphere at a level that would prevent dangerous anthropogenic interference with the climate system. Such a level should be achieved within a time frame sufficient to allow ecosystems to adapt naturally to climate change, to ensure

---

a Titles of articles are included solely to assist the reader.

that food production is not threatened and to enable economic development to proceed in a sustainable manner.

Article 3. Principles

In their actions to achieve the objective of the Convention and to implement its provisions, the Parties shall be guided, *inter alia*, by the following:

1. The Parties should protect the climate system for the benefit of present and future generations of humankind, on the basis of equity and in accordance with their common but differentiated responsibilities and respective capabilities. Accordingly, the developed country Parties should take the lead in combating climate change and the adverse effects thereof.
2. The specific needs and special circumstances of developing country Parties, especially those that are particularly vulnerable to the adverse effects of climate change, and of those Parties, especially developing country Parties, that would have to bear a disproportionate or abnormal burden under the Convention, should be given full consideration.
3. The Parties should take precautionary measures to anticipate, prevent or minimize the causes of climate change and mitigate its adverse effects. Where there are threats of serious or irreversible damage, lack of full scientific certainty should not be used as a reason for postponing such measures, taking into account that policies and measures to deal with climate change should be cost-effective so as to ensure global benefits at the lowest possible cost. To achieve this, such policies and measures should take into account different socio-economic contexts, be comprehensive, cover all relevant sources, sinks and reservoirs of greenhouse gases and adaptation, and comprise all economic sectors. Efforts to address climate change may be carried out cooperatively by interested Parties.
4. The Parties have a right to, and should, promote sustainable development. Policies and measures to protect the climate system against human-induced change should be appropriate for the specific conditions of each Party and should be integrated with national development programmes, taking into account that economic development is essential for adopting measures to address climate change.
5. The Parties should cooperate to promote a supportive and open international economic system that would lead to sustainable economic growth and development in all Parties, particularly developing country Parties, thus enabling them better to address the problems of climate change. Measures taken to combat climate change, including unilateral ones, should not constitute a means of arbitrary or unjustifiable discrimination or a disguised restriction on international trade.

Article 4 Commitments

1. All Parties, taking into account their common but differentiated responsibilities and their specific national and regional development priorities, objectives and circumstances, shall:

(a) Develop, periodically update, publish and make available to the Conference of the Parties, in accordance with Article 12, national inventories of anthropogenic emissions by sources and removals by sinks of all greenhouse gases not controlled by the Montreal Protocol, using comparable methodologies to be agreed upon by the Conference of the Parties;

(b) Formulate, implement, publish and regularly update national and, where appropriate, regional programmes containing measures to mitigate climate change by addressing anthropogenic emissions by sources and removals by sinks of all greenhouse gases not controlled by the Montreal Protocol, and measures to facilitate adequate adaptation to climate change;

(c) Promote and cooperate in the development, application and diffusion, including transfer, of technologies, practices and processes that control, reduce or prevent anthropogenic emissions of greenhouse gases not controlled by the Montreal Protocol in all relevant sectors, including the energy, transport, industry, agriculture, forestry and waste management sectors;

(d) Promote sustainable management, and promote and cooperate in the conservation and enhancement, as appropriate, of sinks and reservoirs of all greenhouse gases not controlled by the Montreal Protocol, including biomass, forests and oceans as well as other terrestrial, coastal and marine ecosystems;

(e) Cooperate in preparing for adaptation to the impacts of climate change; develop and elaborate appropriate and integrated plans for coastal zone management, water resources and agriculture, and for the protection and rehabilitation of areas, particularly in Africa, affected by drought and desertification, as well as floods;

(f) Take climate change considerations into account, to the extent feasible, in their relevant social, economic and environmental policies and actions, and employ appropriate methods, for example impact assessments, formulated and determined nationally, with a view to minimizing adverse effects on the economy, on public health and on the quality of the environment, of projects or measures undertaken by them to mitigate or adapt to climate change;

(g) Promote and cooperate in scientific, technological, technical, socio-economic and other research, systematic observation and development of data archives related to the climate system and intended to further the understanding and to reduce or eliminate the remaining uncertainties regarding the causes, effects, magnitude and timing of climate change and the economic and social consequences of various response strategies;

(h) Promote and cooperate in the full, open and prompt exchange of relevant scientific, technological, technical, socio-economic and legal information related to the climate system and climate change, and to the economic and social consequences of various response strategies;

(i) Promote and cooperate in education, training and public awareness related to climate change and encourage the widest participation in this process, including that of non-governmental organizations; and

(j) Communicate to the Conference of the Parties information related to implementation, in accordance with Article 12.

2. The developed country Parties and other Parties included in annex I commit themselves specifically as provided for in the following:

(a) Each of these Parties shall adopt national[b] policies and take corresponding measures on the mitigation of climate change, by limiting its anthropogenic emissions of greenhouse gases and protecting and enhancing its greenhouse gas sinks and reservoirs. These policies and measures will demonstrate that developed countries are taking the lead in modifying longer-term trends in anthropogenic emissions consistent with the objective of the Convention, recognizing that the return by the end of the present decade to earlier levels of anthropogenic emissions of carbon dioxide and other greenhouse gases not controlled by the Montreal

b This includes policies and measures adopted by regional economic integration organizations.

Protocol would contribute to such modification, and taking into account the differences in these Parties' starting points and approaches, economic structures and resource bases, the need to maintain strong and sustainable economic growth, available technologies and other individual circumstances, as well as the need for equitable and appropriate contributions by each of these Parties to the global effort regarding that objective. These Parties may implement such policies and measures jointly with other Parties and may assist other Parties in contributing to the achievement of the objective of the Convention and, in particular, that of this subparagraph;

(b) In order to promote progress to this end, each of these Parties shall communicate, within six months of the entry into force of the Convention for it and periodically thereafter, and in accordance with Article 12, detailed information on its policies and measures referred to in subparagraph (a) above, as well as on its resulting projected anthropogenic emissions by sources and removals by sinks of greenhouse gases not controlled by the Montreal Protocol for the period referred to in subparagraph (a), with the aim of returning individually or jointly to their 1990 levels these anthropogenic emissions of carbon dioxide and other greenhouse gases not controlled by the Montreal Protocol. This information will be reviewed by the Conference of the Parties, at its first session and periodically thereafter, in accordance with Article 7;

(c) Calculations of emissions by sources and removals by sinks of greenhouse gases for the purposes of subparagraph (b) above should take into account the best available scientific knowledge, including of the effective capacity of sinks and the respective contributions of such gases to climate change. The Conference of the Parties shall consider and agree on methodologies for these calculations at its first session and review them regularly thereafter;

(d) The Conference of the Parties shall, at its first session, review the adequacy of subparagraphs (a) and (b) above. Such review shall be carried out in the light of the best available scientific information and assessment on climate change and its impacts, as well as relevant technical, social and economic information. Based on this review, the Conference of the Parties shall take appropriate action, which may include the adoption of amendments to the commitments in subparagraphs (a) and (b) above. The Conference of the Parties, at its first session, shall also take decisions regarding criteria for joint implementation as indicated in subparagraph (a) above. A second review of subparagraphs (a) and (b) shall take place not later than 31 December 1998, and thereafter at regular intervals determined by the Conference of the Parties, until the objective of the Convention is met;

(e) Each of these Parties shall:

(i) coordinate as appropriate with other such Parties, relevant economic and administrative instruments developed to achieve the objective of the Convention; and

(ii) identify and periodically review its own policies and practices which encourage activities that lead to greater levels of anthropogenic emissions of greenhouse gases not controlled by the Montreal Protocol than would otherwise occur;

(f) The Conference of the Parties shall review, not later than 31 December 1998, available information with a view to taking decisions regarding such amendments to the lists in annexes I and II as may be appropriate, with the approval of the Party concerned;

(g) Any Party not included in annex I may, in its instrument of ratification, acceptance, approval or accession, or at any time thereafter, notify the Depositary that it intends to be bound by subparagraphs (a) and (b) above. The Depositary shall inform the other signatories and Parties of any such notification.

3. The developed country Parties and other developed Parties included in annex II shall provide new and additional financial resources to meet the agreed full costs incurred by developing country Parties in complying with their obligations under Article 12, paragraph 1. They shall also provide such financial resources, including for the transfer of technology, needed by the developing country Parties to meet the agreed full incremental costs of implementing measures that are covered by paragraph 1 of this Article and that are agreed between a developing country Party and the international entity or entities referred to in Article 11, in accordance with that Article. The implementation of these commitments shall take into account the need for adequacy and predictability in the flow of funds and the importance of appropriate burden sharing among the developed country Parties.

4. The developed country Parties and other developed Parties included in annex II shall also assist the developing country Parties that are particularly vulnerable to the adverse effects of climate change in meeting costs of adaptation to those adverse effects.

5. The developed country Parties and other developed Parties included in annex II shall take all practicable steps to promote, facilitate and finance, as appropriate, the transfer of, or access to, environmentally sound technologies and know-how to other Parties, particularly developing country Parties, to enable them to implement the provisions of the Convention. In this process, the developed country Parties shall support the development and enhancement of endogenous capacities and technologies of developing country Parties. Other Parties and organizations in a position to do so may also assist in facilitating the transfer of such technologies.

6. In the implementation of their commitments under paragraph 2 above, a certain degree of flexibility shall be allowed by the Conference of the Parties to the Parties included in annex I undergoing the process of transition to a market economy, in order to enhance the ability of these Parties to address climate change, including with regard to the historical level of anthropogenic emissions of greenhouse gases not controlled by the Montreal Protocol chosen as a reference.

7. The extent to which developing country Parties will effectively implement their commitments under the Convention will depend on the effective implementation by developed country Parties of their commitments under the Convention related to financial resources and transfer of technology and will take fully into account that economic and social development and poverty eradication are the first and overriding priorities of the developing country Parties.

8. In the implementation of the commitments in this Article, the Parties shall give full consideration to what actions are necessary under the Convention, including actions related to funding, insurance and the transfer of technology, to meet the specific needs and concerns of developing country Parties arising from the adverse effects of climate change and/or the impact of the implementation of response measures, especially on:

(a) Small island countries;
(b) Countries with low-lying coastal areas;
(c) Countries with arid and semi-arid areas, forested areas and areas liable to forest decay;

(d) Countries with areas prone to natural disasters;
(e) Countries with areas liable to drought and desertification;
(f) Countries with areas of high urban atmospheric pollution;
(g) Countries with areas with fragile ecosystems, including mountainous ecosystems;
(h) Countries whose economies are highly dependent on income generated from the production, processing and export, and/or on consumption of fossil fuels and associated energy-intensive products; and
(i) Land-locked and transit countries.

Further, the Conference of the Parties may take actions, as appropriate, with respect to this paragraph.

9. The Parties shall take full account of the specific needs and special situations of the least developed countries in their actions with regard to funding and transfer of technology.

10. The Parties shall, in accordance with Article 10, take into consideration in the implementation of the commitments of the Convention the situation of Parties, particularly developing country Parties, with economies that are vulnerable to the adverse effects of the implementation of measures to respond to climate change. This applies notably to Parties with economies that are highly dependent on income generated from the production, processing and export, and/or consumption of fossil fuels and associated energy-intensive products and/or the use of fossil fuels for which such Parties have serious difficulties in switching to alternatives.

Article 5. Research and Systematic Observation

In carrying out their commitments under Article 4, paragraph 1(g), the Parties shall:

(a) Support and further develop, as appropriate, international and intergovernmental programmemes and networks or organizations aimed at defining, conducting, assessing and financing research, data collection and systematic observation, taking into account the need to minimize duplication of effort;

(b) Support international and intergovernmental efforts to strengthen systematic observation and national scientific and technical research capacities and capabilities, particularly in developing countries, and to promote access to, and the exchange of, data and analyses thereof obtained from areas beyond national jurisdiction; and

(c) Take into account the particular concerns and needs of developing countries and cooperate in improving their endogenous capacities and capabilities to participate in the efforts referred to in subparagraphs (a) and (b) above.

Article 6. Education, Training and Public Awareness

In carrying out their commitments under Article 4, paragraph 1(i), the Parties shall:

(a) Promote and facilitate at the national and, as appropriate, subregional and regional levels, and in accordance with national laws and regulations, and within their respective capacities:

(i) the development and implementation of educational and public awareness programmes on climate change and its effects;
(ii) public access to information on climate change and its effects;
(iii) public participation in addressing climate change and its effects and developing adequate responses; and

(iv) training of scientific, technical and managerial personnel.

(b) Cooperate in and promote, at the international level, and, where appropriate, using existing bodies:

(i) the development and exchange of educational and public awareness material on climate change and its effects; and

(ii) the development and implementation of education and training programmes, including the strengthening of national institutions and the exchange or secondment of personnel to train experts in this field, in particular for developing countries.

Article 7. Conference of the Parties

1. A Conference of the Parties is hereby established.

2. The Conference of the Parties, as the supreme body of this Convention, shall keep under regular review the implementation of the Convention and any related legal instruments that the Conference of the Parties may adopt, and shall make, within its mandate, the decisions necessary to promote the effective implementation of the Convention. To this end, it shall:

(a) Periodically examine the obligations of the Parties and the institutional arrangements under the Convention, in the light of the objective of the Convention, the experience gained in its implementation and the evolution of scientific and technological knowledge;

(b) Promote and facilitate the exchange of information on measures adopted by the Parties to address climate change and its effects, taking into account the differing circumstances, responsibilities and capabilities of the Parties and their respective commitments under the Convention;

(c) Facilitate, at the request of two or more Parties, the coordination of measures adopted by them to address climate change and its effects, taking into account the differing circumstances, responsibilities and capabilities of the Parties and their respective commitments under the Convention;

(d) Promote and guide, in accordance with the objective and provisions of the Convention, the development and periodic refinement of comparable methodologies, to be agreed on by the Conference of the Parties, *inter alia*, for preparing inventories of greenhouse gas emissions by sources and removals by sinks, and for evaluating the effectiveness of measures to limit the emissions and enhance the removals of these gases;

(e) Assess, on the basis of all information made available to it in accordance with the provisions of the Convention, the implementation of the Convention by the Parties, the overall effects of the measures taken pursuant to the Convention, in particular environmental, economic and social effects as well as their cumulative impacts and the extent to which progress towards the objective of the Convention is being achieved;

(f) Consider and adopt regular reports on the implementation of the Convention and ensure their publication;

(g) Make recommendations on any matters necessary for the implementation of the Convention;

(h) Seek to mobilize financial resources in accordance with Article 4, paragraphs 3, 4 and 5, and Article 11;

(i) Establish such subsidiary bodies as are deemed necessary for the implementation of the Convention;

(j) Review reports submitted by its subsidiary bodies and provide guidance to them;

(k) Agree upon and adopt, by consensus, rules of procedure and financial rules for itself and for any subsidiary bodies;

(l) Seek and utilize, where appropriate, the services and cooperation of, and information provided by, competent international organizations and intergovernmental and non-governmental bodies; and

(m) Exercise such other functions as are required for the achievement of the objective of the Convention as well as all other functions assigned to it under the Convention.

3. The Conference of the Parties shall, at its first session, adopt its own rules of procedure as well as those of the subsidiary bodies established by the Convention, which shall include decision-making procedures for matters not already covered by decision-making procedures stipulated in the Convention. Such procedures may include specified majorities required for the adoption of particular decisions.

4. The first session of the Conference of the Parties shall be convened by the interim secretariat referred to in Article 21 and shall take place not later than one year after the date of entry into force of the Convention. Thereafter, ordinary sessions of the Conference of the Parties shall be held every year unless otherwise decided by the Conference of the Parties.

5. Extraordinary sessions of the Conference of the Parties shall be held at such other times as may be deemed necessary by the Conference, or at the written request of any Party, provided that, within six months of the request being communicated to the Parties by the secretariat, it is supported by at least one-third of the Parties.

6. The United Nations, its specialized agencies and the International Atomic Energy Agency, as well as any State member thereof or observers thereto not Party to the Convention, may be represented at sessions of the Conference of the Parties as observers. Any body or agency, whether national or international, governmental or non-governmental, which is qualified in matters covered by the Convention, and which has informed the secretariat of its wish to be represented at a session of the Conference of the Parties as an observer, may be so admitted unless at least one-third of the Parties present object. The admission and participation of observers shall be subject to the rules of procedure adopted by the Conference of the Parties.

Article 8. Secretariat

1. A secretariat is hereby established.

2. The functions of the secretariat shall be:

(a) To make arrangements for sessions of the Conference of the Parties and its subsidiary bodies established under the Convention and to provide them with services as required;

(b) To compile and transmit reports submitted to it;

(c) To facilitate assistance to the Parties, particularly developing country Parties, on request, in the compilation and communication of information required in accordance with the provisions of the Convention;

(d) To prepare reports on its activities and present them to the Conference of the Parties;

(e) To ensure the necessary coordination with the secretariats of other relevant international bodies;

(f) To enter, under the overall guidance of the Conference of the Parties, into such administrative and contractual arrangements as may be required for the effective discharge of its functions; and

(g) To perform the other secretariat functions specified in the Convention and in any of its protocols and such other functions as may be determined by the Conference of the Parties.

3. The Conference of the Parties, at its first session, shall designate a permanent secretariat and make arrangements for its functioning.

Article 9. Subsidiary Body for Scientific and Technological Advice

1. A subsidiary body for scientific and technological advice is hereby established to provide the Conference of the Parties and, as appropriate, its other subsidiary bodies with timely information and advice on scientific and technological matters relating to the Convention. This body shall be open to participation by all Parties and shall be multidisciplinary. It shall comprise government representatives competent in the relevant field of expertise. It shall report regularly to the Conference of the Parties on all aspects of its work.

2. Under the guidance of the Conference of the Parties, and drawing upon existing competent international bodies, this body shall:

(a) Provide assessments of the state of scientific knowledge relating to climate change and its effects;

(b) Prepare scientific assessments on the effects of measures taken in the implementation of the Convention;

(c) Identify innovative, efficient and state-of-the-art technologies and know-how and advise on the ways and means of promoting development and/or transferring such technologies;

(d) Provide advice on scientific programmes, international cooperation in research and development related to climate change, as well as on ways and means of supporting endogenous capacity-building in developing countries; and

(e) Respond to scientific, technological and methodological questions that the Conference of the Parties and its subsidiary bodies may put to the body.

3. The functions and terms of reference of this body may be further elaborated by the Conference of the Parties.

Article 10. Subsidiary Body for Implementation

1. A subsidiary body for implementation is hereby established to assist the Conference of the Parties in the assessment and review of the effective implementation of the Convention. This body shall be open to participation by all Parties and comprise government representatives who are experts on matters related to climate change. It shall report regularly to the Conference of the Parties on all aspects of its work.

2. Under the guidance of the Conference of the Parties, this body shall:

(a) Consider the information communicated in accordance with Article 12, paragraph 1, to assess the overall aggregated effect of the steps taken by the Parties in the light of the latest scientific assessments concerning climate change;

(b) Consider the information communicated in accordance with Article 12, paragraph 2, in order to assist the Conference of the Parties in carrying out the reviews required by Article 4, paragraph 2(d); and

(c) Assist the Conference of the Parties, as appropriate, in the preparation and implementation of its decisions.

Article 11. Financial Mechanism

1. A mechanism for the provision of financial resources on a grant or concessional basis, including for the transfer of technology, is hereby defined. It shall function under the guidance of and be accountable to the Conference of the Parties, which shall decide on its policies, programme priorities and eligibility criteria related to this Convention. Its operation shall be entrusted to one or more existing international entities.
2. The financial mechanism shall have an equitable and balanced representation of all Parties within a transparent system of governance.
3. The Conference of the Parties and the entity or entities entrusted with the operation of the financial mechanism shall agree upon arrangements to give effect to the above paragraphs, which shall include the following:
   (a) Modalities to ensure that the funded projects to address climate change are in conformity with the policies, programme priorities and eligibility criteria established by the Conference of the Parties;
   (b) Modalities by which a particular funding decision may be reconsidered in light of these policies, programme priorities and eligibility criteria;
   (c) Provision by the entity or entities of regular reports to the Conference of the Parties on its funding operations, which is consistent with the requirement for accountability set out in paragraph 1 above; and
   (d) Determination in a predictable and identifiable manner of the amount of funding necessary and available for the implementation of this Convention and the conditions under which that amount shall be periodically reviewed.
4. The Conference of the Parties shall make arrangements to implement the above mentioned provisions at its first session, reviewing and taking into account the interim arrangements referred to in Article 21, paragraph 3, and shall decide whether these interim arrangements shall be maintained. Within four years thereafter, the Conference of the Parties shall review the financial mechanism and take appropriate measures.
5. The developed country Parties may also provide and developing country Parties avail themselves of, financial resources related to the implementation of the Convention through bilateral, regional and other multilateral channels.

Article 12. Communication of Information Related to Implementation

1. In accordance with Article 4, paragraph 1, each Party shall communicate to the Conference of the Parties, through the secretariat, the following elements of information:
   (a) A national inventory of anthropogenic emissions by sources and removals by sinks of all greenhouse gases not controlled by the Montreal Protocol, to the extent its capacities permit, using comparable methodologies to be promoted and agreed upon by the Conference of the Parties;
   (b) A general description of steps taken or envisaged by the Party to implement the Convention; and
   (c) Any other information that the Party considers relevant to the achievement of the objective of the Convention and suitable for inclusion in its communication, including, if feasible, material relevant for calculations of global emission trends.
2. Each developed country Party and each other Party included in annex I shall incorporate in its communication the following elements of information:
   (a) A detailed description of the policies and measures that it has adopted to

implement its commitment under Article 4, paragraphs 2(a) and 2(b); and

(b) A specific estimate of the effects that the policies and measures referred to in subparagraph (a) immediately above will have on anthropogenic emissions by its sources and removals by its sinks of greenhouse gases during the period referred to in Article 4, paragraph 2(a).

3. In addition, each developed country Party and each other developed Party included in annex II shall incorporate details of measures taken in accordance with Article 4, paragraphs 3, 4 and 5.

4. Developing country Parties may, on a voluntary basis, propose projects for financing, including specific technologies, materials, equipment, techniques or practices that would be needed to implement such projects, along with, if possible, an estimate of all incremental costs, of the reductions of emissions and increments of removals of greenhouse gases, as well as an estimate of the consequent benefits.

5. Each developed country Party and each other Party included in annex I shall make its initial communication within six months of the entry into force of the Convention for that Party. Each Party not so listed shall make its initial communication within three years of the entry into force of the Convention for that Party, or of the availability of financial resources in accordance with Article 4, paragraph 3. Parties that are least developed countries may make their initial communication at their discretion. The frequency of subsequent communications by all Parties shall be determined by the Conference of the Parties, taking into account the differentiated timetable set by this paragraph.

6. Information communicated by Parties under this Article shall be transmitted by the secretariat as soon as possible to the Conference of the Parties and to any subsidiary bodies concerned. If necessary, the procedures for the communication of information may be further considered by the Conference of the Parties.

7. From its first session, the Conference of the Parties shall arrange for the provision to developing country Parties of technical and financial support, on request, in compiling and communicating information under this Article, as well as in identifying the technical and financial needs associated with proposed projects and response measures under Article 4. Such support may be provided by other Parties, by competent international organizations and by the secretariat, as appropriate.

8. Any group of Parties may, subject to guidelines adopted by the Conference of the Parties, and to prior notification to the Conference of the Parties, make a joint communication in fulfilment of their obligations under this Article, provided that such a communication includes information on the fulfilment by each of these Parties of its individual obligations under the Convention.

9. Information received by the secretariat that is designated by a Party as confidential, in accordance with criteria to be established by the Conference of the Parties, shall be aggregated by the secretariat to protect its confidentiality before being made available to any of the bodies involved in the communication and review of information.

10. Subject to paragraph 9 above, and without prejudice to the ability of any Party to make public its communication at any time, the secretariat shall make communications by Parties under this Article publicly available at the time they are submitted to the Conference of the Parties.

Article 13. Resolution of Questions Regarding Implementation

The Conference of the Parties shall, at its first session, consider the establishment of a multilateral consultative process, available to Parties on their request, for the resolution of questions regarding the implementation of the Convention.

Article 14. Settlement of Disputes

1. In the event of a dispute between any two or more Parties concerning the interpretation or application of the Convention, the Parties concerned shall seek a settlement of the dispute through negotiation or any other peaceful means of their own choice.

2. When ratifying, accepting, approving or acceding to the Convention, or at any time thereafter, a Party which is not a regional economic integration organization may declare in a written instrument submitted to the Depositary that, in respect of any dispute concerning the interpretation or application of the Convention, it recognizes as compulsory *ipso facto* and without special agreement, in relation to any Party accepting the same obligation:

(a) Submission of the dispute to the International Court of Justice, and/or

(b) Arbitration in accordance with procedures to be adopted by the Conference of the Parties as soon as practicable, in an annex on arbitration.

A Party which is a regional economic integration organization may make a declaration with like effect in relation to arbitration in accordance with the procedures referred to in subparagraph (b) above.

3. A declaration made under paragraph 2 above shall remain in force until it expires in accordance with its terms or until three months after written notice of its revocation has been deposited with the Depositary.

4. A new declaration, a notice of revocation or the expiry of a declaration shall not in any way affect proceedings pending before the International Court of Justice or the arbitral tribunal, unless the parties to the dispute otherwise agree.

5. Subject to the operation of paragraph 2 above, if after twelve months following notification by one Party to another that a dispute exists between them, the Parties concerned have not been able to settle their dispute through the means mentioned in paragraph 1 above, the dispute shall be submitted, at the request of any of the parties to the dispute, to conciliation.

6. A conciliation commission shall be created upon the request of one of the parties to the dispute. The commission shall be composed of an equal number of members appointed by each party concerned and a chairman chosen jointly by the members appointed by each party. The commission shall render a recommendatory award, which the parties shall consider in good faith.

7. Additional procedures relating to conciliation shall be adopted by the Conference of the Parties, as soon as practicable, in an annex on conciliation.

8. The provisions of this Article shall apply to any related legal instrument which the Conference of the Parties may adopt, unless the instrument provides otherwise.

Article 15. Amendments to the Convention

1. Any Party may propose amendments to the Convention.

2. Amendments to the Convention shall be adopted at an ordinary session of the Conference of the Parties. The text of any proposed amendment to the Convention shall be communicated to the Parties by the secretariat at least six months before the meeting at which it is proposed for adoption. The secretariat shall also communicate proposed amendments to the signatories to the Convention and, for information, to the Depositary.

3. The Parties shall make every effort to reach agreement on any proposed amendment to the Convention by consensus. If all efforts at consensus have been exhausted, and no agreement reached, the amendment shall as a last resort be adopted by a three-fourths majority vote of the Parties present and voting at the meeting. The adopted amendment shall be communicated by the secretariat to the Depositary, who shall circulate it to all Parties for their acceptance.
4. Instruments of acceptance in respect of an amendment shall be deposited with the Depositary. An amendment adopted in accordance with paragraph 3 above shall enter into force for those Parties having accepted it on the ninetieth day after the date of receipt by the Depositary of an instrument of acceptance by at least three-fourths of the Parties to the Convention.
5. The amendment shall enter into force for any other Party on the ninetieth day after the date on which that Party deposits with the Depositary its instrument of acceptance of the said amendment.
6. For the purposes of this Article, "Parties present and voting" means Parties present and casting an affirmative or negative vote.

Article 16. Adoption and Amendment of Annexes to the Convention

1. Annexes to the Convention shall form an integral part thereof and, unless otherwise expressly provided, a reference to the Convention constitutes at the same time a reference to any annexes thereto. Without prejudice to the provisions of Article 14, paragraphs 2(b) and 7, such annexes shall be restricted to lists, forms and any other material of a descriptive nature that is of a scientific, technical, procedural or administrative character.
2. Annexes to the Convention shall be proposed and adopted in accordance with the procedure set forth in Article 15, paragraphs 2, 3, and 4.
3. An annex that has been adopted in accordance with paragraph 2 above shall enter into force for all Parties to the Convention six months after the date of the communication by the Depositary to such Parties of the adoption of the annex, except for those Parties that have notified the Depositary, in writing, within that period of their non-acceptance of the annex. The annex shall enter into force for Parties which withdraw their notification of non-acceptance on the ninetieth day after the date on which withdrawal of such notification has been received by the Depositary.
4. The proposal, adoption and entry into force of amendments to annexes to the Convention shall be subject to the same procedure as that for the proposal, adoption and entry into force of annexes to the Convention in accordance with paragraphs 2 and 3 above.
5. If the adoption of an annex or an amendment to an annex involves an amendment to the Convention, that annex or amendment to an annex shall not enter into force until such time as the amendment to the Convention enters into force.

Article 17. Protocols

1. The Conference of the Parties may, at any ordinary session, adopt protocols to the Convention.
2. The text of any proposed protocol shall be communicated to the Parties by the secretariat at least six months before such a session.

3. The requirements for the entry into force of any protocol shall be established by that instrument.
4. Only Parties to the Convention may be Parties to a protocol.
5. Decisions under any protocol shall be taken only by the Parties to the protocol concerned.

Article 18. Right to Vote

1. Each Party to the Convention shall have one vote, except as provided for in paragraph 2 below.
2. Regional economic integration organizations, in matters within their competence, shall exercise their right to vote with a number of votes equal to the number of their member States that are Parties to the Convention. Such an organization shall not exercise its right to vote if any of its member States exercises its right, and vice versa.

Article 19. Depositary

The Secretary-General of the United Nations shall be the Depositary of the Convention and of protocols adopted in accordance with Article 17.

Article 20. Signature

This Convention shall be open for signature by States Members of the United Nations or of any of its specialized agencies or that are Parties to the Statute of the International Court of Justice and by regional economic integration organizations at Rio de Janeiro, during the United Nations Conference on Environment and Development, and thereafter at United Nations Headquarters in New York from 20 June 1992 to 19 June 1993.

Article 21. Interim Arrangements

1. The secretariat functions referred to in Article 8 will be carried out on an interim basis by the secretariat established by the General Assembly of the United Nations in its resolution 45/212 of 21 December 1990, until the completion of the first session of the Conference of the Parties.
2. The head of the interim secretariat referred to in paragraph 1 above will cooperate closely with the Intergovernmental Panel on Climate Change to ensure that the Panel can respond to the need for objective scientific and technical advice. Other relevant scientific bodies could also be consulted.
3. The Global Environment Facility of the United Nations Development Programme, the United Nations Environment Programme and the International Bank for Reconstruction and Development shall be the international entity entrusted with the operation of the financial mechanism referred to in Article 11 on an interim basis. In this connection, the Global Environment Facility should be appropriately restructured and its membership made universal to enable it to fulfil the requirements of Article 11.

Article 22. Ratification, Acceptance, Approval or Accession

1. The Convention shall be subject to ratification, acceptance, approval or accession by States and by regional economic integration organizations. It shall be open for accession from the day after the date on which the Convention is closed for

signature. Instruments of ratification, acceptance, approval or accession shall be deposited with the Depositary.
2. Any regional economic integration organization which becomes a Party to the Convention without any of its member States being a Party shall be bound by all the obligations under the Convention. In the case of such organizations, one or more of whose member States is a Party to the Convention, the organization and its member States shall decide on their respective responsibilities for the performance of their obligations under the Convention. In such cases, the organization and the member States shall not be entitled to exercise rights under the Convention concurrently.
3. In their instruments of ratification, acceptance, approval or accession, regional economic integration organizations shall declare the extent of their competence with respect to the matters governed by the Convention. These organizations shall also inform the Depositary, who shall in turn inform the Parties, of any substantial modification in the extent of their competence.

Article 23. Entry into Force

1. The Convention shall enter into force on the ninetieth day after the date of deposit of the fiftieth instrument of ratification, acceptance, approval or accession.
2. For each State or regional economic integration organization that ratifies, accepts or approves the Convention or accedes thereto after the deposit of the fiftieth instrument of ratification, acceptance, approval or accession, the Convention shall enter into force on the ninetieth day after the date of deposit by such State or regional economic integration organization of its instrument of ratification, acceptance, approval or accession.
3. For the purposes of paragraphs 1 and 2 above, any instrument deposited by a regional economic integration organization shall not be counted as additional to those deposited by States members of the organization.

Article 24. Reservations

No reservations may be made to the Convention.

Article 25. Withdrawal

1. At any time after three years from the date on which the Convention has entered into force for a Party, that Party may withdraw from the Convention by giving written notification to the Depositary.
2. Any such withdrawal shall take effect upon expiry of one year from the date of receipt by the Depositary of the notification of withdrawal, or on such later date as may be specified in the notification of withdrawal.
3. Any Party that withdraws from the Convention shall be considered as also having withdrawn from any protocol to which it is a Party.

Article 26. Authentic Texts

The original of this Convention, of which the Arabic, Chinese, English, French, Russian and Spanish texts are equally authentic, shall be deposited with the Secretary-General of the United Nations.

IN WITNESS WHEREOF the undersigned, being duly authorized to that effect, have signed this Convention.

DONE at New York this ninth day of May one thousand nine hundred and ninety-two.

ANNEX I

Australia, Austria, Belarus[a], Belgium, Bulgaria[a], Canada, Czechoslovakia[a], Denmark, European Community, Estonia[a], Finland, France, Germany, Greece, Hungary[a], Iceland, Ireland, Italy, Japan, Latvia[a], Lithuania[a], Luxembourg, Netherlands, New Zealand, Norway, Poland[a], Portugal, Romania[a], Russian Federation[a], Spain, Sweden, Switzerland, Turkey, Ukraine[a], United Kingdom of Great Britain and Northern Ireland, United States of America.

ANNEX II

Australia, Austria, Belgium, Canada, Denmark, European Community, Finland, France, Germany, Greece, Iceland, Ireland, Italy, Japan, Luxembourg, Netherlands, New Zealand, Norway, Portugal, Spain, Sweden, Switzerland, Turkey, United Kingdom of Great Britain and Northern Ireland, United States of America.

SIGNATORIES OF THE UNITED NATIONS FRAMEWORK CONVENTION ON CLIMATE CHANGE

As of 1 July 1992, there were a total of 156 signatories to the United Nations Framework Convention on Climate Change, including the European Community, as follows:

Afghanistan
Algeria
Angola
Antigua and Barbuda
Argentina
Armenia
Australia
Austria
Azerbaijan
Bahamas
Bahrain
Bangladesh
Barbados
Belarus
Belgium
Belize
Benin
Bhutan
Bolivia
Botswana
Brazil
Bulgaria
Burkina Faso
Burundi
Cameroon
Canada
Cape Verde
Central African Republic
Chad
Chile
China
Colombia
Comoros
Congo
Cook Islands
Costa Rica
Côte d'Ivoire
Croatia
Cuba
Cyprus
Denmark
Djibouti
Dominican Republic
Ecuador
Egypt
El Salvador
Estonia
Ethiopia
European Communities
Finland
France
Gabon
Gambia
Germany
Ghana
Greece
Guatemala
Guinea
Guinea-Bissau
Guyana
Haiti
Honduras
Hungary
Iceland
India
Indonesia
Iran
Ireland
Israel
Italy
Jamaica
Japan
Jordan
Kazakhstan
Kenya
Kiribati
Korea, Democratic People's Republic of
Korea, Republic of
Latvia
Lebanon

a Countries that are undergoing the process of transition to a market economy.

Lesotho
Liberia
Libyan Arab Jamahiriya
Liechtenstein
Lithuania
Luxembourg
Madagascar
Malawi
Maldives
Malta
Marshall Islands
Mauritania
Mauritius
Mexico
Micronesia
Moldova
Monaco
Mongolia
Morocco
Mozambique
Myanmar
Namibia
Nauru
Nepal
Netherlands
New Zealand
Nicaragua
Niger
Nigeria
Norway
Oman
Pakistan
Papua New Guinea
Paraguay
Peru
Philippines
Poland
Portugal
Romania
Russian Federation
Rwanda
Saint Kitts and Nevis
Samoa
San Marino
São Tomé and Principe
Senegal
Seychelles
Singapore
Slovenia
Solomon Islands
Spain
Sri Lanka
Sudan
Suriname
Swaziland
Sweden
Switzerland
Tanzania, United Republic of
Thailand
Togo
Trinidad and Tobago
Tunisia
Tuvalu
Uganda
Ukraine
United Kingdom
United States
Uruguay
Vanuatu
Venezuela
Vietnam
Yemen
Yugoslavia
Zaire
Zambia
Zimbabwe

Index

Index